Molecular Networking

The book builds on the analogy between social groups and assemblies of molecules to introduce the concepts of statistical mechanics, machine learning and data science. Applying a data analytics approach to molecular systems, we show how individual (molecular) features and interactions between molecules, or "communication" processes, allow for the prediction of properties and collective behavior of molecular systems - just as polling and social networking shed light on the behavior of social groups. Applications to systems at the cutting-edge of research for biological, environmental, and energy applications are also presented.

Key features:

- Draws on a data analytics approach of molecular systems.
- Covers hot topics such as artificial intelligence and machine learning of molecular trends.
- Contains applications to systems at the cutting-edge of research for biological, environmental and energy applications.
- Discusses molecular simulation and links with other important, emerging techniques and trends in computational sciences and society.
- Authors have a well-established track record and reputation in the field.

Molecular Networking

Statistical Mechanics in the Age of AI and Machine Learning

Caroline Desgranges

Research Assistant Professor, Department of Physics & Applied Physics, University of Massachusetts Lowell

Jerome Delhommelle

Associate Professor, Department of Chemistry, University of Massachusetts Lowell

CRC Press
Taylor & Francis Group
Boca Raton London New York

CRC Press is an imprint of the
Taylor & Francis Group, an **informa** business

First edition published 2024
by CRC Press
2385 Executive Center Drive, Suite 320, Boca Raton, FL 33431

and by CRC Press
4 Park Square, Milton Park, Abingdon, Oxon, OX14 4RN

ISBN: 978-0-367-43893-7 (hbk)
ISBN: 978-1-032-67081-2 (pbk)
ISBN: 978-1-003-00641-1 (ebk)

DOI: 10.1201/9781003006411

Typeset in Times
by codeMantra

Contents

PART I *Molecular Networking Analytics*

PART II   Static Trends: Equilibrium Statistics

PART III   *Dynamic Trends: Motion Statistics*

Preface

"If the moon, in the act of completing its eternal path round the earth, were gifted with self-consciousness, it would feel thoroughly convinced that it would travel its path on its own."

Albert Einstein,
Einstein-Tagore correspondence [114]

Social networks, machine learning, and artificial intelligence have become part of our daily lives. We live in an era where data analysis is present in all aspects of society and a driving force for many decisions that impact our present and future. While statistics have long played a significant role in numbers-driven domains, the development of novel machine learning algorithms, combined with the increase in computing performance and data storage, has led to a paradigm shift in how we approach and address challenges. For instance, in human health, the concept of precision medicine, which considers the individual features of a patient, has emerged as a promising alternative to one-size-fits-all medical treatments. Similarly, the sampling of opinion through polling methods had been for decades a staple of, for example, commercial and political analyses. It is now complemented by the analysis of data from social networks, which provide a window into human interactions and the interrelation between individual and collective responses.

This interrelation between individual features and collective properties also occurs at the microscopic level. Molecules interact with each other and respond to environmental stimuli, such as changes in external conditions or fields. As a result, they often exhibit a collective response and emergent properties. This is, for instance, the case when a group of molecules undergoes a phase transition, where the properties of the new phase dramatically depart from those of the initial phase. In other words, molecular networking can be thought of as equivalent to social networking for microscopic particles.

In writing this advanced textbook, we aim to provide a modern viewpoint on statistical mechanics. This field, which originated during the second half of the 19th century, is, by essence, data-driven. This is because statistical mechanics draws statistics on the distributions of microscopic quantities to predict macroscopic properties. In the book, we leverage the analogy between social and molecular networking to analyze the connection between individual properties and collective behavior. Part I of the book surveys the essential tools for this probabilistic analysis. This part successively draws concepts from mechanics (both classical and quantum) and statistics, characterizes the "communication processes" that take place at the microscopic scale, introduces the concept of statistical ensembles, discusses how the relation between the micro- and macroscopic worlds, and introduces how machine learning methods can provide further insight on molecular systems. Part II focuses on "polling" molecular systems at equilibrium to determine their properties. After discussing sampling approaches and the Monte Carlo simulation framework, Part II examines the impact of the environment on the nature (adiabatic *vs.* isothermal) of the ensemble before addressing how the onset of collective changes and emergence is captured by a central quantity, known as the partition function. Part II concludes with recent developments in statistical mechanics enabled by machine learning for interaction models, data interpolation and extrapolation, and the elucidation of transition pathways. Part III discusses molecular systems' dynamics and time-dependent behavior, the determination of transport coefficients, and the systems' response to external fields. It then opens on recent results in nonequilibrium statistical mechanics that shed light on the behavior of biological, active, and living systems before introducing how machine learning allows to classify transport behaviors, learn navigation strategies at the microscopic level, and identify new governing equations from data for far-from-equilibrium systems.

We hope this book will provide a valuable resource to all researchers, particularly those starting a scientific journey in this exciting field!

Authors

Dr. Caroline Desgranges received a DEA in Physics in 2005 from the University Paul Sabatier-Toulouse III (France) and a PhD in Chemical Engineering from the University of South Carolina (USA) in 2008. She is currently a Research Assistant Professor in Physics & Applied Physics at the University of Massachusetts Lowell.

Dr. Jerome Delhommelle did his undergraduate studies at the Ecole Normale Superieure Paris-Saclay and received his PhD in Chemistry from the University of Paris-Saclay (France) in 2000. He is currently an Associate Professor in Chemistry at the University of Massachusetts Lowell.

Part I

Molecular Networking Analytics

1 Probabilities, Distributions and Statistics

"So many of the properties of matter, especially when in the gaseous form, can be deduced from the hypothesis that their minute parts are in rapid motion, the velocity increasing with temperature."

James Clerk Maxwell,
Illustrations of the dynamical theory of gases [234]

Individual behaviors are at the origin of any global trend or collective property. This is the case for a social group, for which decisions are often made following a vote, as well as for an assembly of atoms and molecules, for which the characteristics of each constituent, such as their dipole moment or ability, for instance, to form hydrogen bonds, directly impact the properties of the assembly. This makes the fabric of the connection between the two levels: the *microscopic* world of individual atoms and molecules and the *macroscopic* world of bulk phases. In this chapter, we present the analytical and mathematical tools that allow us to characterize these two worlds and understand their interplay.

As stated by Maxwell, matter is closely associated with the concepts of motion and change. All constituents, *i.e.*, atoms and molecules, are small, with a typical length scale of the order of the nanometer and in rapid motion. This means that we need laws and theories, *i.e.*, Mechanics, to account for the motion of atoms and molecules - specifically, classical mechanics and quantum mechanics discussed in the first part of this chapter. This rapid motion has a direct impact on the measurable properties of matter, such as, for instance, its temperature. This is where we need a second theory and set of mathematical relations to characterize the macroscopic properties of matter. This theory is called thermodynamics and is the topic of the second section of this chapter. Just as the group's properties can be deduced from the behavior of the individuals, the connection between microscopic and macroscopic emerges from a statistical analysis of the motion of atoms and molecules. There lies the realm of probabilities, distribution functions, and, ultimately, the micro-macro connection discussed in the third section of this chapter.

1.1 MECHANICS

1.1.1 NEWTON, LAGRANGE, AND HAMILTON

In a first approach, the individual constituents of matter, *i.e.*, atoms and molecules, can be naturally thought of as minute replicas of simple shapes surrounding us. For instance, an atom can be considered as a sphere of mass m that undergoes a rapid motion and collides with neighboring atoms. This motion, and the equations that govern it, have been theorized over the years by several brilliant polymaths, including Newton, Lagrange, and Hamilton (see Figure 1.1). As we will see throughout the chapters, the formalisms they introduce lead to equivalent results. However, these formalisms emphasize different measures of the system's state during motion, making mathematics easier when modeling specific applications. We will start with a brief overview of some of their most powerful findings.

Newton, the famous English polymath, is widely recognized as the founder of classical mechanics, with the publication of *Philosophi Naturalis Principia Mathematica* (Mathematical Principles of Natural Philosophy) in 1687. He proposed the well-known equation $\mathbf{F} = m\mathbf{a}$ that captures the

DOI: 10.1201/9781003006411-2

Figure 1.1   Sir William Rowan Hamilton (1805–1865).

equality between the force exerted upon an object, denoted by the vector $\mathbf{F}$, and the product of the mass m of the object by its acceleration (vector $\mathbf{a}$). The product $m\mathbf{a}$ is also called the rate of change of the momentum $\mathbf{p} = m\mathbf{v}$, in which $\mathbf{v}$ is the velocity

Applying Newton's relation to an atom of mass m and momentum $\mathbf{p}$, we obtain

$$\frac{d\mathbf{p}}{dt} = \dot{\mathbf{p}} = \mathbf{F} \tag{1.1}$$

in which $\dot{\mathbf{p}}$ is a shorthand notation for $\frac{d\mathbf{p}}{dt}$. Since $\mathbf{p} = m\mathbf{v}$ and the acceleration $\mathbf{a}$ is the time-derivative of $\mathbf{v}$, we recover the usual form of Newton's relation $\frac{d\mathbf{p}}{dt} = m\frac{d\mathbf{v}}{dt} = m\mathbf{a}$.

If we now write the coordinates for the two vectors, we can rewrite Newton's relation as

$$\begin{bmatrix} \dot{p}_x \\ \dot{p}_y \\ \dot{p}_z \end{bmatrix} = \begin{bmatrix} F_x \\ F_y \\ F_z \end{bmatrix} = \begin{bmatrix} -\partial U(x,y,z)/\partial x \\ -\partial U(x,y,z)/\partial y \\ -\partial U(x,y,z)/\partial z \end{bmatrix} \tag{1.2}$$

The second equality shows how the force can be calculated as the gradient of the potential energy U. U is often a function of the atom's position. For instance, for a particle with position coordinates (x,y,z) tethered to the point of coordinates $(0,0,0)$ by a spring of spring constant k, we have $U = \frac{1}{2}k(x^2 + y^2 + z^2)$ and thus $(F_x, F_y, F_z) = (-kx, -ky, -kz)$. This means that two sets of coordinates, for a total of six coordinates, will be necessary to characterize the motion of an atom: the momentum coordinates and the position coordinates.

Lagrange, a French-Italian mathematician, proposed another set of equations of motion at the end of the 18th century. These equations yield equivalent results. Their advantage is that they emphasize potential energy rather than forces, which can sometimes be very convenient. For the Lagrangian approach, we start with the definition of K, the kinetic energy of the atom, as

$$K(\dot{x}, \dot{y}, \dot{z}) = \frac{m}{2} \left(\dot{x}^2 + \dot{y}^2 + \dot{z}^2 \right) \tag{1.3}$$

The Lagrangian L is then defined as the difference between the kinetic energy K and the potential energy U

$$L(x,y,z,\dot{x},\dot{y},\dot{z}) = K(\dot{x},\dot{y},\dot{z}) - U(x,y,z) \tag{1.4}$$

L is a function of six variables for each atom, leading to a set of six equations of motion. For example, the first two equations, with respect to $\dot{x}$ and x, are given by

$$\begin{aligned} \frac{\partial L}{\partial \dot{x}} &= \frac{\partial K}{\partial \dot{x}} = m\dot{x} \\ \frac{\partial L}{\partial x} &= -\frac{\partial U}{\partial x} \end{aligned} \tag{1.5}$$

This yields the following Lagrangian form corresponding to Newton's equations

$$\begin{aligned} \frac{d}{dt}\left(\frac{\partial L}{\partial \dot{x}}\right) &= \frac{\partial L}{\partial x} \\ \frac{d}{dt}\left(\frac{\partial L}{\partial \dot{y}}\right) &= \frac{\partial L}{\partial y} \\ \frac{d}{dt}\left(\frac{\partial L}{\partial \dot{z}}\right) &= \frac{\partial L}{\partial z} \end{aligned} \tag{1.6}$$

To obtain a more compact notation, (x,y,z) are often replaced by a set of generic position coordinates (q_1, q_2, q_3), which gives the following equations of motion

$$\frac{d}{dt}\left(\frac{\partial L}{\partial \dot{q}_j}\right) = \frac{\partial L}{\partial q_j} \quad j = 1,2,3 \tag{1.7}$$

During the 19th century, Hamilton, an Irish mathematician, proposed a third approach, which has been instrumental in, among others, the development of quantum mechanics. Rather than focusing on the coordinates q_j and the corresponding velocities $\dot{q}_j$, Hamilton's approach introduces a generalized momentum p_j as the conjugate to q_j according to

$$p_j = \frac{\partial L}{\partial \dot{q}_j} \tag{1.8}$$

For an atom, the Hamiltonian function can be written as

$$H(q_1, q_2, q_3, p_1, p_2, p_3) = \sum_{j=1}^{3} p_j \dot{q}_j - L(\dot{q}_1, \dot{q}_2, \dot{q}_3, q_1, q_2, q_3) \tag{1.9}$$

Since, for an atom, we have $K = \frac{m}{2}\sum_{j=1}^{3}\dot{q}_j^2$, the generalized momentum is given by

$$p_j = \frac{\partial L}{\partial \dot{q}_j} = \frac{\partial K}{\partial \dot{q}_j} = m\dot{q}_j \tag{1.10}$$

and, as a result, $\sum_{j=1}^{3} p_j \dot{q}_j = 2K$. This means that

$$H = \sum_{j=1}^{3} p_j \dot{q}_j - L = 2K - (K - U) = K + U \tag{1.11}$$

and thus that the Hamiltonian H corresponds to the atom's total energy. The Hamiltonian equations of motion can then be written, for $j = 1,2,3$ as

$$\begin{aligned} \left(\frac{\partial H}{\partial p_j}\right) &= \dot{q}_j \\ \left(\frac{\partial H}{\partial q_j}\right) &= -\dot{p}_j \end{aligned} \tag{1.12}$$

leading to six first-order differential equations of motion per atom. This means that, for a system of N atoms, we will need to handle a total of $6N$ first-order differential equations and $6N$ variables, i.e., $3N$ momentum coordinates and $3N$ position coordinates. Given the initial positions and momenta of the N atoms, we can then integrate the $6N$ equations of motion and determine the evolution of the system and its state at any time t.

1.1.2   WAVE FUNCTION AND UNCERTAINTY

As we know, atoms and molecules are composed of sub-atomic particles, such as protons, neutrons, and electrons. By the beginning of the 20th century, scientists had realized that a different type of mechanics, known as quantum mechanics, applied to these sub-atomic particles. A German physicist, Planck, introduced the concept of energy quanta, which means that only finite, well-defined quantities of energy can be absorbed by an atom when light is shined on it. This also implies that electrons in an atom have precise energy or, in other words, occupy discrete energy levels, *i.e.* energy levels separated by gaps. This is dramatically different from classical mechanics, where any amount of energy can be exchanged between a system and its surroundings. This means that, in classical mechanics, the velocity of an object, and thus, the kinetic energy, can take any value. In other words, we can have a continuous, *i.e.*, with absolutely no gap, distribution of velocities for this object. Another significant concept, wave-particle duality, was introduced by the French physicist De Broglie. This led to the realization that, for instance, electrons have a mass, and wave-like properties, such as wavelengths.

Another German scientist, Heisenberg, introduced a new formalism known as matrix mechanics to carry out calculations on quantum systems. Heisenberg also discovered what is now known as the uncertainty principle. He finds that quantum mechanics cannot determine both a particle's position and momentum exactly. This opens the door to having a probabilistic picture of the properties of a particle rather than having a set of exact numbers, as in classical mechanics. He proposes his famous uncertainty principle by stating that

$$\Delta p \times \Delta q \geq \frac{h}{4\pi} \tag{1.13}$$

where Δp is the uncertainty in the determination of the momentum, Δq is the uncertainty in the determination of the position, and $h = 6.62 \times 10^{-34} J.s$ is Planck's constant.

With both the concepts of particle-wave duality and uncertainty, the set is staged to account for a quantum particle's properties. This can be achieved by computing the wave function of a particle $\Psi(\mathbf{q},t)$ at time t and the probability of finding the particle at a given position. Indeed, the probability that the particle is found between position coordinates (q_1, q_2, q_3) and coordinates $(q_1 + dq_1, q_2 + dq_2, q_3 + dq_3)$ is equal to

$$\mathscr{P} = |\Psi(\mathbf{q},t)|^2 dq_1 dq_2 dq_3 \tag{1.14}$$

where $|\Psi(\mathbf{q},t)|^2$ denotes the square modulus for the wave function.

The last piece of the puzzle is finding the quantum equivalent to the classical equations of motion. In 1926, Schrödinger formalized the quantum mechanical equation

$$i\hbar \frac{\partial \Psi(\mathbf{q},t)}{\partial t} = \frac{\hbar^2}{2m} \mathscr{H}(\mathbf{q},t) \tag{1.15}$$

where $\mathscr{H}$ is the Hamiltonian operator, and E is a scalar quantity corresponding to the system's energy. Using the notation (q_1, q_2, q_3) in lieu of (x, y, z), the Hamiltonian operator includes the following two terms

$$\begin{aligned} \mathscr{H} &= -\frac{\hbar^2}{2m} \left(\frac{\partial^2}{\partial q_1^2} + \frac{\partial^2}{\partial q_2^2} + \frac{\partial^2}{\partial q_3^2} \right) + U(q_1, q_2, q_3, t) \\ &= -\frac{\hbar^2}{2m} \nabla^2 + U(q_1, q_2, q_3) \end{aligned} \tag{1.16}$$

where $\hbar$ is $h/2\pi$.

In this equation, the first term corresponds to the kinetic energy of the particle and the second term to the potential energy. We thus recover the total energy, as prescribed originally by Hamilton for classical systems (Eq. 1.11). Both the equations of motion in classical mechanics and the

Schrödinger equation in quantum mechanics are differential equations. This means that exactly like equations of motion can be integrated to determine the evolution of positions and momenta for a classical particle, the Schrödinger equation can be integrated to determine the evolution of the wave function for a quantum particle. We will see in Chapter 2 how to decide which route to take (classical vs. quantum) when modeling atoms and molecules.

1.1.3 QUANTUM ENERGY AND DENSITY OF STATES

Quantum mechanics gives rise to a distribution of discrete energy levels for quantum systems. We examine a simple example corresponding to a quantum particle, such as an electron, trapped in a one-dimensional infinite well. The potential energy of the well can be

$$U(q_1) = 0 \qquad 0 < q_1 < L$$
$$U(q_1) = \infty \quad q_1 < 0 \text{ and } q_1 > L \tag{1.17}$$

Since we are dealing with a one-dimensional situation, a single position coordinate is needed to account for the properties of the quantum particle. Furthermore, the well remains in the same place at all times, which means that the particle's properties are independent of time. The wave function then becomes a function of q_1 only and can be written as $\Psi(q_1)$ (see Figure 1.2).

To determine the distribution of energy levels, we solve the time-independent version of the Schrödinger equation

$$\mathcal{H}\Psi(q_1) = E\Psi(q_1) \tag{1.18}$$

Expanding the Hamiltonian operator, we obtain

$$-\frac{\hbar^2}{2m}\frac{\partial\Psi(q_1)}{\partial q_1^2} = E\Psi(q_1) \tag{1.19}$$

To solve this second-order differential equation, we need to specify the boundary conditions, *i.e.* the value taken by the wave function as the particle reaches the boundaries $q_1 = 0$ and $q_1 = L$. Since the potential energy suddenly becomes infinite at the boundaries, there is zero probability of finding the particle at the boundary and beyond when $U(q_1) = \infty$. Rearranging Eq. 1.19, we obtain

$$\frac{\partial\Psi(q_1)}{\partial q_1^2} + \frac{2mE}{\hbar^2}\Psi(q_1) = 0 \tag{1.20}$$

If we denote by $k = \frac{\sqrt{2mE}}{\hbar}$, the general solution can be written as

$$\Psi(q_1) = A\cos(kq_1) + B\sin(kq_1) \tag{1.21}$$

Applying the boundary conditions ($\Psi(0) = \Psi(L) = 0$), we find $A = 0$ and $kL = n\pi$, with $n = 1, 2, 3, \ldots$ Finally, the factor B is determined by realizing that the total probability of finding the

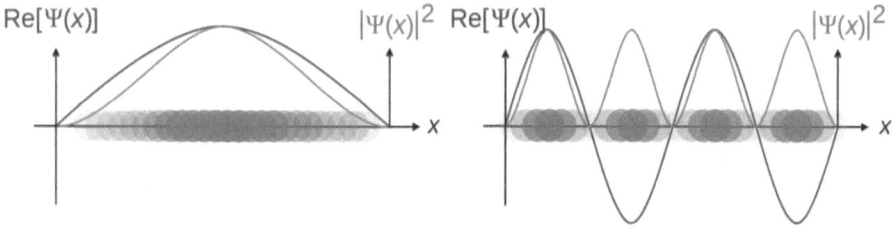

Figure 1.2 Wavefunction for a particle in a box: examples of stationary states.

particle is equal to 1, or, equivalently, that the wave function is normalized as $\int_a^L |\Psi(q_1)|^2 dq_1 = 1$. This yields $B = \sqrt{\frac{2}{L}}$, and the following result for the wave functions $\Psi_n(q_1)$ solutions of Eq. 1.18

$$\Psi_n(q_1) = \frac{\sqrt{2}}{L} \sin\left(\frac{n\pi}{L} q_1\right) \quad n = 1, 2, \ldots \tag{1.22}$$

and the corresponding distribution of energy levels E_n

$$E_n = \frac{h^2 n^2}{8mL^2} \quad n = 1, 2, \ldots \tag{1.23}$$

In other words, an infinity of wave functions associated with an infinite distribution of discrete energy levels of increasing energy satisfy Eq. 1.18.

Generalizing now this result in 3D, and using the notation (q_1, q_2, q_3) rather than (x, y, z), we obtain the following distribution of energy levels

$$E_{n_1, n_2, n_3} = \frac{h^2}{8mL^2} \left(n_1^2 + n_2^2 + n_3^2\right) \quad n_1, n_2, n_3 = 1, 2, 3, \ldots \tag{1.24}$$

Several combinations of microscopic states, or microstates, can have the same energy. Indeed, each microstate is defined by the set of quantum numbers (n_1, n_2, n_3), and we can rapidly see that the set $(1, 1, 3)$ yields the same value as the set $(1, 3, 1)$ for energy. The number of microstates with the same energy is called degeneracy, an important characteristic of a quantum system. The degeneracy increases very quickly with system size. If we now consider a system of N noninteracting particles with no other potential energy than that created by the infinite well, we have $3N$ quantum numbers, and the energy levels are given by

$$E = \frac{h^2}{8mL^2} \sum_{j=1}^{3N} n_j^2 \tag{1.25}$$

The number of states with energy $\leq E$ can be calculated as

$$\Phi(E) = \frac{1}{\Gamma(N+1)\Gamma[(3N/2)+1]} \left(\frac{2\pi ma^2 E}{h^2}\right)^{3N/2} \tag{1.26}$$

where $\Gamma(n)$ here is the gamma function. The number of states between E and $E + \Delta E$ is

$$\Omega(E, \Delta E) = \frac{1}{\Gamma(N+1)\Gamma(3N/2)} \left(\frac{1}{\Gamma(N+1)\Gamma(3N/2)}\right)^{3N/2} E^{(3N/2-1)} \Delta E \tag{1.27}$$

$\Omega(E, \Delta E)$ measures how many states, or energy levels, exist within a very small energy interval ΔE. $\Omega(E, \Delta E)$ also gives access to what is known as the density of states $g(E)$, i.e. $g(E) = \frac{\partial \Phi}{\partial E} = \frac{\Omega(E, \Delta E)}{\Delta E}$. Under ambient conditions, $\Omega(E, \Delta E)$ is of the order of 10^N, which, for a macroscopic system ($N = 6.02 \times 10^{23}$), is gigantic. As we will see in Chapter 3, this has important consequences for macroscopic systems.

1.2 THERMODYNAMICS

1.2.1 PROCESSES, WORK, AND HEAT

We have focused so far on the microscopic level and the computation of the mechanical properties of the "minute" constituents mentioned by Maxwell. We now turn to the properties of the macroscopic systems, i.e., large assemblies of atoms and molecules. A different type of properties, the so-called thermodynamic properties, characterize macroscopic systems. Rather than positions and momenta for individual particles, quantities such as the pressure P, the temperature T, or the volume V of a system define the state of a macroscopic system. There is, of course, a connection between the two, as intuited by Maxwell (*"their minute parts are in rapid motion, the velocity increasing with temperature"*), and as we shall see in the last section of this chapter.

The system's evolution can then be analyzed in terms of changes in its defining quantities or changes in the macroscopic state of the system. Such a change is called a process and can take different forms. A process can occur isothermally, *i.e.*, the temperature remains constant while other quantities, such as V and P, vary. Alternatively, a process can be isobaric, or, in other words, take place at constant pressure, or a process can be isochoric or equivalently take place at constant volume. Another essential feature of macroscopic processes is the physical boundary between the system and its surroundings. Is the system contained in a rigid vessel? Does the vessel have a lid, or is it open? Is it thermally insulated? In other words, does the exchange of matter or energy occur during the process between the system and its surroundings? As we can see, all these considerations are far removed from the discussion of the mechanics of the individual particles in the previous section. Still, they will undoubtedly impact the motion of atoms and molecules.

To start making this connection, we need to quantify how the system is impacted by the processes it undergoes and to define properties to measure its energy. During a process, energy exchange can occur between the system and its surroundings through different mechanisms. The first possibility is heat exchange (see Figure 1.3). The heat absorbed by the system during a process from an initial state (labeled with A) to a final state (labeled with B) can be written as

$$Q = \int_A^B \delta q \tag{1.28}$$

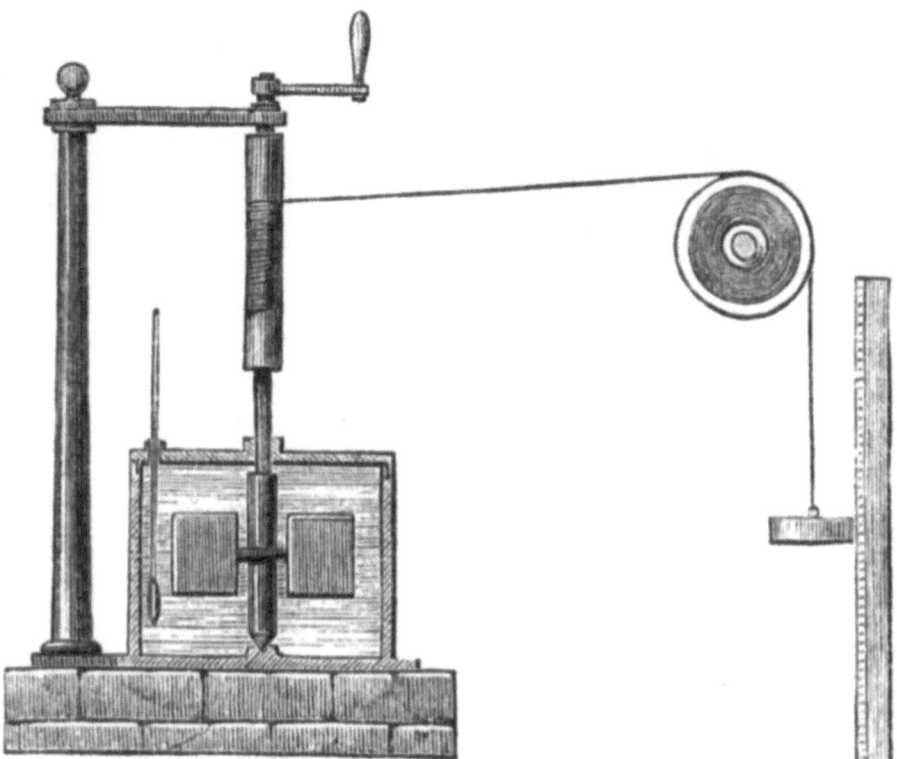

Figure 1.3   Joule's apparatus for measuring the mechanical equivalent of heat.

In practice, the process, and the heat exchange, do not occur instantly but over time. We can imagine splitting it into successive, very small, or infinitesimal steps during which infinitesimal amounts of heat, noted δq, are absorbed. The integral form employed here indicates that Q is the total heat adsorbed during the process. Another possible mechanism for the exchange of energy is through work done on the system by the surroundings over the process

$$W = \int_A^B -PdV \tag{1.29}$$

Here the work results from applying pressure P by the surroundings, which results in a change in V for the system. For instance, subjecting the system to a very large P will result in system compression ($dV < 0$). Of course, the exchange of energy directly depends on the type of process the system undergoes and on the nature of the boundary. For these reasons, Q and W are called path-dependent, meaning their values depend on the path the system follows from state A to state B.

The state of the system, and its energy, changes during the process. Several functions, called state functions, can be used to characterize a state. For instance, the total energy, or internal energy, is noted E and can be obtained by adding the kinetic and potential energy of the system. The enthalpy H takes into account an additional term, PV, that has the same dimension as energy and is written as $H = E + PV$. The entropy S is another state function that correlates with the amount of organization within the system. Finally, free energies are other state functions that often allow us to predict how a system evolves under specific conditions. The Helmholtz free energy A is defined as $A = E - TS$, while the Gibbs free energy G is given by $G = H - TS$. The role of these state functions will be discussed in the next section. Finally, since state functions are characteristic of states only, their variations during a process are not path-dependent. For instance, the change in internal energy will be $\Delta E(A \to B) = E(B) - E(A)$ for any process undergone by the system to go from state A to state B.

1.2.2   FIRST, SECOND AND THIRD LAWS

As with any phenomenon, group, community, or game, there are rules in place that govern how a phenomenon, or thermodynamic process in our case, takes place. A set rule of rule often starts with the simple observation that "we get out what we put in", or, in other words, a conservation law. This is the basis of the first law of thermodynamics, for which we apply this reasoning to energy. Here, the energy that we "get out" is the internal energy $E(B)$ at the end of the process (state B), and energy that we "put in" includes all kinds of energy that we provide to the system during the process, which are the internal energy $E(A)$ at the beginning of the process (state A), the amount of heat Q absorbed by the system during the process and the work W done on the system during the process. Overall we obtain

$$E(B) = E(A) + Q + W \tag{1.30}$$

Rearranging this equation to have all internal energy terms on the left-hand-side, we obtain the first law of thermodynamics

$$\Delta E = E(B) - E(A) = Q + W \tag{1.31}$$

Of course, we need more than one law to predict the complete evolution of a macroscopic system. This is the role played by the second law of thermodynamics. This law involves a state function we mentioned above, the entropy S, and has intrigued generations of scientists since its initial introduction by a German thermodynamicist named Clausius during the 19th century (see Figure 1.4).

We saw that entropy could be related to a change in the system's organization. Still, another physical interpretation relates entropy to the arrow of time and irreversibility - the fact that a process occurs a certain way or along a certain path and cannot take place along the reverse path. For instance, a hot cup of coffee will cool down by releasing heat into the colder surroundings. The process of releasing heat is called irreversible because the reverse process is not physically possible:

Figure 1.4   Rudolf Clausius (1822–1888).

the cup will not heat up even more by absorbing heat from the colder surroundings. The concept of irreversibility is captured by the second law of thermodynamics, which states that

$$\Delta S \geq \frac{Q}{T} \tag{1.32}$$

For an irreversible process, the change in entropy will be strictly greater than the amount of heat over the temperature at which the process occurs (*i.e.*, at room temperature in the example of the cup of coffee). If we are now considering a reversible process, then the second law becomes an equality, and the change in entropy is equal to the ratio of Q over T.

At this point, we have a law for conservation, the first law, and a law for evolution, the second law; we only need a law of reference to specify a starting point or an origin. This is especially important since the first and second laws are only concerned with changes in state functions but do not provide any guidance on what absolute value the entropy should take. This is the purpose of the third law of thermodynamics, proposed by a German scientist, called Nernst, and later refined by two American physical chemists, Lewis and Randall [224] as

"If the entropy of each element in some crystalline state is taken as zero at the absolute zero of temperature, every substance has a positive finite entropy; but at the absolute zero of temperature the entropy may become zero and does so become in the case of perfect crystalline substances."

This gives an origin for the entropy, and the absolute value for the entropy for any conditions can then be obtained by, for instance, subjecting the system to a reversible process, measuring the heat exchanged, and integrating the second law of thermodynamics over the path.

1.2.3   CHANGING CONDITIONS: LEGENDRE TRANSFORMATIONS

We introduced the second law of thermodynamics as a law of evolution. Let us specify what happens to the internal energy during a process. We start from the first law of thermodynamics $\Delta E = Q + W$

and use the second law to quantify Q. We can rapidly obtain that $Q \leq T\Delta S$ from the second law. This leads to

$$\Delta E \leq T\Delta S + W \tag{1.33}$$

If we now replace W by the integral of Eq. 1.29, we obtain

$$\Delta E \leq T\Delta S - \int_A^B P dV \tag{1.34}$$

In other words, E always decreases during a process until it reaches a minimum when the system attains equilibrium where no net change in any quantity occurs (ΔS and dV are equal to 0). This provides a direction for the evolution of the system. Let us delve into this equation and make two assumptions: (i) we assume the process to be reversible and (ii) we now focus on very small, infinitesimal changes. This means that we can write Eq. 1.34 as an equality and for infinitesimal changes in the state function (d notation) rather than changes over the whole process (Δ and $\int_A^B$ symbols). We obtain

$$dE = TdS - PdV \tag{1.35}$$

Here we use the d notation for state functions, such as E and S, and thermodynamic variables, like V and P, rather than the δ symbol for W and Q. This emphasizes that, unlike, for instance, dV, which will take the same value ($V_B - V_A$) regardless of the type of process that connects states A and B, the latter two (δW and δQ) are path-dependent and can take very different values depending on the process. Writing this equation in terms of the partial derivatives of the internal energy E, we have

$$dE = \left(\frac{\partial E}{\partial S}\right)_V dS + \left(\frac{\partial E}{\partial V}\right)_S dV \tag{1.36}$$

which, by identification of the terms factors of dS and dV, respectively, leads to

$$\left(\frac{\partial E}{\partial S}\right)_V = T \text{ and } \left(\frac{\partial E}{\partial V}\right)_S = -P \tag{1.37}$$

where the subscript V in the first derivative, and the subscript S in the second derivative, are held constant.

So far, we have obtained a criterion for the evolution of a system and how it reaches equilibrium as a function of two variables S and V. In essence, E is a function of the two variables, S and V, or, in other words, $E(S,V)$. While "mechanical" properties, like V, P, and T, are straightforward to measure, a thermodynamic function like S is more challenging to evaluate. This prompts the following question: can we write a criterion for the evolution as a function of two mechanical properties such as V and T?

The concept of Legendre transformation allows us to do so. Let us consider a 1D example to show how a Legendre transformation works. If we have a function $f(x)$ of the variable x, the slope g can be calculated for any value of x as

$$g = \frac{f(x) - h}{x - 0} \tag{1.38}$$

where h is the intercept for $f(x)$. Rearranging this equation, we obtain the following function form for h

$$h(g) = f(x) - gx \tag{1.39}$$

Through a Legendre transformation, we have thus changed a function of x, $f(x)$, into an equivalent function of g, $h(g)$. Let us now apply this concept to internal energy. From Eq. 1.37, we already know a "slope" that is a function of the "mechanical" property T as $T = \left(\frac{\partial E}{\partial S}\right)_V$. Since V is held constant in this derivative, we can apply the 1D-Legendre transformation to $E(S,V)$ and write

$$T = \frac{E(S,V) - A(T,V)}{S - 0} \tag{1.40}$$

which yields the Helmholtz free energy $A(T,V)$ as

$$A(T,V) = E(S,V) - TS \tag{1.41}$$

What is the evolution criterion in terms of the two mechanical variables V and T, then? To find it, we differentiate $A(T,V)$ to obtain

$$dA(T,V) = dE(S,V) - d(TS) = dE(S,V) - TdS - SdT \tag{1.42}$$

and replace dE by the inequality provided by Eq. 1.34

$$dA(T,V) \le S\Delta T - PdV \tag{1.43}$$

meaning that the Helmholtz free energy will decrease ($dA < 0$) until it reaches a minimum ($dA = 0$) at equilibrium for a system at constant T and V.

A similar criterion can be established for a system at constant T and P by defining two additional state functions, the enthalpy $H = E + PV$ and the Gibbs free energy $G = H - TS$. Differentiating G yields the following inequality

$$dG(T,P) \le S\Delta T + VdP \tag{1.44}$$

This shows that, for a system at constant T and P, the Gibbs free energy will decrease as the system evolves until G reaches a minimum when the system is at equilibrium.

1.3　STATISTICS AND DISTRIBUTIONS

1.3.1　MAXWELL-BOLTZMANN DISTRIBUTION

Now that we have mathematical equations to follow the motion of individuals and criteria for the evolution of the entire group, we need to develop a way to connect the two worlds. This is the role played by statistics, where the properties of individuals are collected and averaged to yield some global property of the system.

Maxwell was one of the first scientists to factor in the concepts of distribution and probability to determine the properties of macroscopic systems, such as gases. Indeed, Maxwell postulated that not all particles in a gas have the same velocity and that a distribution of these velocities should be used to model such a system. Boltzmann built upon the Maxwell distribution to propose what is now known as the Maxwell-Boltzmann distribution (see Figure 1.5). This distribution is the basis for the kinetic theory of gases, which allows for the calculation of the thermodynamic properties of the system from the mechanical properties of the particles composing the system.

Preceding theories captured the connection between the motion of the particles through their average kinetic energy and temperature. If we look at a single degree of freedom of a particle with velocity v_x along the x axis and evaluate ε_x the average kinetic energy along that direction, we have

$$\varepsilon_x = \frac{1}{2}m < v_x^2 > = \frac{1}{2}k_B T \tag{1.45}$$

where $< v_x^2 >$ is the average square velocity along x. Rearranging this equation provides a direct connection between the temperature T, the mass m of the particle and $< v_x^2 >$ through

$$< v_x^2 > = \frac{k_B T}{m} \tag{1.46}$$

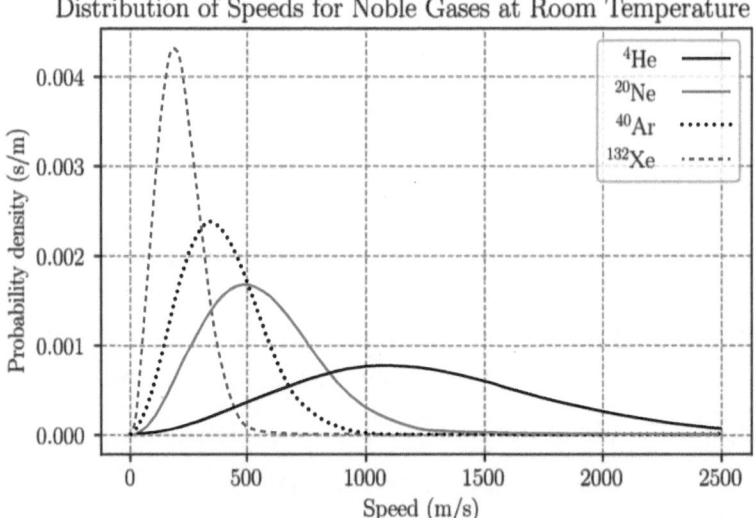

Figure 1.5 Maxwell-Boltzmann distribution functions for the velocity of several monoatomic gases at a temperature of 298.15 K.

However, the lack of knowledge about the exact nature of the velocities limited these theories from providing a complete picture of the system properties. This is the key advance enabled by the Maxwell-Boltzmann (MB) distribution. The MB distribution gives access to the calculation of a wide range of properties because it provides an equation for the full velocity distribution,

From a mathematical standpoint, the MB distribution is a continuous probability distribution, called a χ distribution, with three degrees of freedom, *i.e.* the velocities in the three directions of space. The MB distribution in 3D is given by

$$f(\mathbf{v}) = f(v_x, v_y, v_z) = f(v_x)f(v_y)f(v_z) \tag{1.47}$$

and since it models all possible velocities of the gas particles, we have the total probability

$$\int_{-\infty}^{\infty} f(\mathbf{v})d\mathbf{v} = 1 \tag{1.48}$$

The distribution is thus said to be normalized.

Equation 1.47 also implies that we need to evaluate what happens along 1D, for instance, along x, extend it to the other two dimensions, and calculate the product of the three to obtain the full equation for the distribution. In 1D, the MB distribution $f(v_x)$ is given by

$$f(v_x) = \left(\frac{m}{2\pi kT}\right)^{1/2} e^{-mv_x^2/2kT} \tag{1.49}$$

This distribution is normalized, and we have

$$\int_{-\infty}^{\infty} f(v_x)dv_x = 1 \tag{1.50}$$

Once we have this distribution, we can now calculate different averages. First, the average velocity $< v_x >$ is given by

$$< v_x > = \int_{-\infty}^{\infty} v_x f(v_x)dv_x = 0 \tag{1.51}$$

This is to be expected since the particles are free and equally likely, to move in opposite directions for any velocity value.

Let us now turn to the average of the squared velocities. To this end, we evaluate the following integral

$$< v_x^2 >= \int_{-\infty}^{\infty} v_x^2 f(v_x) dv_x \qquad (1.52)$$

This yields the following equation

$$< v_x^2 >= \left(\frac{m}{2\pi kT}\right)^{1/2} \int_{-\infty}^{\infty} e^{-mv_x^2/2kT} v_x^2 dv_x \qquad (1.53)$$

Calculating the integral, we obtain

$$< v_x^2 >= \left(\frac{m}{2\pi kT}\right)^{1/2} \frac{\pi^{1/2}}{2(m/2kT)^{3/2}} \qquad (1.54)$$

which yields the expected connection between the average of the squared velocities, the temperature, and the mass of the particle

$$< v_x^2 >= \frac{kT}{m} \qquad (1.55)$$

In 3D, following Eq. 1.47, the MB distribution becomes

$$f(v_x, v_y, v_z) = \left(\frac{m}{2\pi kT}\right)^{3/2} \exp\left[-\frac{m}{2kT}(v_x^2 + v_y^2 + v_z^2)\right] \qquad (1.56)$$

To obtain the MB distribution $F(v)$ based on the norm of the velocity vector $v = |\mathbf{v}| = \sqrt{v_x^2 + v_y^2 + v_z^2}$, we switch from cartesian coordinates to spherical coordinates

$$\begin{aligned} v_x &= v \sin\theta \cos\phi \\ v_y &= v \sin d\theta \sin\phi \\ v_z &= v \cos\theta \end{aligned} \qquad (1.57)$$

We differentiate the velocity coordinates to obtain the following product

$$dv_x dv_y dv_z = v^2 dv \sin\theta d\theta d\phi \qquad (1.58)$$

and then integrate over all possible angles

$$F(v)dv = \int_0^\pi d\theta \int_0^{2\pi} d\phi f(v_x, v_y, v_z) \sin\theta v^2 dv \qquad (1.59)$$

This yields the MB distribution $F(v)$

$$F(v) = 4\pi v^2 \left(\frac{m}{2\pi kT}\right)^{3/2} \exp\left(-\frac{mv^2}{2kT}\right) \qquad (1.60)$$

Now that $F(v)$ is known, we have access to all properties for the system. For instance, the average velocity $<v>$ can be obtained through

$$<v> = \int_{-\infty}^{\infty} v F(v) dv \qquad (1.61)$$

$$<v> = 4\pi \left(\frac{m}{2\pi kT}\right)^{3/2} \int_0^{\infty} \exp\left(-\frac{mv^2}{kT}\right) v^3 dv \qquad (1.62)$$

$$<v> = \left(\frac{8kT}{\pi m}\right)^{1/2} = \left(\frac{8RT}{\pi M}\right)^{1/2} \tag{1.63}$$

Similarly, the average squared velocity $<v^2>$ can be calculated as

$$<v^2> = \int_0^\infty v^2 F(v) dv \tag{1.64}$$

yielding

$$<v^2> = \left(\frac{3kT}{m}\right)^{1/2} \tag{1.65}$$

i.e. the result from Eq. 1.55 multiplied by 3, since each particle has 3 degrees of freedom.

Finally, we can also determine the most probable velocity v_{mp} from the MB distribution by differentiating $F(v)$ since $F(v)$ exhibits a maximum for $v = v_{mp}$. We start by calculating

$$\frac{dF(v)}{dv} = \left(\frac{m}{2\pi kT}\right)^{3/2} e^{-mv^2/2kT} \left[8\pi v + 4\pi v^2 \left(-\frac{mv}{kT}\right)\right] \tag{1.66}$$

For $v = v_{mp}$, $\left[\frac{dF(v)}{dv}\right]_{v_{mp}} = 0$ which leads to

$$v_{mp} = \left(\frac{2kT}{m}\right)^{1/2} \tag{1.67}$$

This set of results shows the power of knowing the underlying distribution. All sorts of properties and information can be derived quickly and allow for a complete and rapid system characterization.

1.3.2 PHASE SPACE AND PROBABILITY DISTRIBUTION

The MB distribution focuses on the velocities of the particles of the system. However, when we integrate the equations of motion of the N particles in a system, we have access to $6N$ quantities, standing for the position coordinates $q_1, q_2, \ldots, q_{3N}$ and the momenta coordinates $p_1, p_2, \ldots, p_{3N}$. These $6N$ quantities completely characterize the mechanical state of the N-particles system. Together with the equations of motion, these $6N$ quantities ultimately determine the future and past evolution of the system.

Let us imagine now an Euclidean space with dimension $6N$, with axes marked by each of these $6N$ quantities $q_1, q_2, \ldots, q_{3N}, p_1, p_2, \ldots, p_{3N}$. This space is called the *phase space* for the system. At any given time t, the state of the system can be exactly defined by a set of values for the $6N$ coordinates $(q_1, q_2, \ldots, q_{3N}, p_1, p_2, \ldots, p_{3N})$, which is called a phase point.

Given a set of initial conditions $(q_1(t=0), q_2(t=0), \ldots, q_{3N}(t=0), p_1(t=0), p_2(t=0), \ldots, p_{3N}(t=0))$, the evolution of the system can be obtained by integrating the $6N$ equations of motion proposed by Hamilton to yield a trajectory for the system, *i.e.* a series of phase space points $(q_1(t), q_2(t), \ldots, q_{3N}(t), p_1(t), p_2(t), \ldots, p_{3N}(t))$ that captures the dynamics of the system with t. Now let us consider a system of N particles within a volume V, and that the system is isolated, or, in other words, that its energy E is constant. If we consider a large number of replicas $\mathcal{N}$ of the system, provide different initial conditions to these replicas, and integrate the equations of motion to obtain the phase space trajectories for each of the replicas, we now have an ensemble of trajectories that look like a cloud of phase space points in phase space that all correspond to systems with the same number of particles N, volume V and energy E. Each of these trajectories is independent of the others since the systems are all isolated and share the same probability or are all equally likely to happen since the systems are replicas of one another with the same N, V, and E. We introduce a number density $f(q, p, t)$ to study this cloud of phase space points. This means that the number of systems that have phase points within the interval $(dq, dp) = (dq_1, \ldots, dq_{3N}, dp_1, \ldots, dp_{3N})$ about the point $(q, p) = (q_1, \ldots, q_{3N}, dp_1, \ldots, dq_{3N})$ at time t is $f(q, p, t) dq dp$.

$$\int, \ldots, \int f(q, p) dp dq = \mathcal{N} \tag{1.68}$$

The average of a property $\omega(q, p)$ over this ensemble of trajectories can be calculated as

$$\bar{\omega} = \frac{1}{\mathcal{N}} \int, \ldots \int \omega(q, p) f(q, p, t) dp dq \tag{1.69}$$

Exactly as the phase space trajectories can be obtained by integrating the equations of motion, we can determine the evolution of the number density $f(q, p, t)$ from the laws of mechanics. It can be shown that the time dependence of $f(q, p, t)$ is given by

$$\frac{\partial f}{\partial t} = -\sum_{j=1}^{3N} \left(\frac{\partial f}{\partial q_j} \dot{q}_j + \frac{\partial f}{\partial p_j} \dot{p}_j \right) \tag{1.70}$$

Equivalently, we can write that

$$\frac{df}{dt} = \frac{\partial f}{\partial t} + \sum_{j=1}^{3N} \left(\frac{\partial f}{\partial q_j} \dot{q}_j + \frac{\partial f}{\partial p_j} \dot{p}_j \right) = 0 \tag{1.71}$$

In other words, the number density of the cloud of phase space points is conserved. This equation is known as the Liouville equation and is the basis of what Gibbs called the principle of conservation of density in phase space. If we plug in Hamilton's equations, we obtain

$$\frac{\partial f}{\partial t} = -\sum_{j=1}^{3N} \left(\frac{\partial H}{\partial p_j} \frac{\partial f}{\partial q_j} - \frac{\partial H}{\partial q_j} \frac{\partial f}{\partial p_j} \right) \tag{1.72}$$

which, in Cartesian coordinates, becomes, for a set of N particles

$$\frac{\partial f}{\partial t} + \sum_{j=1}^{N} \frac{\mathbf{p}_j}{m_j} . \nabla_{\mathbf{r}_j} f + \sum_{j=1}^{N} \mathbf{F}_j . \nabla_{\mathbf{p}_j} f = 0 \tag{1.73}$$

where $\nabla_{\mathbf{r}_j}$ is the gradient with respect to the position variables, $\nabla_{\mathbf{p}_j}$ the gradient with respect to the momentum variables, and $\mathbf{F}_j$ is the total force on the $j$the particle. The Liouville equation, as well as the averaging procedure of Eq. 1.69, provides a path towards the calculation of properties over an ensemble of trajectories and a first way to derive information for a group of systems rather than for a single individual system. More broadly, as we will see later, the Liouville equation is essential for modeling nonequilibrium systems.

1.3.3 MICRO-MACRO CONNECTION

We have seen how the mechanical properties of microscopic particles can be obtained and how distribution probabilities provide a path toward determining averages of mechanical properties over groups of particles and ensembles of systems. How do we connect these averages at the micro-level to the properties we can measure at the macro level?

The first connection we have seen is the relation between temperature and the average kinetic energy. From the MB distribution for a group of N particles with 3 degrees of freedom each (corresponding to motions in the three directions of space), we find the total kinetic energy to be

$$\sum_{i=1}^{N} \frac{1}{2} m < v_i^2 > = \frac{3}{2} N k_B T \tag{1.74}$$

A second micro-macro connection can be made for pressure via the kinetic theory of gases. Indeed, we start by making a few assumptions to define a simple model for the particles in a gas. We hypothesize that the particles are point particles that do not interact with one another, go through elastic collisions, and follow the MB distribution. The particles are enclosed within walls, and we can determine pressure from the collisions undergone by the particles with the confining walls and the change of momentum that occurs when particles bounce back on walls. Specifically, if the particle hits the wall with velocity v_x along, *e.g.* the x axis, it will bounce back with a velocity $-v_x$ (elastic collision). The change in momentum is equal to: $\Delta \mathbf{p} = \mathbf{p}_f - \mathbf{p}_i = -2mv_x$. Pressure can be calculated from this change in momentum. Indeed, the general definition for the pressure P

$$P = \frac{|\mathbf{F}|}{A} \tag{1.75}$$

where $|\mathbf{F}|$ is the norm of the force exerted here upon the wall and A is the area over which the force is exerted, here, the area of the wall (see Figure 1.6).

Furthermore, $|\mathbf{F}|$ can be obtained from the change in momentum as

$$\mathbf{F} = -\frac{2mv_x}{\Delta t} \tag{1.76}$$

where Δt is the period over which the change in momentum takes place. To obtain Δt, we evaluate the distance traveled by the atom of $2L$ on its way to the wall, and then back after bouncing off the wall at the velocity v_x. This gives

$$v_x = 2L/\Delta t \tag{1.77}$$

Since the volume V is given by $V = AL$, we obtain $P = mv_x^2/V$ along a single dimension. Generalizing this result to $3D$ and N particles, we obtain

$$P = \frac{Nm\mathbf{v}^2}{3V} \tag{1.78}$$

which, after using the MB result for the average squared velocity ($|\mathbf{v}|^2 = 3kT/m$), gives the well-known ideal gas law $PV = Nk_BT$.

Several other micro-macro connections can be made for gas properties through the kinetic theory of gases, including, *e.g.*, calculating transport coefficients such as diffusion coefficients or viscosity. However, for condensed phases or particles whose characteristics depart from the assumptions made above, we need to resort to another set of theories and methods that account more accurately for the particles' properties. This is the motivation for the algorithms, numerical methods, computer simulations, artificial intelligence approaches, and machine learning models we discuss in the following chapters.

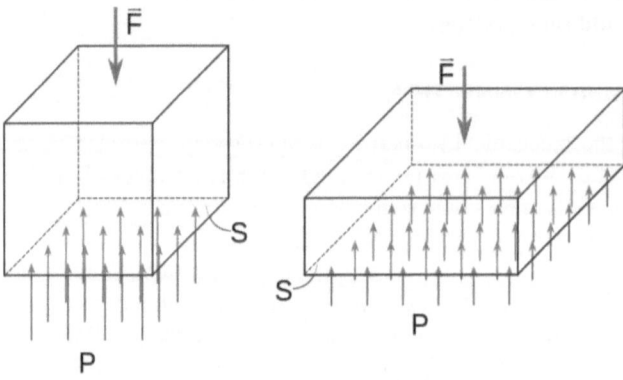

Figure 1.6 Pressure P arises when a force $\mathbf{F}$ is exerted on a surface S.

2 Communication Rules in Molecular Systems

> "Moreover, just as the quark is a symbol of the physical laws that, once discovered, come into full view before the mind's analytical eye, so the jaguar is, for me at least, a possible metaphor for the elusive complex adaptative system."
>
> Murray Gell-Mann,
> *The Quark and the Jaguar* [156]

How do individuals sense their environment, be it their surroundings or, *e.g.*, the motion of other individuals? How do they respond to it? Answering these two questions undoubtedly goes a long way toward having a model that accounts for the behavior of individuals and the collective properties and features of the group. In a social group, "sensing" what others think or intend to do, *i.e.*, gauging moods, opinions, or intentions, routinely occurs through a wide range of "communication" processes. Human beings generally rely on their senses and human interactions. For instance, opinions are assessed through conversations, acts are measured through observations, and intents are estimated through an ensemble of conversations and observations. Since they are deprived of our senses, how do molecular systems proceed to "sense", communicate, and organize as a group? This is the topic of this chapter, with the characterization of the interactions between molecules and between molecules and their environment.

A word of caution before we proceed further. We have made tremendous progress in understanding and modeling molecular interactions over the past century. However, as we constantly look to expand the range of systems we fully understand, we come across new challenges. Perhaps, as in Gell-Mann's metaphor, we are facing a sort of a continuum between the world of objects and phenomena thought to be fully understood - the quark metaphor - and the realm of beings and events for which we have yet to discover a widely accepted model - the jaguar metaphor. Molecular systems are much closer to the quark than the jaguar in terms of complexity. However, does this mean we have a complete picture of every process occurring at the molecular scale? As we will see in this chapter, a full understanding of many microscopic processes still eludes us. How systems consider environmental cues and what can be broadly defined as "sensing" are still very active research areas.

2.1 COMMUNICATION AND INTERACTIONS

2.1.1 INTERACTIONS IN A QUANTUM WORLD

Molecules, and all matter for that purpose, are made of atoms. As a result, in a first approximation, these can be modeled as composed of two types of charged sub-atomic particles, *e.g.*, nuclei, and electrons. To model such systems, we can resort to quantum mechanics and invoke the Schrödinger equation introduced in Section 1.1.3. This time, however, the Hamiltonian operator needs to be written for a system with a total of, say, N_n nuclei and N_e electrons, with N_n being the sum of all nuclei from all atoms present in the system and N_e corresponding to the sum of all electrons from all atoms. The Hamiltonian becomes for such a system

$$\mathscr{H} = -\sum_{i=1}^{N_n} \frac{\hbar^2}{2M} \nabla_i^2 - \sum_{i=1}^{N_e} \frac{\hbar^2}{2m} \nabla_i^2 + U(q_{i_n}, q_{i_e}) \qquad (2.1)$$

DOI: 10.1201/9781003006411-3

in which M denotes the mass of a nucleus, m is the mass of an electron, and q_{i_n} and q_{i_e} denote the position of nucleus i_n and electron i_e. The Hamiltonian operator for this multi-particle system involves a much greater number of coordinates and momenta. As a result, the Schrödinger equation has become much more complex to solve analytically than the example of the electron in a box provided in Section 1.1.3. This is only partly due to the increase in the number of variables. An additional challenge lies in the form of the potential energy operator, and is known as "electronic correlation". Indeed, a closer inspection of what $U(q_{i_n}, q_{i_e})$ entails reveals that

$$U(q_{i_n}, q_{i_e}) = \sum_{a=1}^{N_n} \sum_{b \neq i} \frac{Z^2 e^2}{R_{ab}} - \sum_{a=1}^{N_n} \sum_{i=1}^{N_e} \frac{Ze^2}{R_{ai}} + \sum_{i=1}^{N_e} \sum_{j \neq i} \frac{e^2}{r_{ij}} \tag{2.2}$$

where running indices a and b refer to nuclei, running indices i and j correspond to electrons, Ze is the positive charge carried by each of the nuclei, $-e$ is the negative charge carried by each of the electrons, R_{ab} denotes an inter-nuclei distance, R_{ai} a distance between a nucleus and an electron, and r_{ij} a distance between two electrons. The $1/R$ and $1/r$ terms each involve two different particles, which means that the differential equations for these two particles are coupled and that the motions of the two particles are interdependent or correlated. This makes solving these equations much more challenging.

To address this challenge, several approximations have been introduced. One of these approximations is known as the Born–Oppenheimer approximation. It stems from the large difference between the mass of electrons and the mass of the particles, protons, and neutrons, that make up the nuclei. Indeed, the mass of a proton is three orders of magnitude larger than that of an electron. In turn, nuclei move extremely slowly compared to the much lighter electrons. They can be assumed to remain immobile when solving the Schrödinger equation for the electrons of the system. This approximation, which is also at the heart of the Franck–Condon spectroscopy principle, simplifies the problem considerably. Indeed, in the equation for $U(q_{i_n}, q_{i_e})$, the first term $\sum_{a=1}^{N_n} \sum_{b \neq i} \frac{Z^2 e^2}{R_{ab}}$ becomes a constant, whereas the second term $-\sum_{a=1}^{N_n} \sum_{i=1}^{N_e} \frac{Ze^2}{R_{ai}}$ effectively becomes decorrelated since R_{ai} only depends on the position of electron i, now that nucleus a stays immobile. The challenge presented by the last term, $\sum_{i=1}^{N_e} \sum_{j \neq i} \frac{e^2}{r_{ij}}$, still remains. Indeed, this term involves the relative position of two electrons, which have the same mass and move *a priori* at similar speeds, and whose motions are correlated. It requires a much more detailed mathematical treatment, which is outside this book's scope. Several approaches have been developed to tackle this issue, such as, for instance, perturbation methods in *ab initio* calculations and quantum density functional theory (DFT) methods.

There are two important conclusions we can draw from this discussion. First, putting a theory to the test and modeling a system requires several pragmatic decisions. This is the case with the first approximation we have come across, the Born-Oppenheimer approximation, to reduce the complexity of the equations and enable us to solve them. Second, this example emphasizes the need to employ a model whose level of detail is aligned with the predicted property type. This is the reasoning behind the Franck-Condon principle (see Figure 2.1), which assumes that internuclear distances will not vary during electronic transitions and that no additional insight will be gained by considering the infinitesimal displacement of these nuclei. Will solving the full Schrödinger equation, rather than applying other approximations and, for instance, solving the equations of classical mechanics, always be warranted? In the following sections, we will see that it can be advantageous to resort to approximate methods if the process under study does not involve electronic transitions or extensive bond-breaking and bond-forming across the system.

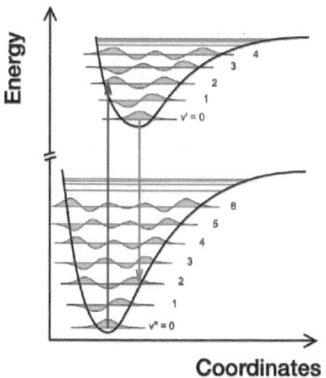

Coordinates

Figure 2.1 Franck-Condon principle: electronic transitions occur along a vertical line without changes in internuclear distance.

2.1.2 COARSE-GRAINING: TIGHT-BINDING

We have seen a first approximation arising from the difference in masses and velocities between two types of quantum, sub-atomic particles. Do we need to solve the full Schrödinger equation for all electrons in all cases? For instance, is it necessary to consider all electrons when a bond forms between two atoms? Indeed, the commonly used picture for bond formation involves the sharing of two electrons. In addition, these two electrons belong to the valence shell, which contains the highest energy electrons of the atom or, in other words, among the last to be listed when building the atom's electron configuration. The core electrons with lower energy, such as, for instance, the $1s$ electrons in a carbon atom, have little impact when forming carbon-carbon or carbon-hydrogen bonds. This has led scientists to lump together the overall contribution from these core electrons in an "effective" way and only consider the valence electrons explicitly when solving the Schrödinger equation. One such example is the tight-binding approach developed in the 1970s.

Tight-binding schemes [49,60,178,294,321] split the energy into an electronic part and a phenomenological, effective, short-ranged repulsive part, which accounts for the fact that particles cannot overlap. The electronic part is obtained by solving the Schrödinger equation and by summing the single-electron energy eigenvalues of this equation. The repulsive energy groups in a single term the repulsion between the atomic core electrons and nuclei. The overall potential energy U is given by

$$U = U_{\text{TB}} + U_R$$
$$= \sum_n 2 \langle \Psi_n | H_{\text{TB}} | \Psi_n \rangle + U_R \tag{2.3}$$

In Eq. 2.3, the repulsive energy U_R is a function of the position of atoms only and is assumed to be independent of their electronic states. U_{TB} is the electronic energy obtained from the lowest eigenvalues of the tight-binding Hamiltonian H_{TB}. In the latter, the factor of two accounts for the two possible spin states that exist for each energy level, spins up and down. The matrix elements for the Hamiltonian H_{TB} are obtained from the Slater–Koster formalism [335]. In this formalism, we assume that bonding results from a coupling between pairs of atoms that are close to one another, giving rise to what is known as the two-center approximation. This is from this approximation that the "tight-binding" name stems since, according to this model, electrons are tightly tethered to the atom they originate from or, in other words, are localized close to the nucleus. If we now consider the example of a silicon atom [83,94], there are four valence atomic orbitals (s, p_x, p_y, p_z) on the $n = 3$ shell. These four valence orbitals constitute a basis set for each atom. A single silicon atom is thus represented by a 4×4 matrix with the following form

$$\begin{pmatrix} \varepsilon_s & 0 & 0 & 0 \\ 0 & \varepsilon_p & 0 & 0 \\ 0 & 0 & \varepsilon_p & 0 \\ 0 & 0 & 0 & \varepsilon_p \end{pmatrix} \tag{2.4}$$

in which ε_s and ε_p correspond to the energies for the s and p orbitals.

This silicon atom's orbitals can overlap with another atom's orbitals, provided that this second atom is close by. This second atom also has four valence orbitals and is associated with a 4×4 matrix with the same form as in Eq. 2.4. This means that the overall H_{TB} matrix will now have an 8×8 dimension and can be described as composed of 4 (4×4) blocks. There are two diagonal blocks, which contain the information about the orbital energies for the two atoms of the system and have the form given in Eq. 2.4, and two off-diagonal blocks that contain the information about the overlap between the orbitals of the two atoms. These off-diagonal blocks contain hopping matrix elements, calculated from the distance-dependent tight-binding overlaps $h_\alpha(r_{ij})$ between the two atoms labeled with the indices i and j. Here α denotes the type of overlap that can occur either between two s orbitals (one on each atom), a s from one atom with a p from the other atom, two p orbitals (one on each atom) along the silicon-silicon bond axis (σ symmetry) or two p orbitals (one on each atom) perpendicularly to the silicon-silicon bond (π symmetry). In "tight-binding" notations, α takes the form of either $ss\sigma$, $sp\sigma$, $pp\sigma$ and $pp\pi$). The off-diagonal blocks for the pair of atoms are given by

$$\begin{pmatrix} h^f_{ss\sigma} & d_x h^f_{sp\sigma} & d_y h^f_{sp\sigma} & d_z h^f_{sp\sigma} \\ -d_x h^f_{sp\sigma} & d_x^2 h^f_{pp\sigma} + (1-d_x^2)h^f_{pp\pi} & d_x d_y (h^f_{sp\sigma} - h^f_{sp\pi}) & d_x d_z (h^f_{sp\sigma} - h^f_{sp\pi}) \\ -d_y h^f_{sp\sigma} & d_y d_x (h^f_{sp\sigma} - h^f_{sp\pi}) & d_y^2 h^f_{pp\sigma} + (1-d_y^2)h^f_{pp\pi} & d_y d_z (h^f_{sp\sigma} - h^f_{sp\pi}) \\ -d_z h^f_{sp\sigma} & d_z d_x (h^f_{sp\sigma} - h^f_{sp\pi}) & d_z d_y (h^f_{sp\sigma} - h^f_{sp\pi}) & d_z^2 h^f_{pp\sigma} + (1-d_z^2)h^f_{pp\pi} \end{pmatrix} \tag{2.5}$$

where $h^f_{ss\sigma}$, $h^f_{sp\sigma}$, $h^f_{pp\sigma}$ and $h^f_{pp\sigma}$ are the hopping functions. These hopping functions are a function of the distance between the two atoms i and j. They can take the form of short-range scaling functions [167,214] as

$$h_\alpha(r_{ij}) = h_\alpha(r_0) \times \left(\frac{r_0}{r_{ij}} \right)^n \times \exp\left[n \left(-\left(\frac{r_{ij}}{r_{c\alpha}} \right)^{n_{c\alpha}} + \left(\frac{r_0}{r_{c\alpha}} \right)^{n_{c\alpha}} \right) \right] \tag{2.6}$$

where r_0, $r_{c\alpha}$, $n_{c\alpha}$ and n are potential parameters [214]. The repulsive energy is calculated as a sum of a functional of a repulsive pair potential $\phi(r_{ij})$

$$U_R = \sum_i f \left[\sum_j \phi(r_{ij}) \right]$$
$$f(x) = C_1 x + C_2 x^2 + C_3 x^3 + C_4 x^4 \tag{2.7}$$
$$\phi(r_{ij}) = \left(\frac{r_0}{r_{ij}} \right)^m \times \exp\left[m \left(-\left(\frac{r_{ij}}{d_c} \right)^{m_c} + \left(\frac{r_0}{d_c} \right)^{m_c} \right) \right]$$

where C_n ($n = 1, 2, 3$ or 4), r_0, d_c, m and m_c are also potential parameters [214]. The energy of this system of two silicon atoms can now be calculated by adding the tight-binding and repulsive energies. The tight-binding energy is obtained by diagonalizing the 8×8 tight-binding matrix and taking the four lowest eigenvalues. This is because each silicon atom has four valence electrons, leading to eight electrons that can be accommodated by the four orbitals associated with these four lowest energies or, equivalently, the four lowest eigenvalues. Generalizing this approach to a system of N silicon atoms, we obtain a $4N \times 4N$ matrix with the same form for the diagonal and off-diagonal blocks. The system's energy can then be calculated by diagonalizing the matrix and adding the repulsive energy for the entire system. The tight-binding approximation has been highly efficient for the modeling of carbon- or silicon-based systems and has led to a deeper understanding of carbon nanostructures and of semi-conducting devices such as computer chips, which often rely on silicon,

2.1.3 FURTHER COARSE-GRAINING: A CLASSICAL WORLD

What happens now if, during our study of a system, bonds between atoms are neither formed nor broken? Do we need to solve the Schrödinger equation for the system? Alternatively, can we take further the "effective" approach we have started to examine in the previous section? Let us consider, for instance, a system containing rare gas atoms, such as argon atoms. The electron configuration for these atoms takes the following form for the valence shell $(ns)^2(np)^6$, which means that all valence shell orbitals are fully occupied. This, however, does not provide any means for two neighboring argon atoms to share electrons and form a bond and accounts for the virtually non-existent reactivity of noble gases. This means that, when accounting for the properties of a system of argon atoms, the energy can be calculated from a simple, functional form of the distance between atoms, and solving the Schrödinger equation does not bring any additional insight. This is the basis for an effective and classical approach to determining energy in atomic and molecular systems. Indeed, in the 1920s, the British physicist Lennard-Jones proposed to evaluate the interaction between atoms as the sum of two inverse power laws of the distance between the atoms [223]. As in the previous section, the first of the two power laws is negative and accounts for attractive interactions. In contrast, the second is positive and is linked to short-range repulsive interactions. The Lennard-Jones potential is given by

$$\phi(r_{ij}) = 4\varepsilon_{ij}\left[\left(\frac{\sigma_{ij}}{r_{ij}}\right)^{12} - \left(\frac{\sigma_{ij}}{r_{ij}}\right)^{6}\right] \tag{2.8}$$

where r_{ij} is the distance between two atoms i and j, σ_{ij} is the exclusion diameter for the pair of atoms i and j, and ε_{ij} is the depth of the interaction well for the pair of atoms. The σ and ε parameters depend on the element the potential models. For instance, a suitable parameter set for an argon atom is $\sigma = 3.4$ Angstroms and $\varepsilon/k_B = 119.8$ K.

We now examine why σ is the exclusion parameter and why ε is termed the well depth. For this purpose, we start by calculating the distance for which the interaction potential takes a value of 0. Solving for $\phi(r_{ij} = 0$ leads to

$$\left[\left(\frac{\sigma_{ij}}{r_{ij}}\right)^{12} - \left(\frac{\sigma_{ij}}{r_{ij}}\right)^{6}\right]_{\phi=0} = 0 \tag{2.9}$$

and thus

$$[r_{ij}]_{\phi=0} = \sigma_{ij} \tag{2.10}$$

This shows that the parameter σ corresponds to the distance for which the attractive and repulsive interactions cancel out. In other words, for any distance shorter than σ, the potential energy will be positive, and repulsion will dominate. This gives σ its name of exclusion diameter. On the other hand, for any distance greater than σ, the potential energy will become negative, and the two atoms will attract one another. If we now look for the extremum of the Lennard-Jones function, we need to differentiate Eq. 2.8. This gives

$$\frac{d\phi(r_{ij})}{dr_{ij}} = 4\varepsilon_{ij}\left[-\frac{12\sigma_{ij}^{12}}{r_{ij}^{13}} + \frac{6\sigma_{ij}^{6}}{r_{ij}^{7}}\right] = 0 \tag{2.11}$$

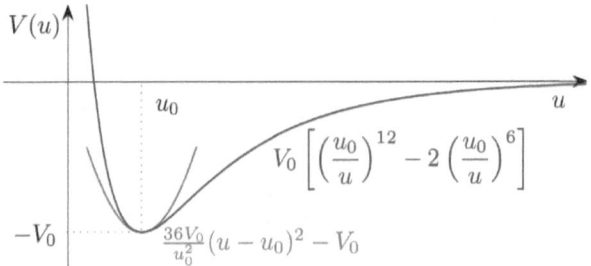

Figure 2.2 Example of a Lennard-Jones potential $V(u)$ as a function of the distance u with the depth of the interaction well noted as V_0 and the exclusion parameter noted as u_0. The parabolic curve shows the results of a harmonic approximation near the minimum.

which leads to

$$[\phi(r_{ij})]_{\left(\frac{d\phi}{dr}=0\right)} = 2^{1/6}\sigma_{ij} \tag{2.12}$$

For this distance, the potential takes its minimum value of

$$\phi(r_{ij})_{[r_{ij}=2^{1/6}\sigma_{ij}]} = -\varepsilon_{ij} \tag{2.13}$$

This shows why ε_{ij} is the depth of the interaction well for the Lennard-Jones potential. The Lennard-Jones potential is an example of a coarse-grained model for interatomic interactions (see Figure 2.2). The σ and ε parameters are atom-dependent and take, for instance, the value of $\sigma = 3.4$ Angstroms and $\varepsilon/k_B = 120$ K for Argon, which lead to an excellent agreement with the experimental data for the thermodynamic properties of Argon. Very interestingly, the $1/r^6$ dependence of the attractive part of the potential reflects the range of interactions commonly termed as *van der Waals interactions* present in any molecular system.

2.2 INTERACTIONS BETWEEN MOLECULES

2.2.1 MOLECULAR PROPERTIES AND INTERACTIONS

We have just seen the connection between a coarse-grained classical potential, like the Lennard-Jones potential, and interactions, like the van der Waals interactions, that govern the properties of a molecular system and its response to external changes. This opens the door to introducing additional coarse-grained potential functions for specific, molecule-dependent interactions. What happens if, for instance, a molecule bears an electric dipole? This is the case of many molecules, such as water, in which the imbalance between the intrinsic properties of O and H atoms, such as their difference in electronegativity, leads to the existence of a permanent dipole moment **M** (see Figure 2.3). Upon applying an electric field **E**, water molecules reorient such that their dipole moments align with the electric field to minimize the potential energy $E_f = -\mathbf{M}.\mathbf{E}$. We need to introduce another coarse-grained model that reflects the anisotropy of the H_2O molecules to model this behavior. A popular solution, one of many possibilities, is to use a distribution of point charges, often placed on the molecule's atoms, to account for this molecular property. In the SPC/E model for water [20], each water molecule is modeled with a single Lennard-Jones "site" (placed on the oxygen atom) and three point charges (one on each atom too). Two water molecules a and b will give rise to the following interaction energy

$$U_{ab} = 4\varepsilon\left[\left(\frac{\sigma}{r_{O_aO_b}}\right)^{12} - \left(\frac{\sigma}{r_{O_aO_b}}\right)^6\right] + \sum_{i=1}^{3}\sum_{j=1}^{3}\left[\frac{q_iq_j}{4\pi\varepsilon_0 r_{ij}}\right] \tag{2.14}$$

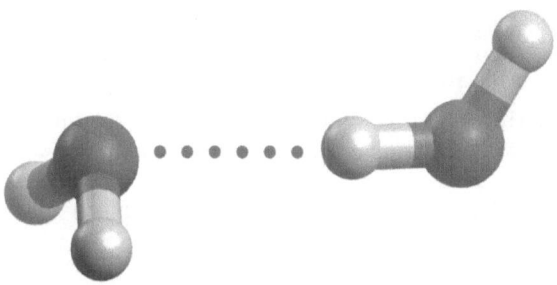

Figure 2.3 Dimer of H_2O molecules with, for each molecule, the O atom shown as a red sphere and the H atoms as white spheres. The distribution of point charges (negative on O and positive on H) allows for the onset of a dipole moment and hydrogen bonding, shown as a dashed line.

in which i denotes an atom of molecule a and j an atom of molecule b. As we choose the Lennard-Jones parameters to reproduce the thermodynamic properties of a given molecule, the model uses a set of charges that mimics an effective dipole moment for a water molecule in the liquid phase. In practice, this effective dipole has a larger magnitude than the molecule's dipole in the gas phase. This is due to a property known as polarizability. If we focus on a water molecule in the gas phase, the water molecule bears a dipole moment of 1.85 D (Debye). Now let us immerse this molecule in a condensed phase of water, either a liquid or a solid phase. The water molecules surrounding the molecule we focus on exert an electric field on this molecule. This, in turn, gives rise to an induced dipole moment that adds to the 1.85 D, leading to an effective value of about 2.1 D.

Of course, choosing an *ad hoc* value for this effective dipole moment is equivalent to making assumptions about the environment surrounding the water molecule. It is assumed that it is immersed in liquid water and its environment will not change. What happens if the temperature increases, and liquid water becomes a vapor? The field created by the surrounding water molecules becomes much weaker, and, as such, the effective dipole moment specified by the SPC/E model overestimates the actual induced dipole on the water molecules. What happens now if ions are added to liquid water? In this case, the field exerted by the environment on the water molecule is stronger, and the SPC/E effective dipole underestimates the induced dipole. Adapting the value of the point charges to the environment in which water is immersed is then required, *e.g.*, by explicitly including the effect of polarizability. This can be performed by setting the value of the point charges to yield the gas phase dipole moment, adding a polarizable site on the molecule with a polarizability α, and calculating the field exerted by the surroundings $\mathbf{E}$. This gives an additional potential energy per polarizable site of

$$U = \frac{1}{2}\alpha\mathbf{E}^2 \tag{2.15}$$

This example illustrates the power of coarse-graining, as it considerably simplifies the problem at hand and the need to examine carefully the assumptions made upon coarse-graining. Coarse-graining generally assumes a fixed setting and environment. This, in turn, limits the ability of the model to adapt to any change in the environment.

2.2.2 TWO-BODY VS. MANY-BODY POTENTIALS

Effective models can also be designed to account for complex spatial correlations between atoms. These correlations depend not only on distances between atoms but also on the angles between vectors joining neighboring atoms. Such a situation can occur even in the absence of chemical bonds such as, for instance, in the case of semi-conducting materials like Silicon. In Silicon, there exists a high degree of tetrahedral ordering that leads to the formation, in the solid phase, of a solid phase with a diamond-like structure. Effective models require that, in addition to distances between silicon atoms, the angles between vectors joining neighboring silicon atoms be considered in the model definition. Doing so introduces an effective three-body interaction accompanying the two-body, or pair, interaction. Two examples of such models are the Stillinger–Weber [344] and Tersoff [358] potentials. The Stillinger-Weber potential U_{SW} is the sum of a two-body term and a three-body term u_3. The two-body term between two atoms i and j is given by

$$u_2(r_{ij}) = A\varepsilon(B(r_{ij}/\sigma)^{-p} - (r_{ij}/\sigma)^{-q})\exp\left[((r_{ij}/\sigma - a)^{-1}\right], (r_{ij}/\sigma) < a$$
$$= 0, (r_{ij}/\sigma) \geq a \tag{2.16}$$

where ε, σ, A, B, p and a are potential parameters [344]. The three-body term between three atoms i, j and k is given by

$$u_3(\mathbf{r}_i, \mathbf{r}_j, \mathbf{r}_k) = \varepsilon\left[h(r_{ij}, r_{ik}, \theta_{jik}) + h(r_{ji}, r_{jk}, \theta_{ijk}) + h(r_{ki}, r_{kj}, \theta_{ikj})\right] \tag{2.17}$$

where the h function is defined for $r < a$ as e.g. in the case of $h(r_{ij}, r_{ik}, \theta_{jik})$

$$h(r_{ij}, r_{ik}, \theta_{jik}) = \lambda \exp\left[\gamma(r_{ij}/\sigma - a)^{-1} + \gamma(r_{ik}/\sigma - a)^{-1}\right] \times \left(\cos\theta_{jik} + 1/3\right)^2 \tag{2.18}$$

where θ_{jik} denotes the angle between vectors $\mathbf{r}_{ij}$ and $\mathbf{r}_{ik}$, subtended by vertex i, and where λ and γ are potential parameters [344]. The Tersoff potential [358] is based on a bond-order potential description of the interactions. In this model, the interactions between two atoms i and j are given by

$$V(r_{ij}) = f_c(r_{ij})\left[A\exp(-\lambda r_{ij}) - B\exp(-\mu r_{ij})b_{ij}\right]$$
$$f_c(r_{ij}) = \frac{1}{2}\left[1 + \cos\left(\frac{r_{ij} - R}{S - R}\right)\right] \tag{2.19}$$

where b_{ij} is the bond order parameter, which is a many-body term that depends on the strength of the interaction between atoms i and j, $f_c(r_{ij})$ is a cutoff function, and A, B, λ, μ, S and R are potential parameters. The bond order parameter directly depends on the bond geometry according to

$$b_{ij} = \left(1 + \beta^n \zeta_{ij}^n\right)^{-1/2n}$$
$$\zeta_{ij} = \sum_{k \neq i,j} f_c(r_{ik})g(\theta_{ijk})$$
$$g(\theta_{ijk}) = 1 + \frac{c^2}{d^2} - \frac{c^2}{\left[d^2 + \left(h - \cos\theta_{ijk}\right)\right]} \tag{2.20}$$

where β, c, d and h are potential parameters and θ_{jik} denotes the angle between vectors $\mathbf{r}_{ij}$ and $\mathbf{r}_{ik}$.

Another example of a many-body model is the embedded-atom potentials (EAM) [68,138,256, 346] approach that has been developed for metallic systems. The idea here is that the interactions, and thus the properties, are functions of the local density in the system. In recent years, the quantum-corrected Sutton-Chen [231] embedded atoms (qSC-EAM) potential has emerged as a very effective way to model interactions in single-component, as well as multi-component, metallic systems. In this approach, the potential energy U of a system containing N atoms is defined as the sum of a two-body contribution and a many-body contribution

$$U = \frac{1}{2}\sum_{i=1}^{N}\sum_{j \neq i}^{} \varepsilon\left(\frac{a}{r_{ij}}\right)^n - \varepsilon C\sum_{i=1}^{N}\sqrt{\rho_i} \tag{2.21}$$

in which r_{ij} is the distance between two atoms i and j and the density term ρ_i is given by

$$\rho_i = \sum_{j \neq i}\left(\frac{a}{r_{ij}}\right)^m \tag{2.22}$$

2.2.3 TOWARD MACRO- AND BIO-MOLECULES

The previous sections have focused on strategies to provide effective models for the interactions in atomic systems. How do these strategies evolve for molecules, polymers (see Figure 2.4) or macromolecules, and biological systems, such as membranes and proteins? In such cases, chemical bonds join atoms to one another. As we have seen previously, the energy of these compounds can be calculated by solving the Schrödinger equation. However, if chemical bonds are neither formed nor broken during the system's evolution, the system can be modeled with an effective, purely classical model. For instance, intramolecular vibrations, such as stretching or bending modes, can be accounted for by harmonic springs. For the stretching mode for a bond between atoms a and b, the corresponding potential energy is given by

$$U_{\text{stretch}}(r) = \frac{1}{2}k_s(r - r^o)^2 \tag{2.23}$$

where k_s is a spring constant, r the distance between a and b and r^o the equilibrium bond length. Similarly, for bending, $i.e.$, the vibration of the angle θ between the bond connecting the atoms a and b, and the bond connecting the atoms b and c, the potential energy can be written as

$$U_{\text{bend}}(\theta) = \frac{1}{2}k_b(\theta - \theta^o)^2 \tag{2.24}$$

in which k_b is a spring constant and θ^o the equilibrium bond angle. A third type of intramolecular vibration, known as torsion, also occurs in molecular systems between three consecutive chemical bonds $a - b - c - d$. Torsion corresponds to out-of-plane deformation of the molecules or, in other words, changes in the molecular conformation. Torsion potentials generally involve a Fourier series that is a function of the so-called torsion, or dihedral, angle, $i.e.$ the angle ϕ between the plane defined by atoms (a, b, c) and the plane defined by atoms (b, c, d). The torsion potential energy can be calculated as follows

$$U_{\text{tors}}(\phi) = \sum_n V_n \cos^n \phi \tag{2.25}$$

Different mathematical functions and series can be substituted to those listed in Eqs. 2.23–2.25. Furthermore, potential energy functions can be added to achieve specific objectives that depend on the nature of the molecule to be modeled. This is the case of functions that maintain the planarity

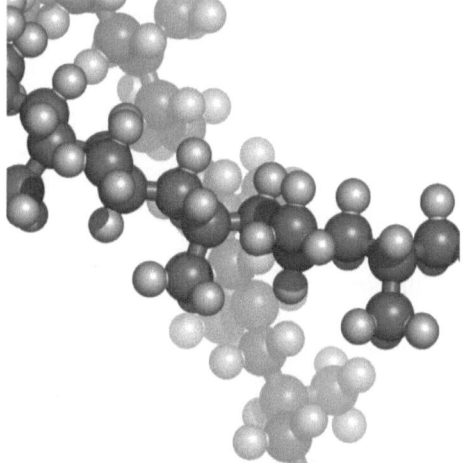

Figure 2.4 Molecular structure of a polymer with the example of syndiotactic polystyrene. The flexibility of the polymer chain is accounted for by the stretching, bending, and torsion potentials discussed in the text.

of a molecule, *e.g.*, for an aromatic compound can be used and are termed as improper torsion potentials.

Adding all the contributions we have seen so far provides an effective force field with the versatility to model a wide range of systems. This is the approach taken by commonly used force fields for biological systems. For instance, the CHARMM (Chemistry at HARvard Macromolecular Mechanics) force field [36], developed in Martin Karplus' group, models many biological systems, including peptides, proteins, lipids, and carbohydrates. The overall CHARMM potential energy can be summarized as

$$U_{\text{CHARMM}} = \sum_{\text{bonds}} k_b (b - b_o)^2 + \sum_{\text{angles}} k_\theta (\theta - \theta_0)^2 + \sum_{\text{Urey-Bradley}} k_{UB} (S - S^o)$$
$$+ \sum_{\text{dihedrals}} k_\chi (1 + \cos(n\chi - \delta)) + \sum_{\text{impropers}} k_i (\psi - \psi_o)$$
$$+ \sum_{\text{nonbonded}} \varepsilon_{ij} \left[\left(\frac{R_{\min,ij}}{r_{ij}} \right)^{12} - 2 \left(\frac{R_{\min,ij}}{r_{ij}} \right)^6 \right] + \frac{q_i q_j}{r_{ij}} \tag{2.26}$$

in which k_b, k_θ, k_{UB}, k_χ and k_i denote the bond, bending, Urey-Bradley, torsion, and improper torsion force constants, b, θ, S, χ and ψ are the bond length, bond angle, Urey-Bradley 1,3-distance, dihedral angle, improper dihedral angle, and the superscript ^{o} denote the equilibrium value. In the Fourier series for the torsion potential, n is the multiplicity, and δ is the phase angle. Here, slightly modified potential energy functions have been used to account for the various types of intramolecular and intermolecular (last two terms: Lennard-Jones and Coulombic potential terms) contributions to the overall potential energy.

More recently, effective force fields have even been able to account for the effects of chemical reactions through the definition of reactive force fields [374], in which functional forms model the bond breaking and forming. This means we can model highly complex biological systems with effective force fields. If, for instance, we sequence and break down the atomic formula of each system, including, *e.g.*, cells and their surroundings, we could, in principle, model any cellular process. However, this also means considering many atoms and calculating their evolution over extremely long periods. Indeed, cells have dimensions of the order of 1 to 100 $1 - 100 \, \mu m$ ($1 \times 10^{-6} m - 1 \times 10^{-4} m$), which far exceeds the dimension of an atom, which are of the order of a few ($10^{-10} m$). In comparison, biological processes can take days (about $10^5 s$), which also considerably exceeds the characteristic times of atomic processes ($10^{-15} s$). This calls for the understanding of the various types of communication processes that arise in such systems and how we can effectively model them.

2.3 BEYOND INTERACTIONS

2.3.1 SIGNALING

We have seen how correlations and communication occur at the atomic level through interactions. If we are now to consider more extensive, complex systems, such as those found in biology, we need to have a broader picture of the communication processes that take place on this larger scale. Biological cells communicate with one another, with their surroundings, and communication even within cells themselves through processes often called signaling (see Figure 2.5). Signaling is a very general term encompassing communication processes in their broadest sense. Indeed, signals from the surroundings can have a physical origin and consist of a change in mechanical pressure or temperature or have a chemical basis and result from a change in concentration in a chemical compound. The latter phenomenon is called chemical signaling. It is the mechanism cells employ for intercellular communication and when the communication occurs within the cell itself or intracellular signaling. Chemical signals are generally molecules or proteins produced and "emitted" toward the target cell, often termed ligands. This classification can be further refined depending on how

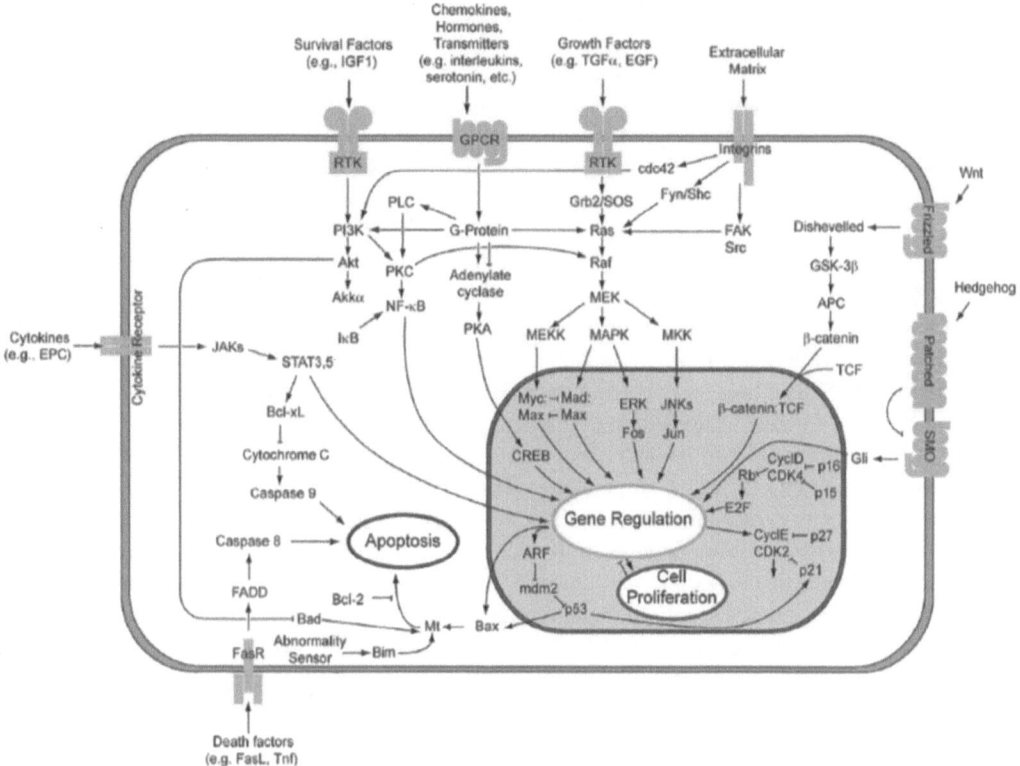

Figure 2.5   The complex processes underlying cell signaling: signal transduction pathways.

far the chemical signal travels, with autocrine signaling occurring when a cell targets itself during the communication process, direct signaling when the signal travels through a junction connecting two neighboring cells, paracrine signaling when the signal travels to a nearby cell and endocrine signaling when the signal travels through the bloodstream to a remote target cell.

For the communication process to occur, target cells must be able to detect signals or, in other words, be equipped with a receptor that can selectively decipher the chemical signal sent to the cell. To be read by the cellular receptor, these ligands need to bind to the protein that serves as a receptor on the target cell. This means that the receptor is often located on the cell's surface to detect an external stimulus. Examples include G-protein coupled receptors or receptors with intrinsic enzymatic activity. Furthermore, the binding process is a particular interaction. Any receptor on any given cell will not be able to read the chemical signal, as it involves the recognition, and subsequent attachment, by the protein receptor of the ligand. Once the ligand-receptor binding step has been completed, there is generally a conformational change in the receptor, leading to the transmission of the signal inside the cell or, in other words, to the cytoplasmic side. This triggers a series of events in the cell as the signal gets transmitted to effector molecules down the signaling pathway by the passage of conformation changes from one protein to the next. For instance, the binding of a ligand to the G protein-coupled receptor results in a conformation change in the receptor that activates a G protein which, in turn, triggers an effector protein that generates another messenger. Conformation changes often occur through chemical reactions known as phosphorylations, catalyzed by enzymes known as kinases, and dephosphorylations, catalyzed by phosphatases. This sequence of events finally triggers an action such as *e.g.*, the activation of gene transcription, or the opening of an ion channel. This sequence of events can be understood as a transduction process, in which the incoming signal is transformed into another, here a chemical signal resulting in a physical process, *i.e.*, the opening of a channel.

2.3.2 PHORESIS AND ACTIVE MATTER

The next step is understanding how biological systems, such as cells and bacteria, respond to an external stimulus. This happens when natural microorganisms move in a fluid; for instance, bacteria move according to nutrient gradient to improve their odds of survival (see Figure 2.6). In such systems, explored by Purcell, the biological systems move in a fluid of high viscosity, which means that inertia is negligible and that a characteristic of these systems, known as their Reynolds number, is very low. The Reynolds number is defined as $R_e = \rho v_0 L_0 / \eta$, in which ρ is the fluid density, η the fluid viscosity, v_0 a characteristic velocity of the moving (swimming) biological object, and L_0 a characteristic dimension for the biological swimmer. As a result of their small size and slow velocities, the Reynolds number of natural swimmers with a size of the order of the μm is well below 1 less than unity, which is well below the Reynolds number for a human swimming in a pool $R_e = 10^4$. Building on Gray, Hancock, Berg, and Anderson's experimental work on biological swimmers, Purcell showed in his "scallop theorem" that propulsion at low R_e results from time-reversal symmetry-breaking swimming moves. This has drawn considerable interest in these systems in recent years. Although, in practice, swimming is a very challenging task at low R_e, the limit of low R_e numbers dramatically simplifies the study and analysis of the motion and behavior of microswimmers. Microswimmers generally employ cilia or flagella for propulsion and periodic changes in their body shape. Various strategies for the design of synthetic swimmers have been proposed. This includes phoresis. Inhomogeneous catalysis of chemical reactions (diffusiophoresis), thermal gradients (thermophoresis), and biology-inspired concepts, such as rotating helices. Such machines provide the basis for multi-functional and highly responsive artificial materials, which exhibit emergent behavior and the ability to perform specific tasks in response to signals from each other and the environment.

A model often used for active matter is the two-dimensional Active Brownian Particles (ABP) model. In this case, an active fluid is composed of disks of radius a and with a packing fraction $\phi = \frac{N \pi a^2}{L^2}$, where L is the simulation cell length. Each colloidal disk is described by the position $\mathbf{r}_i$ of its center and the orientation θ_i of a polar axis $\hat{\mathbf{n}}_i = (\cos \theta_i, \sin \theta_i)$. The interactions between two disks are given by

$$\Phi(r_{ij}) = \begin{cases} \frac{k}{2}(2a - r_{ij})^2 & r_{ij} < 2a \\ 0 & r_{ij} \geq 2a \end{cases}$$

in which r_{ij} denotes the distance between the disks i and j, and the equations of motion are as follows

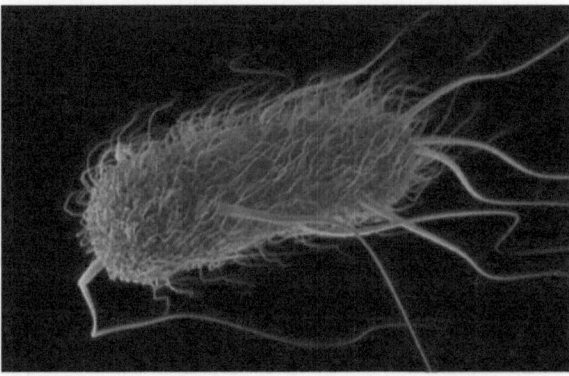

Figure 2.6 Microorganisms, such as the *Escherichia coli* bacteria, have the ability to beat their flagella and move in response to chemical gradients, *i.e.* changes in the local concentration in nutrients.

$$\dot{\mathbf{r}}_i = v\hat{\mathbf{n}}_i + \mu \sum_j \mathbf{F}_{ij} + \sqrt{2D_T}\eta_i^T$$
$$\dot{\theta} = \sqrt{2D_R}\eta_i^R \tag{2.27}$$

where v the single-particle self-propulsion speed, $\mu = \frac{D_T}{k_B T}$ the mobility, D_T and D_R the translational and rotational diffusion constants, which in the low-Reynolds number regime are related through $D_R = \frac{3D_T}{4a^2}$. The η_i are Gaussian white noise with zero mean and correlations $< \eta_i(t)\eta_j(t') > = \delta_{ij}\delta(t-t')$.

Recent experimental work has shown that light-driven active colloidal particles can be designed by grafting oxide catalysts such as hematite or titania on a surfactant droplet and immersing the resulting colloidal particles into a H_2O_2 matrix. Then, the system is subjected to light, thereby inducing the onset of chemical gradients in O_2 and H_2O_2 through the reaction of decomposition of H_2O_2 catalyzed by the oxide. This, in turn, leads to the self-propulsion of the colloidal particles through diffusiophoretic effects. Thus, specifying a spatial pattern for applying light to a system of self-propelled particles provides a unique avenue for realizing and controlling active assemblies.

2.3.3 CHEMOTAXIS

Chemotaxis denotes the motion of an organism in response to a chemical stimulus, such as spatial gradients in the concentration of specific compounds. This is an essential step in the operation of any biological system, as, for instance, bacteria develop survival strategies by swimming toward regions where the concentration of a given nutrient is the highest or swimming away from the areas where the toxin concentration is the highest. Similarly, chemotaxis is crucial for the development of an organism, *e.g.*, through the migration of neurons and for its regular operation, such as during the motion of leukocytes or immune cells in response to an infection. In turn, anomalous cell migration, and the underlying anomalous chemotactic behavior, can lead to diseases, including inflammation and cancer metastasis. Biological active matter routinely uses chemotaxis for its survival and normal function. For example, starving Dictyostelium cells use signaling to survive. These cells produce specific chemicals when starving. This attracts other cells, which exhibit a chemotactic behavior by swimming toward the starving cells. This results in the aggregation of cells, with the newly formed multicellular aggregate able to sustain long starvation periods.

Focusing now on the case of synthetic microswimmers, their propulsion mechanisms involve the production of various chemical compounds. The resulting spatial gradients in these chemicals create phoretic fields, which bias the swimming direction toward (or away from) producing microswimmers. Synthetic microswimmers, therefore, provide a synthetic analog to motile microorganisms interacting by chemotaxis toward (or away from) self-produced chemical fields. This can lead to ensembles of synthetic motile particles, which assemble into clusters, even at low particle density. Such clusters either proceed toward macrophase separation, resembling Dictyostelium aggregation, or to dynamic clusters of self-limited size (dynamic clustering). This considerably extends the range of means by which microscopic particles can communicate with one another, assemble and respond to environmental cues.

3 Ensemble Approach
Finding Descriptors and Reducing Dimensions

"We may imagine a great number of systems of the same nature, but differing in the configurations and velocities which they have at a given instant."

Josiah Willard Gibbs,
Elementary Principles in Statistical Mechanics [159]

Now that we have defined how members of the molecular group interact and communicate with one another, we need to identify means to characterize, label and interpret the collective properties exhibited by the group. The starting point we saw in the first chapter is the ubiquity of statistics and probability distributions. There is a very large number of molecules in a system (or molecular group), typically of the order of 10^{23}, and the velocities of individual molecules are captured by the Maxwell-Boltzmann probability distribution. Using the Maxwell-Boltzmann distribution, we can calculate their average over the group and determine its relation with temperature. What we will see in this chapter is that a similar probabilistic reasoning applies to the group itself. Indeed, every property, be it individual or collective, can be accessed by performing averages. The thought experiment that Gibbs (1839–1903), recipient of the first American doctorate in Engineering (see Figure 3.1) and whom Einstein referred to as "the greatest American mind in history", proposes above is to imagine many different realizations of the system, with different positions and velocities for the individual molecules of the group. The set of configurations is what Gibbs calls an ensemble.

This set constitutes a series of samples, over which measurements, or calculations, can be performed and averages can be calculated to yield collective properties. If the system consists, *e.g.*,

Figure 3.1 Josiah Willard Gibbs (1839–1903).

DOI: 10.1201/9781003006411-4

18 g of water under ambient conditions (1 mole or 10^{23} molecules of water at $T = 293$ K and $P = 1$ atm), an ensemble can be "created" by pouring the desired quantity of water in a very large number of beakers. Then, if we had a camera that could capture the positions and velocities of each water molecule, we could take pictures of each of these beakers and then start computing averages and predicting properties. As stated by Gibbs, this is different from following a particular system, *i.e.*, taking a large number of pictures over time of a single beaker filled and calculating averages over the time-series of data. In other words, an ensemble of configurations does not have to be a time-series of configurations. Fortunately, we will see later on that, because of a property known as ergodicity, performing an average over an ensemble or over a long enough time-series yields the same results.

3.1 COLLECTIONS AND ENSEMBLES

3.1.1 MAKING SENSE OF THE MICROSCOPIC BIG DATA

What defines a macroscopic system? Consider our example of a mole of water. We quickly find that we need very few parameters to describe this system entirely. These can be, for instance, its volume, density, and temperature, and will serve as *macroscopic descriptors*. This is very different from what is required once we look at the mole of water with a microscopic lens. At the microscopic scale, this system can take many configurations, atom positions, and atom velocities for many atoms. Another way to put this is to recall that for a microscopic (quantum) system with N particles, the degeneracy (see Figure 3.2), or number of states within an infinitesimal energy interval ΔE of E, $\Omega(E, \Delta E)$ (Eq. 1.27) is of the order of 10^N, which is, once again, huge for a macroscopic system ($N \sim 10^{23}$). Hence, there is, at the microscopic level, a tremendously large dataset for our example system. The idea behind statistical mechanics is to connect these two scales. For instance, through averages, we can extract the macroscopic properties of a group of molecules from the exceedingly large number of data (positions and momenta) for the individual molecules. We first need to address a couple of

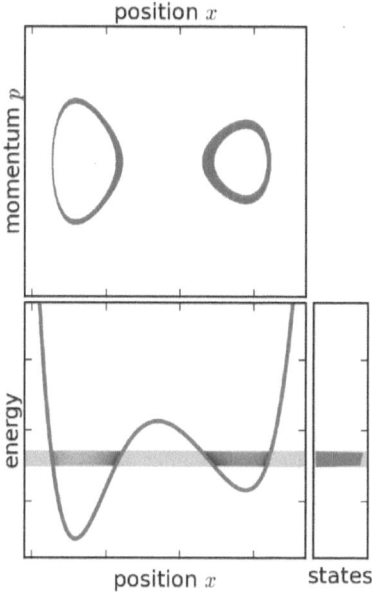

Figure 3.2 States within a narrow interval of energy (shown as a band in the bottom plot) for a one-particle system in 1D. The range of positions (x) and momenta (p) is shown on the top plot, while the density of states is plotted on the right.

questions to apply this approach. Which types of properties are accessible through averaging? And how should we perform this average in practice?

The properties we collect at the microscopic level are mechanical properties. We can locate the atoms and determine their positions. We can measure their velocities and obtain their momenta. Given, *e.g.*, an interatomic interaction potential, such as the Lennard-Jones potential from Chapter 2, we can also calculate their potential energy. This means that macroscopic mechanical properties, such as potential energy, kinetic energy, and pressure, can quickly be evaluated for the entire system through a direct, arithmetic average. On the other hand, nonmechanical properties, including thermodynamic properties such as free energy and entropy, require an extra step. By thermodynamic properties, we mean here properties whose definition relies on temperature. In this case, the connection between a macroscopic thermodynamic property and the microscopic mechanical properties will require making an assumption, or postulate, that the thermodynamic property will correspond to a specific average mechanical property. For instance, we will postulate that the average total energy, calculated as the sum of potential and kinetic energies, equals the thermodynamic internal energy. Similarly, we will assume that the average mechanical pressure is the same as the thermodynamic pressure.

Next, we will talk about how the averaging needs to be performed. This is where Gibbs' contribution becomes especially significant. Gibbs proposes calculating the average of microscopic mechanical properties for every state consistent with the few parameters necessary to define a macroscopic system, for instance, for a given density, volume, and total energy. This set of three parameters specifies the volume occupied by the system V, the total number of particles in the system N (since the density N/V has been determined), and the total energy E. The average is then calculated according to the principle of equal *a priori* probabilities, *i.e.*, by giving each microscopic (quantum) state the same weight. We saw that the degeneracy, or number of microstates, means that each of the $\Omega(E, \Delta E)$ microscopic states is given the same significance in the averaging process and appears the same number of times in the sum. This principle can be understood as the consequence that we do not have any information that would lead us to consider any of the $\Omega(E, \Delta E)$ microscopic states to be of greater significance than another.

3.1.2 DEFINING ENSEMBLES

We are starting to see the basic principles of an ensemble emerge. What is called a statistical ensemble is, in fact, an ensemble of systems or, in other words, a collection of a vast number of systems that all have the following point in common: they all are realizations of a thermodynamic state characterized by a few fixed parameters, or macroscopic descriptors. The first example we saw above, in which each system contains a set number of particles N, a fixed volume V, and a fixed energy E, is called the *microcanonical* ensemble is denoted by the following triplet (N, V, E). Another possible and popular choice of fixed variables is to keep constant the number of particles N, the volume V, and the temperature T (see Figure 3.3). This constitutes the *canonical* ensemble, denoted by the triplet (N, V, T). The microcanonical ensemble corresponds to an ensemble of isolated systems since the energy of each system E is kept constant, and there is no energy exchanged between each system and its surroundings. Such an ensemble is termed *adiabatic* ensemble, with the (N, V, E) ensemble being only one of the existing adiabatic ensembles, as we will discuss later. On the other hand, in the canonical ensemble, energy exchanges occur between each of the systems and their surroundings. Heat flows in and out of each system to ensure that the temperature of each system is constant. The canonical ensemble is called an isothermal ensemble, as T is kept constant. Exactly like the microcanonical ensemble for adiabatic ensembles, the canonical ensemble is only one of the possible choices of isothermal ensembles.

We can thus define the critical characteristics of the ensemble from the data we have on each of the systems that are members of the ensemble. Considering the microcanonical ensemble (N, V, E), as we have seen, each system, or element, of the ensemble, contains a given number of molecules

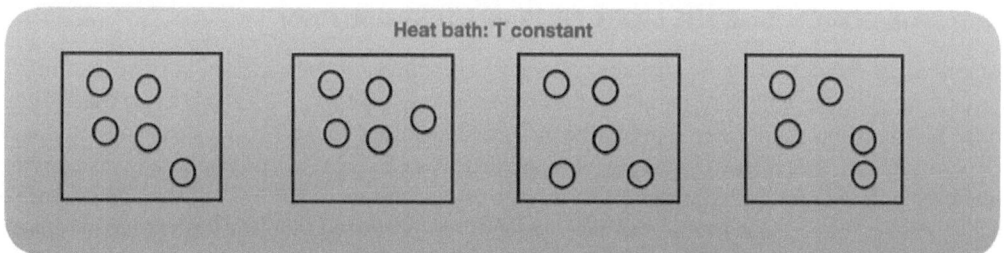

Figure 3.3 A few members of a canonical ensemble. All systems contain the same number of particles N, have the same volume V, and are at the same temperature T as they are immersed in a heat bath that ensures the temperature remains constant.

N, has a given volume V, and a given energy E. Suppose we now define as $\tilde{N}$ the vast number of elements in the ensemble. In that case, the total energy of the ensemble is then $\tilde{E} = \tilde{N}E$, the total volume occupied by the ensemble is $\tilde{N}V$, and the total number of molecules contained in the ensemble is $\tilde{N}N$. Similarly, in the canonical ensemble (N, V, T), the total volume occupied by the ensemble is $\tilde{N}V$, and the total number of molecules contained in the ensemble is $\tilde{N}N$. However, the system's total energy differs from $\tilde{N}E$ since it is the temperature T that is fixed for each system in the canonical ensemble and not the energy E as in the case of the microcanonical ensemble. This means that the energy of each system can take a different value. Of course, physical laws, such as those from quantum mechanics, apply. The energy of each system is provided by a solution, or energy eigenvalue E_j, of the Schrödinger equation, with the index j serving as a label for this energy level of degeneracy $\Omega(E_j)$. Considering now the ensemble, if, for each energy level E_j, we find n_j elements of energy E_j in the ensemble, then we can write that

$$\tilde{N} = \sum_j n_j \tag{3.1}$$

$$\tilde{E} = \sum_j n_j E_j \tag{3.2}$$

Using the vocabulary of quantum mechanics, each n_j can be interpreted as an occupation number, *i.e.*, the number of times a system of energy E_j appears in the ensemble. The set of numbers $\mathbf{n} \equiv (n_1, n_2, \ldots)$ constitutes what is called a distribution, which, as we discuss below, plays a significant role in the ensemble averaging process.

The final steps consist of (i) defining the concept of ensemble average and (ii) making the connection between mechanics and thermodynamics. First, we calculate the ensemble average of a mechanical property as the average value of this property over all ensemble members by applying the principle of equal *a priori* probabilities. Second, we make the following assumption to determine a thermodynamic property from a mechanical property. We postulate that the ensemble average of a mechanical property equals a corresponding thermodynamic property. We will discuss these points, carry out these calculations in the following few sections and shed light on the central enigma in statistical mechanics, as famously put by Schrödinger [323]: "There is, essentially, only one problem in statistical thermodynamics, the distribution of a given amount of energy E over identical systems. Or perhaps better, to determine the distribution of an assembly of identical systems over the possible states in which the system can find itself, given that the energy of the assembly is a constant E." Schrödinger goes on to state that [323]:

"To determine the distribution' means in principle to make oneself familiar with any possible distribution-of-the-energy (or state-of-the-assembly), to classify them suitably, *i.e.* in the way suiting the purpose in question and to count the numbers in the classes, to be able to judge of the probability of certain features or characteristics turning up in the assembly."

3.1.3 THE CONCEPT OF THE MOST PROBABLE DISTRIBUTION

We now focus on addressing the question raised by Schrödinger on the nature of the distributions. To this end, we continue with the example of the canonical ensemble (N, V, T) and provide more details on the physical constraints that apply to the ensemble members. A wall bounds each system that is not permeable to molecules to ensure that the number of molecules N in a system remains constant. On the other hand, the wall allows heat to flow in or out of the system to keep the temperature of each system constant. This means that all $\tilde{N}$ systems are immersed in a heat bath at temperature T and that when equilibrium has been reached, the entire ensemble is at a uniform temperature T. If we now enclose the whole ensemble into a thermally insulating envelope, the ensemble is isolated, and the total energy $\tilde{E} = \sum_j n_j E_j$ is constant. We are now ready to apply the principle of equal *a priori* probabilities to the canonical ensemble. This means that every distribution $(n_1, n_2, \ldots)$, as defined in Eqs. 3.1 and 3.2, is equally probable and has the same weight as any other distribution in the calculation of ensemble averages.

The systems all share the same properties, so we cannot tell one from another, just as in the picture of an ensemble of water-filled beakers we started the chapter with. The systems are thus indistinguishable, which means that there is, in fact, a large number of ways to distribute n_1 systems in a first group at energy E_1, n_2 systems in a second group at energy $E_1, \ldots$ and obtain a specific distribution. Mathematically, this number of ways is denoted by $W(\mathbf{n}) \equiv W(a_1, a_2, a_3, \ldots)$ and is given by

$$W(\mathbf{n}) = \frac{\tilde{N}!}{n_1! n_2! n_3! \ldots} = \frac{\tilde{N}!}{\prod_k n_k!} \tag{3.3}$$

For a given distribution, the probability of finding a system with an energy E_j is given by the ratio $\frac{n_j}{\tilde{N}}$. If we now consider all possible distributions, this probability becomes equal to the ratio of the average number $< n_j >$ of systems with energy E_j to the total number of systems and can be written as

$$P_j = \frac{< n_j >}{\tilde{N}} = \frac{1}{\tilde{N}} \frac{\sum_{\mathbf{n}} W(\mathbf{n}) n_j(\mathbf{n})}{\sum_{\mathbf{n}} W(\mathbf{n})} \tag{3.4}$$

in which $n_j(\mathbf{n})$ stands for the number of systems with energy E_j for a given distribution $\mathbf{n}$.

Knowing these probabilities would allow us to calculate ensemble averages in the canonical ensemble. For any mechanical property B, we can evaluate its ensemble average $< B >$ according to the following equation

$$< B > = \sum_k P_k B_k \tag{3.5}$$

in which B_k is the value taken by the property B in the state of energy E_j. This is where the fact that we have a huge number of systems in the ensemble becomes extremely useful. For very large n_k and thus $\tilde{N}$, $W(\mathbf{n})$ becomes sharply peaked around a specific distribution which we call $\mathbf{n}^* = (n_1^*, n_2^*, \ldots)$. This property exhibited by the multinomial coefficient $W(\mathbf{n})$ can be illustrated through a plot of the behavior of the binomial coefficient $W_2 = \frac{\tilde{N}!}{N_1! N_2!}$ (where we consider only two levels $\tilde{N} = N_1 + N_2$) for increasing values of $\tilde{N}$. W_2 can be written as a function of N_1 only $W_2 = \frac{\tilde{N}!}{N_1!(\tilde{N}-N_1)!}$ and graphed on a 2D-plot (see Figure 3.4).

The last step consists of identifying this specific distribution $\mathbf{n}^*$ around which $W(\mathbf{n})$ is sharply peaked, *i.e.*, the central, or most probable, distribution. To do this, we rewrite the probability P_j to take into account that only the central distribution $\mathbf{n}^*$ makes a significant contribution

$$P_j = \frac{< n_j >}{\tilde{N}} = \frac{1}{\tilde{N}} \frac{\sum_{\mathbf{n}} W(\mathbf{n}) n_j(\mathbf{n})}{\sum_{\mathbf{n}} W(\mathbf{n})} = \frac{1}{\tilde{N}} \frac{W(\mathbf{n}^*) n_j^*}{W(\mathbf{n}^*)} = \frac{n_j^*}{\tilde{N}} \tag{3.6}$$

To identify $\mathbf{n}^*$, we need to maximize the function $W(\mathbf{n})$ or, equivalently, its logarithm $\ln W(\mathbf{n})$, under the two constraints that the distributions $\mathbf{n}$ have to satisfy. We employ a method known as

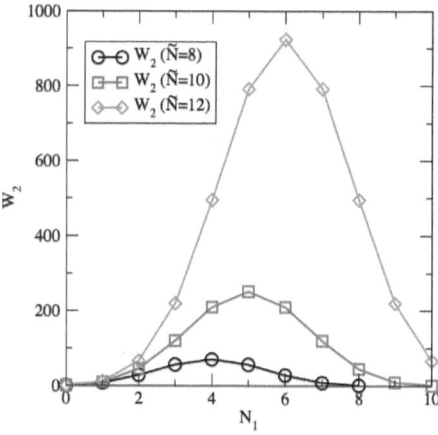

Figure 3.4 Binomial coefficient $W_2 = \frac{\tilde{N}!}{N_1!(\tilde{N}-N_1)!}$ as a function of N_1. As $\tilde{N}$ increases, W_2 becomes more and more sharply peaked.

the method of Lagrange multipliers. In practice, we rewrite Eqs. 3.1 and 3.2 as the two following constraints $g_1(n_1, n_2, \ldots)$ and $g_2(n_1, n_2, \ldots)$ as

$$\begin{aligned} g_1(n_1, n_2, \ldots) &= \textstyle\sum_j n_j - \tilde{N} = 0 \\ g_2(n_1, n_2, \ldots) &= \textstyle\sum_j n_j E_j - \tilde{E} = 0 \end{aligned} \tag{3.7}$$

and then maximize $\ln W(n_1, n_2, \ldots)$ subject to these two constraints by writing that, for all n_j, we have at the maximum

$$\frac{\partial}{\partial n_j} \ln W(n_1, n_2, \ldots) - \frac{\partial}{\partial n_j} g_1(n_1, n_2, \ldots) - \frac{\partial}{\partial n_j} g_2(n_1, n_2, \ldots) = 0 \tag{3.8}$$

Since $\tilde{N}$ and $\tilde{E}$ are constant, this gives

$$\frac{\partial}{\partial n_j} \ln W(n_1, n_2, \ldots) - \frac{\partial}{\partial n_j} \alpha \sum_k n_k - \frac{\partial}{\partial n_j} \beta \sum_k n_k E_k = 0 \tag{3.9}$$

The last two terms on the left-hand-side can be easily differentiated to yield

$$\frac{\partial}{\partial n_j} \ln W(n_1, n_2, \ldots) - \alpha - \beta E_j = 0 \tag{3.10}$$

We now use Stirling's approximation to evaluate the first term (see Figure 3.5).

This approximation states that, for any large number, such as $\tilde{N}$, we can estimate that $\ln \tilde{N}! = \tilde{N} \ln \tilde{N} - \tilde{N}$. We recall that $W(n_1, n_2, \ldots) = \frac{\tilde{N}}{\prod_k n_k!}$ and that since both $\tilde{N}$ and all n_j are very large, we can apply Stirling's approximation to obtain

$$\begin{aligned} \ln W(n_1, n_2, \ldots) &= \ln \tilde{N}! - \sum_k \ln n_k! \\ &= \tilde{N} \ln \tilde{N} - \tilde{N} - \sum_k (n_k \ln n_k! - n_k) \\ &= \tilde{N} \ln \tilde{N} - \sum_k (n_k \ln n_k!) \end{aligned} \tag{3.11}$$

For the distribution $\mathbf{n}^* = (n_1^*, n_2^*, \ldots)$ that maximize $\ln W$, we thus have for all n_j^*

$$-\ln n_j^* - 1 - \alpha - \beta E_j = 0 \tag{3.12}$$

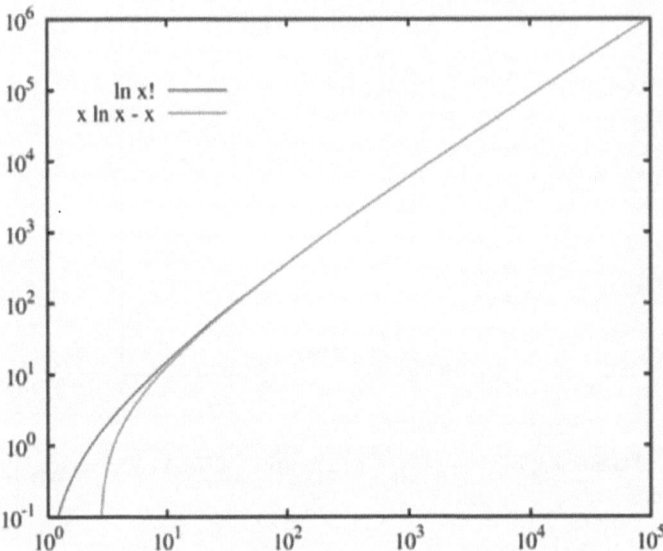

Figure 3.5 Stirling's approximation. Here $\ln(x!)$ and $x\ln(x) - x$ are both plotted against x. The two functions become equivalent for large x values.

and finally obtain the most probable distribution $\mathbf{n}^* \equiv (n_1^*, n_2^*, \ldots)$ with each n_j^* given by

$$n_j^* = \exp(-\alpha - 1)\exp(-\beta E_j) \tag{3.13}$$

3.2   INDIVIDUALS IN AN ISOTHERMAL WORLD: THE CANONICAL ENSEMBLE

3.2.1   KEY PARAMETERS AND MULTIPLIERS

Now that we have identified how we determine the most probable distribution, we need to clarify the value of the two multipliers α and β. Like any other distribution, the most probable distribution $\mathbf{n}^*$ satisfies Eq. 3.1. Summing Eq. 3.13 over all occupation numbers n_j^*, we obtain

$$\sum_j n_j^* = \exp(-\alpha - 1)\sum_j \exp(-\beta E_j) = \tilde{N} \tag{3.14}$$

From the last two terms, we obtain the following equation for $\exp(\alpha + 1)$

$$\exp(\alpha + 1) = \frac{1}{\tilde{N}}\sum_j \exp(-\beta E_j) \tag{3.15}$$

We are now almost ready to evaluate ensemble averages. To this end, we use the equation for α and plug it into the definition for the probability P_j (Eq. 3.6) to obtain

$$P_j = \frac{n_j^*}{\tilde{N}} = \frac{\exp(-\alpha - 1)\exp(-\beta E_j)}{\tilde{N}} = \frac{\exp(-\beta E_j)}{\sum_j \exp(-\beta E_j)} \tag{3.16}$$

This means that the average in the canonical ensemble of any property B can be evaluated as

$$<B> = \sum_j P_j B_j = \frac{\sum_j B_j \exp(-\beta E_j)}{\sum_j \exp(-\beta E_j)} \tag{3.17}$$

In particular, if we focus on the total energy of the system $B \equiv E$, we have that

$$< E >= \frac{\sum_j E_j \exp(-\beta E_j)}{\sum_j \exp(-\beta E_j)} \tag{3.18}$$

and postulate that the average of the total energy $< E >$ is equal to its thermodynamic counterpart, *i.e.*, the internal energy.

The mechanical pressure can be evaluated through the work done on the system when the volume increases by, *e.g.*, dV, and the number of molecules N remains constant. For a given state j of energy E_j, the work is given by $\delta W = -p_j dV$ and thus, the corresponding pressure is $p_j = -\left(\frac{\partial E_j}{\partial V}\right)_N$. This leads to the canonical average of the mechanical pressure $< p >$ as

$$< p >= \sum_j p_j P_j = -\frac{\sum_j \left(\frac{\partial E_j}{\partial V}\right) e^{-\beta E_j}}{\sum_j e^{-\beta E_j}} \tag{3.19}$$

and postulate that this average is equal to the thermodynamic pressure.

We aim to determine the defining equation for β from the mechanical-thermodynamic equivalency we have just postulated. If we differentiate the average energy $< E >$ (Eq. 3.18) with respect to V, while keeping the other variables N and β constant, we obtain

$$\left(\frac{\partial < E >}{\partial V}\right)_{N,\beta} = - < p > + \beta < Ep > - \beta < E >< p > \tag{3.20}$$

in which $< Ep >$ denotes the average of the product Ep, *i.e.*, $\sum_j E_j p_j P_j$.

If we now differentiate the average pressure $< p >$ with respect to β, while keeping the other variables N and V constant, we can show that

$$\left(\frac{\partial < p >}{\partial \beta}\right)_{N,V} = < E >< p > - < Ep > \tag{3.21}$$

Combining the previous two equations thus gives

$$\left(\frac{\partial < E >}{\partial V}\right)_{N,\beta} + \beta \left(\frac{\partial < p >}{\partial \beta}\right)_{N,V} = - < p > \tag{3.22}$$

which we compare with the following relation from macroscopic (classical) thermodynamics

$$\left(\frac{\partial E}{\partial V}\right)_{N,1/T} + \frac{1}{T}\left(\frac{\partial p}{\partial 1/T}\right)_{N,V} = -p \tag{3.23}$$

Therefore, the postulate that canonical averages of mechanical properties are equal to their thermodynamic counterparts implies that the two equations are identical. This, in turn, means that $\beta = \frac{1}{k_B T}$, in which k_B is the constant known as Boltzmann's constant ($k_B = 1.38 \times 10^{-23}$ J/K).

3.2.2   THE CENTRAL PARTITION FUNCTION

We saw that the sum of $\exp(-\beta E_j)$ over all states j plays a special role in ensemble averages. This sum, denoted by Q, is called the partition function and, in the case of the canonical ensemble, is given by

$$Q(N,V,T) = \sum_j e^{-E_j(N,V)/k_B T} \tag{3.24}$$

In this equation, we can see that Q is a function of the three critical parameters of the canonical ensemble, *i.e.*, the number of molecules N, volume V, and temperature T. On the other hand, the energy associated with each of the j states is a function of N and V (see, for instance, the example of the quantum energy of N noninteracting particles of Eq. 1.25). Q captures the connection between mechanics and thermodynamics by encapsulating the dependence of the system's properties on mechanical energy and temperature in a single function. As discussed in Chapter 2, depending on the system and accuracy level required, the mechanical energy $E_j(N,V)$ will be either calculated from quantum mechanics or evaluated as a sum of contributions from two-body, three-body, or many-body classical terms.

We now focus on the relation between the partition function of the ensemble and the properties of the particles that compose the system. To achieve this, we start with the case of a system composed of N atoms in a fixed volume V and constant temperature T. We now assume that the atoms do not interact and behave independently from one another, which is typical of atoms in an ideal gas. Then, we start by assuming that the atoms are distinguishable and label each atom with an index i ($i = 1, 2, \ldots, N$). The energy of the state E_j is thus equal to the sum of the energies $\varepsilon_{i,j}$ of all atoms i composing the system. This means that the partition function can now be expressed as

$$
\begin{aligned}
Q(N,V,T) &= \sum_j e^{-E_j(N,V)/k_B T} \\
&= \sum_j e^{-(\sum_i \varepsilon_{i,j})/k_B T} \\
&= \sum_j \Pi_i e^{-\varepsilon_{i,j}/k_B T} \\
&= \Pi_i \sum_j e^{-\varepsilon_{i,j}/k_B T} \\
&= \Pi_i q_i(N,V,T)
\end{aligned}
\tag{3.25}
$$

Thus, provided that we can split the total energy, or Hamiltonian, of the system into a sum of independent terms, we will be able to write the partition function of the N distinguishable particles as the product of the individual, atomic, partition functions $q_i = \sum_j e^{-\varepsilon_{i,j}/k_B T}$. This significant result considerably simplifies the statistical mechanics of atomic and molecular systems.

To obtain the system's actual partition function, we now consider that atoms are identical, and we cannot distinguish atoms from one another. This means that we can interchange labels in the sum $\sum_i \varepsilon_i, j$ or, in other words, that we took into account $N!$ times the same terms in the sum. It implies that the partition function for the system is equal to

$$
Q(N,V,T) = \frac{q^N}{N!}
\tag{3.26}
$$

with the atomic partition function $q = \sum_j e^{-\varepsilon_j/k_B T}$ taking the same value for all atoms composing the system. We will discuss in the next chapter how the individual particle's partition function is impacted by the nature of the particles (atom vs. diatomic molecule vs. polyatomic molecule) composing the system.

3.2.3 PARTITION FUNCTION AND THERMODYNAMICS

And the partition function enables the calculation of all thermodynamic properties of a system. It can be shown that thermodynamic properties are all partial derivatives, or linear combinations of, the logarithm of Q. To show this, we start by differentiating with respect to β (and at constant N and V), the logarithm of Q. We obtain

$$\left(\frac{\partial \ln Q}{\partial \beta}\right)_{N,V} = \frac{\left(\frac{\partial Q}{\partial \beta}\right)_{N,V}}{Q} \tag{3.27}$$

with the quantity in the numerator being equal to

$$\left(\frac{\partial Q}{\partial \beta}\right)_{N,V} = -\sum_j E_j \exp(-\beta E_j) \tag{3.28}$$

If we now compare with the average energy $<E>$ from Eq. 3.18, we have

$$<E> = \frac{\sum_j E_j \exp(-\beta E_j)}{\sum_j \exp(-\beta E_j)} = -\frac{\left(\frac{\partial Q}{\partial \beta}\right)_{N,V}}{Q} = -\left(\frac{\partial \ln Q}{\partial \beta}\right)_{N,V} \tag{3.29}$$

To make the T-dependence appear, we write that

$$-\left(\frac{\partial \ln Q}{\partial \beta}\right)_{N,V} = -\left(\frac{\partial \ln Q}{\partial T}\right)_{N,V} \times \left(\frac{\partial T}{\partial \beta}\right)_{N,V} \tag{3.30}$$

Since $\beta = 1/k_B T$, or equivalently, $T = 1/k_B \beta$, we have

$$-\left(\frac{\partial \ln Q}{\partial T}\right)_{N,V} = \frac{1}{k_B \beta^2}\left(\frac{\partial \ln Q}{\partial T}\right)_{N,V} = k_B T^2 \left(\frac{\partial \ln Q}{\partial T}\right)_{N,V} \tag{3.31}$$

This gives us the first thermodynamic average property, the internal energy, as a function of Q and the parameter for the canonical ensemble N, V, and T as

$$<E> = k_B T^2 \left(\frac{\partial \ln Q}{\partial T}\right)_{N,V} \tag{3.32}$$

A similar approach can be used to derive all other thermodynamic properties from $\ln Q$, including the thermodynamic pressure $<p>$ from Eq. 3.19 as

$$<p> = k_B T \left(\frac{\partial \ln Q}{\partial V}\right)_{N,T} \tag{3.33}$$

The connection with the other thermodynamic properties can be made by invoking the fundamental relations of thermodynamics. For instance, the Helmholtz free energy $<A>$, which is given by the difference between the internal energy and the product of temperature by entropy, is also related to pressure via the following equation

$$<p> = -\left(\frac{\partial <A>}{\partial V}\right)_{N,T} \tag{3.34}$$

Comparing Eqs. 3.33 and 3.34 leads us to identify the following equation for $<A>$

$$<A> = -k_B T \ln Q \tag{3.35}$$

with the integration constant set to 0. Setting this constant to 0, or the origin for the free energy scale to 0, does not impact the connection with thermodynamics as thermodynamics deals with energy changes and changes in $\Delta <A>$, for which the constant drops out of the calculations. Interestingly, as we will see in the case of other isothermal ensembles, corresponding to a set of fixed parameters other than the canonical set of (N,V,T), the thermodynamic potential (here $<A>$ for the canonical ensemble) will always be given by the product of $-k_B T$ by the logarithm of the partition function of the ensemble. We finally give the expression of entropy $< S >$

$$< S >= \frac{< E > - < A >}{T} = k_B T \left(\frac{\partial \ln Q}{\partial T} \right)_{N,V} + k_B \ln Q \tag{3.36}$$

The entropy is obtained through the thermodynamic relation between Helmholtz and internal energies. Similarly, all other thermodynamic properties can be derived using classical thermodynamics equations and results.

3.3 INDIVIDUALS IN ISOLATION: THE MICROCANONICAL ENSEMBLE

3.3.1 NUMBER AND DENSITY OF STATES

In the microcanonical ensemble, the key parameters are the number of particles N, the volume V, and the energy E. Since an ensemble is a collection of replicas, or microstates, that share the same N, V, and E, the knowledge of the number of states $\Omega(E)$ with the same energy E provides all the information necessary to calculate properties in the microcanonical ensemble and serves as the partition function for this ensemble. In other words, $Q(N,V,E) = \Omega(E)$.

We saw in Chapter 1 how we could evaluate the number of states $\Omega(E)$ between a given energy E and an infinitesimally large energy $E + \Delta E$ on the example of an electron trapped in a cubic box of length L. The general approach to determining $\Omega(E)$ and the related density of states $g(E)$ follows the same idea and involves considering "energy shells". We start by defining the phase space volume $\Phi(E)$, enclosed by the energy surface E

$$\Phi(E) = \int \int_{H(\mathbf{p},\mathbf{q})<E} d^{3N}q d^{3N}p \tag{3.37}$$

in which q and p denote a position and momentum coordinate, respectively, and $H(\mathbf{p},\mathbf{q})$ is the Hamiltonian of the system.

What is the value taken by this function in a simple case, $i.e.$, for an ideal gas? For such a system, the N particles do not interact, and the Hamiltonian is equal to the kinetic energy of the system: $H(\mathbf{p},\mathbf{q}) = \sum_{i=1}^{3} N \frac{p_i^2}{2m}$, in which m is the mass of a particle. The phase space volume $\Phi(E)$, enclosed by the energy surface E, becomes

$$\Phi(E) = \int \int_{p_i^2 < 2mE} d^{3N}q d^{3N}p \tag{3.38}$$

We first integrate the position coordinates, leading to

$$\Phi(E) = V^N \int_{p_i^2 < 2mE} d^{3N}p \tag{3.39}$$

The integral over the momenta coordinates corresponds to the volume occupied by a 3N-dimensional sphere of radius $\sqrt{2mE}$. This gives the overall result for $\Phi(E)$

$$\Phi(E) = V^N \frac{\pi^{3N/2}}{(3N/2)\Gamma(3N/2)} \sqrt{2mE}^{3N} \tag{3.40}$$

in which Γ is known as the gamma function (for an integer M, $\Gamma(M+1) = M!$).

We now consider an energy shell of infinitesimal thickness ΔE around E and calculate the number of states with energy between E and $E + \Delta E$, $i.e.$, in this energy shell. It is given by

$$\Omega(E) = \Phi(E + \Delta E) - \Phi(E) \tag{3.41}$$

Figure 3.6 Ludwig Boltzmann (1844–1906).

which, by taking the limit $\Delta E \to 0$, gives the following relation between the number of states $\Omega(E)$ and the density of states $g(E)$

$$\Omega(E) = \frac{\partial \Phi(E)}{\partial E} \times \Delta E = g(E) \Delta E \tag{3.42}$$

and the probability distribution $\rho(\mathbf{p}, \mathbf{q})$ associated with a given state, or $(\mathbf{p}, \mathbf{q})$ set, is thus given by

$$\rho(\mathbf{p}, \mathbf{q}) = \frac{1}{\Omega(E)} \tag{3.43}$$

3.3.2 BOLTZMANN'S ENTROPY

We now reach the point where we can invoke the famous Boltzmann formula for entropy. It is actually so famous that it is engraved in Boltzmann's tombstone in Vienna (see Figure 3.6). According to this formula, the entropy S is equal to

$$S(N, V, E) = k_B \ln \Omega(E) \tag{3.44}$$

To determine S, we use the following approximation. Comparing the equations for the logarithm of $\Phi(E)$ and for the logarithm of $\Omega(E)$, we see that they differ by the value taken exponent of E. However, we can write that

$$\begin{aligned} \ln \Phi(E) &\sim 3N/2 \ln E \\ \ln \Omega(E) &\sim (3N/2 - 1) \ln E \end{aligned} \tag{3.45}$$

If N is very large, $3N/2$ and $3N/2 - 1$ can be approximated to be equal. Thus, we can evaluate S as $k_B \ln \left(\frac{\Phi(E)}{n_0} \right)$, in which $n_0 = h^{3N} N!$ is a factor that ensures that the ratio inside the logarithm is dimensionless.

$$S(N, V, E) = k_B \ln \left(\frac{\Phi(E)}{h^{3N} N!} \right) \tag{3.46}$$

Plugging into this equation, the result obtained in Eq. 3.40 gives

$$S(N,V,E) = k_B \ln \frac{V^N \pi^{3N/2} \sqrt{2mE}^{3N}}{(3N/2)\Gamma(3N/2)h^{3N}N!} \tag{3.47}$$

Then, using the properties of the Γ function, the Stirling approximation, and the fact that for an ideal atomic gas, we can relate the energy E to the average temperature $<T>$ through $E = (3/2)Nk_B <T>$, we obtain the following equation for S

$$S(N,V,E) = k_B \ln \left(\frac{\Phi(E)}{h^{3N}N!} \right) = Nk_B \ln \left[\left(\frac{2\pi mk_B <T>}{h^2} \right)^{3/2} \frac{Ve^{5/2}}{N} \right] \tag{3.48}$$

This equation is known as the Sackur-Tetrode formula. Before examining how we obtain other thermodynamic properties in the microcanonical ensemble, we briefly comment on how the Sackur-Tetrode formula for entropy predicts the expected behavior for entropy in macroscopic systems. To this end, we consider a generic process that takes the system from a macroscopic state A characterized by (N_A, V_A, T_A) to a macroscopic state B with (N_B, V_B, T_B). The entropy change is given by

$$\Delta S = S_B - S_A$$
$$= N_B k_B \ln \left[\left(\frac{2\pi mk_B T_B}{h^2} \right)^{3/2} \frac{V_B e^{5/2}}{N_B} \right] - N_A k_B \ln \left[\left(\frac{2\pi mk_B T_A}{h^2} \right)^{3/2} \frac{V_A e^{5/2}}{N_A} \right] \tag{3.49}$$

If the system undergoes an expansion, then $V_B > V_A$ and the other parameters are constant ($N_A = N_B = N$ and $T_A = T_B = T$). Plugging this into Eq. 3.49 yields

$$\Delta S = Nk_B \ln \left[\frac{V_B}{V_A} \right] > 0 \tag{3.50}$$

which corresponds exactly to the classical thermodynamics relation. In addition, $\Delta S > 0$, which is what we expect from an increase in molar volume such as, *e.g.* when a system goes from liquid$\rightarrow$vapor and the increase, at the microscopic level, in the number of available energy states and, equivalently, of possible spatial arrangements of atoms in a larger volume V_B. Similarly, if the system sees its temperature increase $T_B > T_A$ and the other parameters are constant ($V_A = V_B = V$ and $N_A = N_B = N$) which yields

$$\Delta S = \frac{3N}{2} Nk_B \ln \left[\frac{T_B}{T_A} \right] > 0 \tag{3.51}$$

This is the result expected from classical thermodynamics and the experiment.

3.3.3 THERMODYNAMIC FUNCTIONS

We have seen for isothermal systems on the example of the canonical ensemble that we can evaluate all thermodynamic properties of the system once we have the partition function for the ensemble. The same reasoning applies to adiabatic ensembles like the microcanonical ensemble, with the partition function $Q(N,V,E) = \Omega(N,V,E)$. Thus, precisely as we have seen in the previous section for determining the entropy, all properties are functions of the logarithm of $\Omega(N,V,E)$ and its partial derivatives.

To determine thermodynamic properties, we resort to the usual thermodynamic relations and start by differentiating the entropy to obtain

$$
\begin{aligned}
dS &= \left(\frac{\partial S}{\partial E}\right)_{N,V} dE + \left(\frac{\partial S}{\partial V}\right)_{N,E} dV + \left(\frac{\partial S}{\partial N}\right)_{V,E} dN \\
&= \frac{1}{T} dE + \frac{p}{T} dV - \frac{\mu}{T} dN
\end{aligned}
\tag{3.52}
$$

We now identify the partial derivatives of S and thus of the logarithm of $\Omega(N,V,E)$ to the average temperature $<T>$, average pressure $<p>$ and chemical potential μ to obtain

$$
\begin{aligned}
\frac{1}{k_B <T>} &= \left(\frac{\partial \ln \Omega(N,V,E)}{\partial E}\right)_{N,V} \\
<p> &= k_B <T> \left(\frac{\partial \ln \Omega(N,V,E)}{\partial V}\right)_{N,E} \\
\mu &= k_B <T> \left(\frac{\partial \ln \Omega(N,V,E)}{\partial N}\right)_{V,E} dN
\end{aligned}
\tag{3.53}
$$

Combined with the Sackur-Tetrode equation (Eq. 3.49), the equation for $<p>$ gives the well-known equation

$$
\begin{aligned}
<p> &= k_B <T> \left(\frac{\partial S}{\partial V}\right)_{N,E} \\
&= \frac{N k_B <T>}{V}
\end{aligned}
\tag{3.54}
$$

and allows us to recover the well-known ideal gas law.

The equations for $<T>$ and $<p>$ then provide access to other thermodynamic functions such as the Helmholtz free energy A, enthalpy H, and Gibbs free energy G through Legendre transformations

$$
\begin{aligned}
A &= E - <T> S \\
H &= E + <p> V \\
G &= E + <p> V - <T> S
\end{aligned}
\tag{3.55}
$$

4 Accounting for Individual Features and Changes

"The forces between atoms are due to the motion and distribution of electrons and must be calculated, therefore, using quantum mechanics (...). It must be possible, therefore, to calculate the reaction rates by the methods of statistical mechanics."

Henry Eyring,
The activated complex in chemical reactions [130]

We have examined in Chapter 3 the general policies that govern the behavior of the members of the molecular group. This has allowed us to identify the role played, for instance, in the canonical ensemble, by energy levels (or, more specifically, by $\exp(-E_j)$ for any state j) as weighting factors in ensemble averages. These weighing factors also appear in the partition function for the ensemble and, in turn, dictate the value taken by any group property. We now propose to focus on the properties of the individual and examine how they impact the group's properties. To this end, we need to consider that fluids and solids may be composed of molecules, *i.e.*, of atoms connected by chemical bonds rather than composed of particles moving independently. As discussed by Eyring in the opening statement, this interdependence can be evaluated through the equations of quantum mechanics. Furthermore, it also gives rise to several events, such as bond forming and breaking, at the heart of chemical reactions and can be captured by statistical mechanics. We discuss in the following how we factor in individual and molecular features in the determination of the properties of the molecular group and how we account for the identity changes that occur for these individuals as a result of the chemical reactions.

4.1 MOLECULES IN A CANONICAL WORLD

4.1.1 FEATURES AND CONSEQUENCES

The first step consists in taking into account the nature of the molecules in the group. One way to approach this would be to start from scratch and work on solving the Schrödinger equation. What does this equation look like now that we have a system composed of molecules? Since this equation accounts for the entire system's energy, the Hamiltonian includes all terms that are function of the position of all the atoms (*i.e.*, all nuclei and electrons) composing the system. Upon closer inspection, we realize that the Hamiltonian for a molecular system takes the same form as the Hamiltonian for a system containing the same atoms. This implies that bonding is accounted for through some of the energy levels, or the Hamiltonian eigenvalues, obtained by solving the Schrödinger equation. Since the Hamiltonian is the same for atomic and molecular systems, provided that the system's composition remains the same, we can directly start from some of the results we obtained in Chapter 3. More specifically, we previously discussed that, for a monoatomic ideal gas in the canonical (N,V,T) ensemble, the partition function is given by $Q(N,V,T) = \left[q^N/N!\right]$. So far, since we were focusing on atomic systems with a single type of atom, or element, in the system, the individual partition function q corresponded to the partition function of a single atom $q = q_a$. Now that the individual is a molecule, we must replace the atomic partition function q_a with its molecular counterpart q_m. Once we have identified the molecular partition function q_m, we will be able to calculate the partition function of the ensemble as $Q(N,V,T) = \left[q_m^N/N!\right]$ for a system of N molecules in a volume V and at temperature T. What is q_a then for an atom? To answer this question, we continue

DOI: 10.1201/9781003006411-5

with our example of a monoatomic ideal gas. As with any partition function, everything starts with the energy levels. The energy of each atom is obtained by solving the Schrödinger equation with the following Hamiltonian

$$\mathscr{H} = \mathscr{H}_{\text{trans}} + \mathscr{H}_{\text{elec}} + \mathscr{H}_{\text{nucl}} \tag{4.1}$$

in which we have written the Hamiltonian as a sum of three terms. The first term corresponds to the kinetic energy and the 3 degrees of freedom for the translation, as each atom can move along the three directions of space x, y, z. The other two terms are associated with electronic and nuclear energy. The overall energy is equal to the sum of the eigenvalues for each of the three sub-Hamiltonians

$$\varepsilon = \varepsilon_{\text{trans}} + \varepsilon_{\text{elec}} + \varepsilon_{\text{nucl}} \tag{4.2}$$

In practice, atoms will all be in their ground electronic and nuclear states. This is because the gap in energy between the ground state and the excited states is enormous. Thus, if we set the origin for the electronic and the nuclear energy to 0, we have $\varepsilon_{\text{elec}} = 0$ and $\varepsilon_{\text{nucl}} = 0$, the two partition functions become $q_{\text{elec}} = \exp(-\beta \varepsilon_{\text{elec}}) = 1$ and $q_{\text{nucl}} = 1$. However, the gaps between energy levels for the translation are much smaller, and the exact approximation cannot be made for the translational partition function.

As previously discussed, in a cubic box of edge a, and thus of volume $V = a^3$, the energy states for the translation (or eigenvalues for H_{trans}) are given by

$$\varepsilon(n_x, n_y, n_z) = \frac{h^2}{8ma^2}(n_x^2 + n_y^2 + n_z^2) \tag{4.3}$$

in which n_x, n_y, and n_z are positive integers. This gives q_{trans} as

$$q_{\text{trans}} = \sum_{(n_x, n_y, n_z)=1}^{\infty} e^{-\beta \varepsilon(n_x, n_y, n_z)} \tag{4.4}$$

$$q_{\text{trans}} = \sum_{n_x=1}^{\infty} \exp\left(-\frac{\beta h^2 n_x^2}{8ma^2}\right) \sum_{n_y=1}^{\infty} \exp\left(-\frac{\beta h^2 n_y^2}{8ma^2}\right) \sum_{n_z=1}^{\infty} \exp\left(-\frac{\beta h^2 n_z^2}{8ma^2}\right) \tag{4.5}$$

With $n^2 = n_x^2 + n_y^2 + n_z^2$, we can write

$$q_{\text{trans}} = \sum_{n=1}^{\infty}\left(\exp\left(-\frac{\beta h^2 n^2}{8ma^2}\right)\right)^3 \tag{4.6}$$

Replacing the infinite sum by an integral, we obtain

$$q_{\text{trans}} = \left(\int_0^{\infty} e^{-\beta h^2 n^2/8ma^2} dn\right)^3 = \left(\frac{2\pi mkT}{h^2}\right)^{3/2} V \tag{4.7}$$

This leaves us with the atomic partition function $q_a = q_{\text{trans}} q_{\text{elec}} q_{\text{nucl}}$. Since $q_{\text{elec}} = q_{\text{nucl}} = 1$, we have $q_a = \left(\frac{2\pi mkT}{h^2}\right)^{3/2} V$ in the case of a monoatomic ideal gas. The next step is identifying how the molecular partition function q_m differs from the atomic partition function q_a. We now gradually increase the complexity of the molecule in the next two sections, starting with the case of diatomic molecules before examining polyatomic molecules.

4.1.2 THE CASE OF DIATOMIC MOLECULES

In a diatomic molecule, we are now considering an individual with a Hamiltonian that is a function of twice as many variables as before. Indeed, there are now two sets of coordinates $\mathbf{r}_1$ and $\mathbf{r}_2$ and momenta for the two atoms, as well as two masses (m_1 and m_2), etc. Rather than handling the two atoms separately, the idea is to perform a change of variables and recast the motion of the two atoms into an equivalent physical problem akin to the relative motion of a planet (the Earth) and its satellite (the Moon). This amounts to replacing the study of two atoms of coordinates and masses $(\mathbf{r}_1, m_1)$ and $(\mathbf{r}_2, m_2)$ by the study of an "atom" of mass $M = m_1 + m_2$ positioned at the center-of-mass $(m_1 \times \mathbf{r}_1 + m_2 \times \mathbf{r}_2)/M$ plus a second "atom" with a reduced mass $\mu = (m_1 \times m_2)/M$ orbiting, and vibrating, around the center-of-mass and positioned at $\mathbf{r} = \mathbf{r}_2 - \mathbf{r}_1$ from the center-of-mass (see Figure 4.1). This allows for the separation of the motion between the translational degrees of freedom (motion of the center of mass) and the internal degrees of freedom (orbiting of the reduced-mass particle).

The Hamiltonian can thus be written as

$$\mathcal{H} = \mathcal{H}_{\text{trans}} + \mathcal{H}_{\text{int}} \tag{4.8}$$

and the corresponding energies, or eigenvalues of the Hamiltonian, are

$$\varepsilon = \varepsilon_{\text{trans}} + \varepsilon_{\text{int}} \tag{4.9}$$

This means that the partition function for the molecule is

$$q_m = q_{\text{trans},m} q_{\text{int},m} \tag{4.10}$$

The translational partition function for the central "atom" with total mass $m_1 + m_2$ can be written as

$$q_{\text{trans},m} = \left[\frac{2\pi(m_1 + m_2)kT}{h^2} \right]^{3/2} V \tag{4.11}$$

The internal partition function consists of the orbiting and vibrating of the reduced mass. The amplitude of the vibration is very small, giving rise to a separation of the motions of rotation and vibration or, in other words, to the rigid rotor-harmonic oscillator approximation. This leads to

$$\mathcal{H}_{\text{int},\updownarrow} = \mathcal{H}_{\text{rot},m} + \mathcal{H}_{\text{vib},m} \, \varepsilon_{\text{int},m} = \varepsilon_{\text{rot},m} + \varepsilon_{\text{vib},m} q_{\text{int},m} = q_{\text{rot},m} q_{\text{vib},m} \tag{4.12}$$

The next step involves solving the equations for the rotation and the vibration to find the corresponding energies (eigenvalues) and then the partition functions for the corresponding motions. For the rotation, we obtain

$$q_{\text{rot},m} = \frac{8\pi^2 I k_B T}{\sigma h^2} \tag{4.13}$$

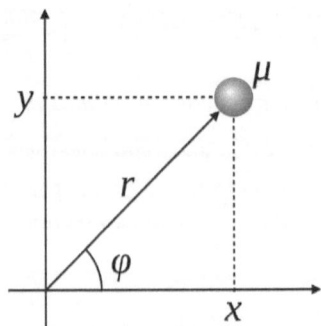

Figure 4.1 The study of the relative motion of the two atoms, rotating and vibrating around the center-of-mass, is replaced by the study of the motion of the reduced mass μ of coordinate $\mathbf{r}$.

in which I denotes the moment of inertia ($I = \sum_i m_i r_i^2$ for $i = 1, 2$). In this equation, σ denotes a symmetry number. This number correlates with molecular symmetry. For a homonuclear diatomic molecule like H_2, it is equal to 2, while $\sigma = 1$ for a heteronuclear diatomic molecule like CO. For the vibration, we have

$$q_{vib,m} = e^{-\beta h\nu/2}(1 - e^{-\beta h\nu})^{-1} \tag{4.14}$$

in which ν denotes the frequency with which the bond between the two atoms vibrates.

Taking the product of all partition functions, the molecular partition function for a diatomic molecule in its ground state is thus given by

$$q_m = \left(\frac{2\pi(m_1 + m_2)kT}{h^2}\right)^{3/2} V \frac{8\pi^2 I k_B T}{\sigma h^2} e^{-\beta h\nu/2}(1 - e^{-\beta h\nu})^{-1} \tag{4.15}$$

4.1.3 MOLECULAR SYMMETRY AND POLYATOMIC MOLECULES

What happens now if there are many more atoms in the molecule? This will result in many more internal degrees of freedom and, *e.g.*, quite a few additional vibrations. The good news is that we can extend the rigid rotor-harmonic oscillator approximation to polyatomic molecules. Simply put, we will have to work within a very similar framework

$$q_m = q_{trans,m} q_{int,m} \tag{4.16}$$

with

$$q_{trans,m} = \left[\frac{2\pi m k_B T}{h^2}\right]^{3/2} V \tag{4.17}$$

in which m denotes this time the total mass of the molecule and, for the internal part, the same combination as before, *i.e.*,

$$q_{int,m} = q_{rot,m} q_{vib,m} \tag{4.18}$$

Taking the general case of a nonlinear polyatomic molecule, we find the rotational partition function

$$q_{rot,m} = \frac{\pi^{1/2}}{\sigma} \left(\frac{T^3}{\Theta_A \Theta_B \Theta_C}\right)^{1/2} \tag{4.19}$$

σ denotes the symmetry number, *i.e.*, the number of physically indistinguishable configurations realized during a complete molecule rotation. Θ_A, Θ_B, and Θ_C are the so-called rotational temperatures. Here A, B, and C stand for the three axes of inertia of the molecule (with moments of inertia I_A, I_B, and I_C), and the rotational temperatures are defined as, for instance, for Θ_A as $\Theta_A = h^2/(8\pi^2 I_A k_B)$.

For the vibrational partition function, we find

$$q_{vib,m} = \prod_{j=1}^{3n-6} \frac{e^{-\Theta_{\nu j}/2T}}{(1 - e^{-\Theta_{\nu j}/T})} \tag{4.20}$$

in which the vibrational temperatures $\Theta_{\nu j}$ are defined as $\Theta_{\nu j} = h\nu_j/k_B$ (ν_j is the frequency of vibration of a bond labeled by the index j, $j = 1, 2, \ldots$).

Taking the product of all partition functions, we obtain the following general expression for the molecular partition function of a nonlinear polyatomic molecule

$$q_m = \left(\frac{2\pi M k_B T}{h^2}\right)^{3/2} V \frac{\pi^{1/2}}{\sigma} \left(\frac{T^3}{\Theta_A \Theta_B \Theta_C}\right)^{1/2} \left\{\prod_{j=1}^{3n-6} \frac{e^{-\Theta_{\nu j}/2T}}{(1 - e^{-\Theta_{\nu j}/T})}\right\} \tag{4.21}$$

Figure 4.2 llustration of William of Ockham, from a 1341 manuscript of Ockham's Summa Logicae.

4.2 CONNECTING WITH THE MACROSCOPIC WORLD

4.2.1 ARE ALL FEATURES ESSENTIAL?

This is the crucial question at the center of the building of any model. What level of detail should we take into account? Should we explicitly account for the ability of chemical bonds to break or form if we are merely looking to determine the density of an inert chemical compound? In other words, how many features should our model include to be deemed efficient and reliable? With the advent of ever-increasing computing power, a tempting answer would be to have as many features as possible (*e.g.*, as our computing resources allow), with the most advanced level of accuracy available to us. However, this approach has two main drawbacks. First, it only sometimes allows for an efficient use of our resources. Second, handling too many features may obscure the interpretation of the results, as too much complexity in the model and too many features accounted for might obscure the understanding of the results. It is, therefore, often desirable to use a law of economy or parsimony when building a model for the individual molecules and the group. This principle was stated many centuries ago and is known as Ockham's razor. The scholastic philosopher William of Ockham (1285–1347/49) said that *pluralitas non est ponenda sine necessitate*, which means that "plurality should not be posited without necessity". In other words, when two models or theories accurately describe the facts, we should always choose the simpler model over the more complex one. It is believed that Ockham (see Figure 4.2) illustrated this principle so sharply that it became known as "Ockham's razor". This principle summarizes the concept of efficient causality. It is a guiding design principle for a model that should be complex enough to yield accurate results yet remain simple enough to provide a clear understanding of the feature-result relationship.

4.2.2 MODEL-PARTITION FUNCTION INTERPLAY

The choice of a specific model for a given system strongly impacts how we evaluate its properties. Suppose we consider a molecule of, for instance, methane, or CH_4. It can be pictured as a tetrahedron with a large atom at its center, the C atom, with four smaller H atoms at the corners of the tetrahedron, each placed at a distance of 0.11 nm from the central C atom. A periodic chart shows that a C atom has a van der Waals radius of 0.17 nm while a H atom has a van der Waals radius of 0.11 nm. This means that the CH_4 molecule appears as a large sphere (the C atom), with four bits of H extending beyond the carbon sphere (see Figure 4.3).

If we are interested in basic thermodynamic properties that primarily hinge on the "space" occupied by the molecule (*e.g.*, density and phase properties), then it makes a lot of sense to approximate the molecule of CH_4 to a sphere, and essentially model it as a monoatomic ideal gas. In this case,

Figure 4.3 Space-filling model of the CH_4 molecule.

the partition function for the molecule of CH_4 is precisely that of an atom, and thus we have

$$q_{CH_4} = \left(\frac{2\pi MkT}{h^2}\right)^{3/2} V \tag{4.22}$$

in which $M = (m_C + 4m_H)$ is the total mass of the molecule. The partition function in the canonical ensemble, *i.e.*, for a system of N_{CH_4} molecules in a volume V and at a temperature T is given by

$$Q(N_{CH_4}, V, T) = \frac{q_{CH_4}^{N_{CH_4}}}{N_{CH_4}!} \tag{4.23}$$

and, following the development outlined in Section 3.2.3, we can calculate the properties of this system of CH_4 molecules by taking partial derivatives of the logarithm of $Q(N_{CH_4}, V, T)$. For instance, if we consider internal energy, we see that

$$E = k_B T^2 \left(\frac{\partial \ln Q}{\partial T}\right)_{N_{CH_4}, V} \tag{4.24}$$

Plugging in the expression for Q we have for our system composed of ideal "monoatomic" CH_4 molecules

$$E = \frac{3}{2} N_{CH_4} k_B T \tag{4.25}$$

Similarly, for pressure, we have

$$p = k_B T \left(\frac{\partial \ln Q}{\partial V}\right)_{N_{CH_4}, T} \tag{4.26}$$

which yields the ideal gas law $pV = N_{CH_4} k_B T$. Finally, this partition function allows us to calculate the Helmholtz free energy A and the entropy S of this system of CH_4 molecules as

$$<A> = -k_B T \ln Q(N_{CH_4}, V, T) \tag{4.27}$$

$$S = k_B T \left(\frac{\partial \ln Q}{\partial T} \right)_{N_{CH_4}, V} + k_B \ln Q \tag{4.28}$$

$$\frac{S}{N_{CH_4} k_B} = \ln \left[\frac{2\pi M k_B T}{h^2} \right]^{3/2} \frac{V e^{5/2}}{N_{CH_4}} \tag{4.29}$$

This leaves us with a question: how should we adapt these calculations if we need to take into account the molecular features of the CH_4 molecule, such as, for instance, the rotation of the molecule around its molecular axes or the vibrations of the C-H bonds? This is what we explore in the next section.

4.2.3 THERMODYNAMIC PROPERTIES AND IDEALITY

Now that the partition function of molecules is known, we can adapt the determination of thermodynamic properties in the previous section to obtain the thermodynamic properties for a polyatomic ideal gas. This involves calculating the partial derivatives of the partition function $Q = q_m^N/N!$ (and, more precisely, of its logarithm) with respect to T to obtain the internal energy E or with respect to V to determine p. Starting with the example of a diatomic molecule, we have

$$q_m = \left(\frac{2\pi m k_B T}{h^2} \right)^{3/2} V \frac{8\pi^2 I k_B T}{\sigma h^2} e^{-\beta h\nu/2} (1 - e^{-\beta h\nu})^{-1} \tag{4.30}$$

which, when we follow the steps outlined above, gives us the following equation for the internal energy and pressure

$$\frac{E}{N k_B T} = \frac{5}{2} + \frac{h\nu}{2k_B T} + \frac{h\nu/k_B T}{e^{h\nu/k_B T} - 1} \tag{4.31}$$

$$pV = N k_B T \tag{4.32}$$

and for the entropy

$$\frac{S}{N k_B} = \ln \left[\frac{2\pi(m_1 + m_2) k_B T}{h^2} \right]^{3/2} \frac{V e^{5/2}}{N} + \ln \frac{8\pi^2 I k_B T e}{\sigma h^2}$$
$$+ \frac{h\nu/k_B T}{e^{h\nu/k_B T} - 1} - \ln(1 - e^{-h\nu/k_B T}) \tag{4.33}$$

Adapting this approach to nonlinear polyatomic molecules, we start, in this case, from the following partition function q_m

$$q = \left(\frac{2\pi M k_B T}{h^2} \right)^{3/2} V \frac{\pi^{1/2}}{\sigma} \left(\frac{T^3}{\Theta_A \Theta_B \Theta_C} \right)^{1/2} \prod_{j=1}^{3n-6} \frac{e^{-\Theta_{\nu_j}/2T}}{(1 - e^{-\Theta_{\nu_j}/T})} \tag{4.34}$$

This gives the Helmholtz free energy A

$$-\frac{A}{N k_B T} = \ln \left[\left(\frac{2\pi M k_B T}{h^2} \right)^{3/2} \frac{V e}{N} \right] + \ln \frac{\pi^\pi 1/2}{\sigma} \left(\frac{T^3}{\Theta_A \Theta_B \Theta_C} \right)^{1/2}$$
$$- \sum_{j=1}^{3n-6} \left[\frac{\Theta_{\nu_j}}{2T} + \ln(1 - e^{-\Theta_{\nu_j}/T}) \right] \tag{4.35}$$

and thus, for the internal energy and pressure

$$\frac{E}{N k_B T} = \frac{3}{2} + \frac{3}{2} + \sum_{j=1}^{3n-6} \left(\frac{\Theta_{\nu_j}}{2T} + \frac{\Theta_{\nu_j}/T}{e^{\Theta_{\nu_j}/T} - 1} \right) \tag{4.36}$$

$$pV = Nk_BT \tag{4.37}$$

The entropy S for polyatomic nonlinear molecules can be written as

$$\frac{S}{Nk_B} = \ln \left[\frac{2\pi Mk_BT}{h^2} \right]^{3/2} \frac{Ve^{5/2}}{N} + \ln \frac{\pi^{1/2}e^{3/2}}{\sigma} \left(\frac{T^3}{\Theta_A\Theta_B\Theta_C} \right)^{1/2}$$
$$+ \sum_{j=1}^{3n-6} \left[\frac{\Theta_{v_j}/T}{e^{\Theta_{v_j}/T} - 1} - \ln(1 - e^{-\Theta_{v_j}/T}) \right] \tag{4.38}$$

4.3 CHANGING IDENTITIES: CHEMICAL REACTIONS

4.3.1 REACTION PROPERTIES AND PARAMETERS

In a chemical reaction, a set of chemical compounds transforms into another, and molecules change their identities. These changes result from the formation or breaking of chemical bonds between atoms or, in other words, from modifications in how pairs of electrons are shared between atoms. The molecules that undergo these changes, or reactants, have different properties when compared to the molecules, or products, that are obtained at the end of the chemical reaction. Chemical reactions are often not straightforward and involve a series of steps marked by the formation of intermediates, *i.e.*, compounds that are, for instance, adducts of two reactants. When taken together, this series of steps defines what is known as the reaction mechanism (see Figure 4.4).

Adding up each step, just like mathematical equations, gives rise to an overall balanced chemical equation in which each element appears the same number of times on both sides of the chemical equation. Two main types of properties characterize a chemical reaction. The first type includes thermodynamic properties and allows tracking of how identity changes lead to variations in energy distribution between compounds. For a chemical reaction between two reactants A and B yielding two products C and D, we can write the following chemical equation

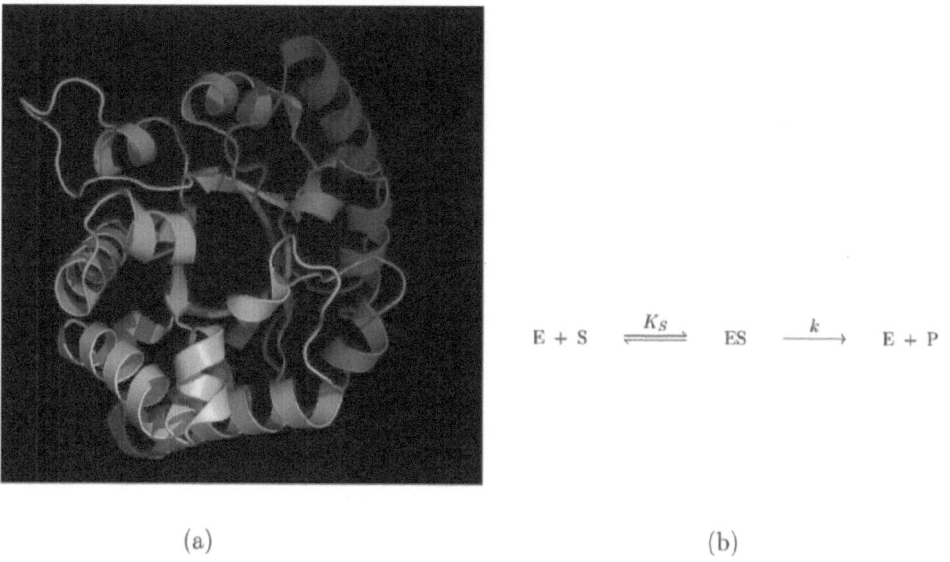

(a)   (b)

Figure 4.4 (a) Ribbon diagram of an enzyme (a triose-phosphate isomerase), and (b) Example of a reaction mechanism for an enzymatic reaction. The enzyme (E) and substrate (S) form an adduct in the first step, which later gives rise to the formation of product (P) and the regeneration of the enzyme in the second step.

$$n_A A + n_B B = n_C C + n_D D \tag{4.39}$$

in which n_A, n_B, n_C, n_D are positive integers, known as the stoichiometric coefficients of the reaction. The $=$ sign stands for the chemical equilibrium that exists between reactants and products as the sum of the forward reaction $n_A A + n_B B \rightarrow n_C C + n_D D$ and of the reverse reaction $n_C C + n_D D \rightarrow n_A A + n_B B$. The relation between the chemical composition of the system and energy is captured by the equilibrium constant K

$$K = \frac{a_C^{n_C} a_D^{n_D}}{a_A^{n_A} a_B^{n_B}} = \exp\left(-\frac{\Delta_r G^0}{RT}\right) \tag{4.40}$$

Here, a_A, a_B, a_C, a_D are known as the activities of each of the compounds and are directly related to the amount of the compounds contained in the reactor. The activity is dimensionless, and the exact expression of the activity depends on the phase in which the compound is. For instance, if the compound A is in solution, we have $a_A = [A]/c_0$, in which $[A]$ denotes the concentration in A (in $mol.L^{-1}$) and $c_0 = 1 \ mol.L^{-1}$ the unit concentration. If A is a gas, $a_A = p_A/p_0$, in which p_A denotes the partial pressure in A (in bar) and $p_0 = 1$ bar. The equilibrium constant, and thus the amounts for the compounds involved in the reaction, is related to the reaction Gibbs free energy $\Delta_r G^0$. $\Delta_r G^0$ depends on temperature and is a characteristic of the energy balance for the reaction. It is obtained from the Gibbs free energy of each of the reactants and product as

$$\Delta_r G^0 = (n_C G_C^0 + n_D G_D^0) - (n_A G_A^0 + n_B G_B^0) \tag{4.41}$$

This means that, once the thermodynamic properties are known (here the standard Gibbs free energy G^0 of a compound that can be found in reference tables), K gives access to the relative amounts of reactants and products and thus to the yield of a chemical reaction.

The second type of properties that characterize chemical reactions are kinetic properties. They allow us to quantify how fast a reaction proceeds and how, for instance, a given concentration or partial pressure in a given compound can impact the rate at which the reaction proceeds. For a simple forward reaction $A \rightarrow B$, we can define the order of reaction r as

$$r = \frac{d[B]}{dt} = -\frac{d[A]}{dt} \tag{4.42}$$

and write r as the following for a reaction occurring in a solution

$$r = k[A]^\alpha \tag{4.43}$$

in which k is a rate constant, $[...]$ denotes the concentration, and α is known as the order of the chemical reaction with respect to reactant A. If the reaction occurs in the gas phase, we have a similar equation for r

$$r = k p_A^\alpha \tag{4.44}$$

with the partial pressure p_A in A replacing the concentration $[A]$ in the equation for r.

Suppose the reaction is known to be an elementary reaction (single steps in multi-step complex reactions are often deemed to be elementary reactions). In that case, the order is equal to the stoichiometric coefficient in the chemical reaction (here $\alpha = 1$). The rate constant k is temperature-dependent and follows the empirical rule proposed by Arrhenius (see Figure 4.5)

$$k = A \exp\left(-\frac{E_a}{RT}\right) \tag{4.45}$$

in which E_a is the activation energy for the reaction. According to Arrhenius' law, an increase in T leads to a greater value for k and thus an increased reaction rate, which we often see in Nature. Furthermore, Arrhenius' law also provides a measure of the impact of a catalyst on a chemical reaction. Indeed, the reaction intermediate for the catalyzed reaction, i.e., the adduct between catalyst and reactants, is associated with a lower activation energy $E_a,$cat that is lower than E_a for the uncatalyzed chemical reaction. This means that $k_{cat} > k$ and, as a result, that the presence of a catalyst accelerates the rate of the chemical reaction.

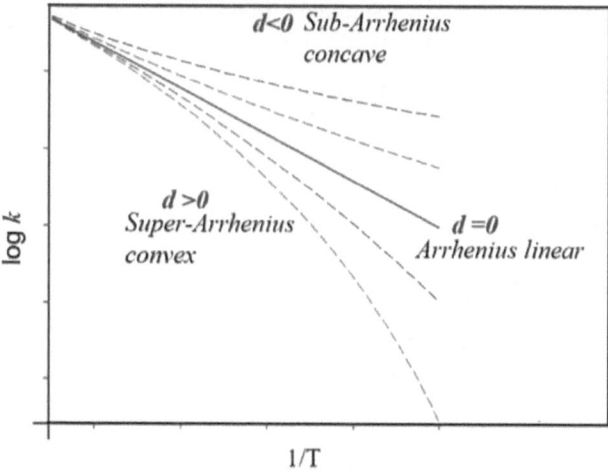

Figure 4.5 Plot of the logarithm of the rate constant k against the reciprocal temperature $1/T$. Arrhenius' law (Eq. 4.45) predicts a linear behavior, with sub-Arrhenius and super-Arrhenius behaviors shown on either side of the Arrhenius line.

4.3.2 PARTITION FUNCTIONS AND EQUILIBRIUM CONSTANTS

The connection between the equilibrium constant k and the Gibbs free energy of the reactants and products provides a direct route to relate K and the partition functions. To illustrate this, we consider the example of a reaction $n_A A + n_B B \rightarrow n_C C + n_D D$ taking place in the gas phase and consider all reactants and products to behave as ideal gases. Starting with the Helmholtz free energy for any compound, we can write that, for instance, for A

$$
\begin{aligned}
A_A &= -k_B T \ln Q_A \\
&= -k_B T \ln \left(\frac{q_A^{N_A}}{N_A!} \right) \\
&= -N_A k_B T \ln (q_A) - N_A k_B T (N_A \ln (N_A) - N_A) \\
&= -N_A k_B T \ln \left(\frac{q_A}{N_A} \right) - N_A k_B T
\end{aligned}
\tag{4.46}
$$

in which we have used the Stirling approximation to approximate $\ln (N_A!)$. Since we have an ideal gas, we can write that $p_A V = N_A k_B T$ leading us to

$$
G_A = A_A + p_A V = -N_A k_B T \ln \left(\frac{q_A}{N_A} \right)
\tag{4.47}
$$

and, for one mole of A, we have $N_A = \mathscr{N}$ and $N_A k_B = R$, which gives

$$
G_A^0 = -RT \ln \left(\frac{q_A}{\mathscr{N}} \right)
\tag{4.48}
$$

We have similar equations for all reactants and products. This gives the following equation for the reaction Gibbs free energy

$$
\begin{aligned}
\Delta_r G^0 &= n_C G_C^0 + n_D G_D^0 - n_A G_A^0 - n_B G_B^0 \\
&= -RT \ln \left[\frac{\left(\frac{q_C}{\mathscr{N}} \right)^{n_C} \left(\frac{q_D}{\mathscr{N}} \right)^{n_D}}{\left(\frac{q_A}{\mathscr{N}} \right)^{n_A} \left(\frac{q_B}{\mathscr{N}} \right)^{n_B}} \right]
\end{aligned}
\tag{4.49}
$$

Since we also have $\Delta_r G^0 = -RT \ln K$, we find the following relation between the equilibrium constant and the molecular partition functions of the reactant and products.

$$K = \left(\frac{q_C^{n_C} q_D^{n_D}}{q_A^{n_A} q_B^{n_B}} \right) \mathcal{N}^{n_A + n_B - n_C - n_D} \tag{4.50}$$

This shows how the amounts of reactants and products, and thus the yield, of a chemical reaction can be predicted from the knowledge of the partition function.

4.3.3 THE ACTIVATED COMPLEX

A chemical reaction often involves the formation of intermediates, with a chemical structure that can be characterized as in-between that of reactants and products. Such an intermediate can form due to a partial rearrangement of reacting molecules or from a partial bonding, or adduct, of two reactants. The idea is that creating this intermediate requires energy. If we visualize the reaction pathway, the system starts from an energy minimum (the reactants), increases in energy as the reaction intermediate starts to form reaches an energy maximum when the formation of the intermediate is complete (transition state), and finally decreases in energy as the intermediate converts into products and reaches a new energy minimum (the products). This picture is similar to the example of a hiker climbing up a hill (reaching a maximum in potential energy at the top) and going down the hill afterward (reaching an energy minimum at the bottom of the hill). Transition state theory developed in the 1930s with significant contributions from Eyring and later by Evans and Polanyi. This has given rise to the concept of activated complex, the critical intermediate during the reaction, and plays a significant role in its kinetics. The International Union of Pure and Applied Chemistry (IUPAC) defines an activated complex as "that assembly of atoms which corresponds to an arbitrary infinitesimally small region at or near the saddle point of a potential energy surface" [253] (see Figure 4.6). In the rest of this section, we use the superscript ‡ commonly used to denote the activated complex.

If we consider the example of the generic reaction

$$A + BC \rightarrow AB + C \tag{4.51}$$

the activated complex $(A - B - C^{\ddagger})$ will take the form of the following intermediate

$$A + BC \rightleftharpoons A - B - C^{\ddagger} \rightarrow AB + C \tag{4.52}$$

The overall reaction rate r can be measured through the formation rate of each product AB or C. This gives

$$r = \frac{d[AB]}{dt} = k \left[A - B - C^{\ddagger} \right] \tag{4.53}$$

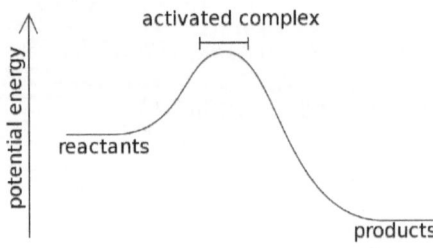

Figure 4.6 Projection of the potential energy surface on a 2D plot. The 2D plot shows the formation of the activated complex at the saddle point as the reaction takes place and reactants (left) are converted into products (right).

in which k is the rate constant for the reaction of dissociation of the activated complex. In terms of partition functions, k is inversely proportional to the partition function of the activated complex. The main difference between the products and the activated complex is the existence of an additional vibration in the activated complex. As discussed above, the vibrational partition function for the extra vibration in the activated complex can be written as

$$q_{\text{vib}} = \frac{1}{1 - \exp\left(\frac{h v^{\ddagger}}{k_B T}\right)} \tag{4.54}$$

Since the activated complex is an adduct and no strong chemical bond has formed, the vibration is loose, and $v^{\ddagger}$ is small. In turn, this allows us to make the following approximation

$$q_{\text{vib}} \sim \frac{h v^{\ddagger}}{k_B T} \tag{4.55}$$

and to obtain the following result for the rate constant k

$$k = \frac{k_B T}{h v^{\ddagger}} \tag{4.56}$$

Now we take into account the fact that the activated complex is involved in a chemical equilibrium with the two reactants A and BC, we can relate $\left[A - B - C^{\ddagger}\right]$ to $[A]$, $[BC]$, and to the equilibrium constant $K^{\ddagger}$

$$\left[A - B - C^{\ddagger}\right] = \frac{K^{\ddagger}[A][BC]}{c_0} \tag{4.57}$$

which gives the overall reaction rate

$$r = \frac{k_B T}{h v^{\ddagger}} \frac{K^{\ddagger}[A][BC]}{c_0} \tag{4.58}$$

Introducing the reaction Gibbs free energy for the formation of the activated complex $\Delta_r G^{\ddagger}$, we have

$$K^{\ddagger} = \exp\left(-\frac{\Delta_r G^{\ddagger}}{RT}\right) \tag{4.59}$$

and find for the reaction rate the following law

$$r = \frac{k_B T}{h v^{\ddagger} c_0} \exp\left(-\frac{\Delta_r G^{\ddagger}}{RT}\right)[A][BC] \tag{4.60}$$

One of the successes of the activated complex theory is that it enables us to recover Arrhenius' law ($k = A\exp(-E_a/RT)$). Indeed, we have obtained an overall rate constant k_o, factor of the concentrations of the two reactants $[A]$ and $[BC]$, as

$$k_o = \frac{k_B T}{h v^{\ddagger} c_0} \exp\left(-\frac{\Delta_r G^{\ddagger}}{RT}\right) \tag{4.61}$$

It provides a physical interpretation of what the activation energy E_a is (here, the reaction free energy for the formation of the activated complex), of the temperature dependence of the exponential, as well as a physical interpretation of the underpinnings of the pre-factor A.

5 Machine Learning and Molecular Systems

"Because of the 'all-or-none' character of nervous activity, neural events and the relations among them can be treated by means of propositional logic. It is found that the behavior of every net can be described in these terms."

Warren S. McCulloch and Walter Pitts,
A logical calculus of the ideas immanent in nervous activity [251]

How does the human brain operate? This extremely complex question remains an outstanding challenge and has fascinated philosophers, scientists, and neurologists for many years. It has given rise, among others, to a new field that draws inspiration from many scientific, engineering, medical and social science disciplines and has experienced, in recent years, tremendous developments: the field of Artificial Intelligence (AI). As discussed by Russell and Norvig [319], AI focuses not only on understanding but also building intelligent entities, *i.e.*, machines that can operate effectively and make rational decisions to achieve their goals. Much knowledge and experience can be gleaned by observing and mimicking how the brain learns from examples, builds models to predict outcomes, and takes rational actions, *i.e.* take the actions that achieve a goal. This sequence of operations is at the basis of what is now called Machine Learning (ML), which, together with Knowledge Representation, Automated Reasoning, Robotics, Natural Language Processing, and Computer Vision, are the components of AI. Over the 20th century, researchers have started to build a mathematical model of how the brain operates, as Warren McCulloch and Walter Pitts discussed in the opening citation of this chapter. They have identified that the information gets coded into signals (*e.g.*, electrical impulses) that are sent to neurons (*i.e.*, the cells that process information in the brain), and that the decoding and decision-making process is achieved via a collection of interconnected neurons. This has given rise to one of the most famous methods used in ML, known as artificial neural networks. These advances were followed by many others, including the design of the Turing test by the well-known mathematician Alan Turing (1912–1954) to assess if one interacts with a machine or a human, the proposal of the foundations of AI based on representation and reasoning by Marvin Minsky (Turing Award winner in 1969) and John McCarthy (Turing Award winner in 1971) and, more recently, the development of deep learning approaches by Yoshua Bengio, Geoffrey Hinton, and Yann LeCun (Turing Award winners in 2018). Before discussing how these approaches can be applied to understand the behavior of molecular groups and, eventually, to program them, we start in this chapter by introducing the main ML methods and algorithms.

5.1 DISTINGUISHING FROM THE MOLECULAR CROWD

5.1.1 LABELS AND CLASSES

ML methods allow us to make predictions and draw conclusions from data. When we look at the data, we can start to see that there are different types of data (numbers, images, etc.) and that there are also different ways of organizing the data. We can sometimes have data that includes a category label. In other words, we know, in that case, what we are looking at as we have the relation between input (cause) and output (consequences) laid out. To illustrate this, let us look at two examples, one based on a dataset of numbers and the other on a dataset of images. The first dataset lists for 1000 configurations of our molecular system two properties (number density N/V and enthalpy -

DOI: 10.1201/9781003006411-6

the input) and the physical state corresponding to each configuration (solid or vapor - the output). We thus have 1000 entries with, for each entry, three pieces of information (two input and one output). The second dataset provides for each of these 1000 configurations a snapshot of the system (a picture - the input) and the physical state corresponding to each configuration (solid or vapor - the output). In this case, we have 1000 entries with, for each entry, two pieces of information (one input and one output). In both cases, the input-output relation is built into the data, with the physical state as the "category label" (solid or vapor). Is that, however, always the case? Preparing labeled data generally requires a human expert to step in. Because our brain has been trained, we can identify a configuration corresponds to the solid phase (with atoms packed within a tight space) from a vapor configuration (with atoms all located far away). We can train ML models to map this input-output, image-state, relation and replicate our human expertise. This will lead to what is known as supervised learning. What happens now if the human expert is unavailable and we have to work with data without category labels, *i.e.*, unlabeled data? Let us imagine, for instance, that we only have access, in the first example, to the numbers for density and enthalpy and, in the second example, to the images of the configurations without knowing which state of matter they are associated with. We can learn a lot from unlabeled data, and that will form the basis for what is known as unsupervised learning. We can discern patterns in the data and define categories by examining the data. For instance, if we look at the density for our 1000 configurations, we will notice that there are indeed two main subsets, one with very low numbers for the density (typically less than 0.1) and the other with large densities (typically larger than 0.3). Similarly, we will likely see two subsets for enthalpy, with a subgroup composed of positive enthalpies and the other subset with negative enthalpies. This means that, even though we did not get any help from a human expert, we immediately spot that the data can be organized into two classes, delimited by a threshold (0.3 for the density and 0 for the enthalpy in our example). There are, of course, several ML methods that are much more systematic and accurate than the rough estimate we have just determined between the "solid" and "vapor" classes to determine thresholds and achieve classification tasks. This is what we explore in the next section.

5.1.2   IDENTIFYING AND HANDLING PATTERNS

The first learning style we discuss is unsupervised learning. This learning style applies when we have unlabeled data to work from. In unsupervised learning, the ML algorithm allows learning patterns in the input without explicit feedback. Clustering is arguably the most common unsupervised learning method. It consists in organizing the input data in subgroups with similar features and, as a result, leads to the detection of multiple categories in the data (see Figure 5.1).

To illustrate how clustering works, we consider the following example and set the number of clusters to be identified to $k = 2$. Our goal is thus to identify two clusters from a list of numbers $(2, 3, 4, 10, 11, 12, 20, 25, 30)$. The implementation of the k-means clustering approach leads to the following steps:

1. We start with two numbers from the list as potential centroids C_1 and C_2 or, in other words, centers for our two clusters: $C_1 = 4$ and $C_2 = 12$ (the initial choice of centroid values does not impact the final result). This allows us to split the list into two initial clusters composed of the numbers closest to a given centroid (with the lowest Euclidean distance from the centroid). This gives a first list $K_1 = (2, 3, 4)$ for the numbers closest to C_1 and a second list $K_2 = (10, 11, 12, 20, 25, 30)$ for the numbers closet to C_2.
2. We then refine our estimates for the two centroids C_1 and C_2 by taking the means of the lists K_1 and K_2, respectively. This gives us new estimates of $C_1 = \frac{2+3+4}{3} = 3$ and of $C_2 = \frac{10+11+12+20+25+30}{6} = 18$. We then update our lists for the two clusters to obtain $K_1 = (2, 3, 4, 10)$ and $K_2 = (11, 12, 20, 25, 30)$.

k-Means Clustering

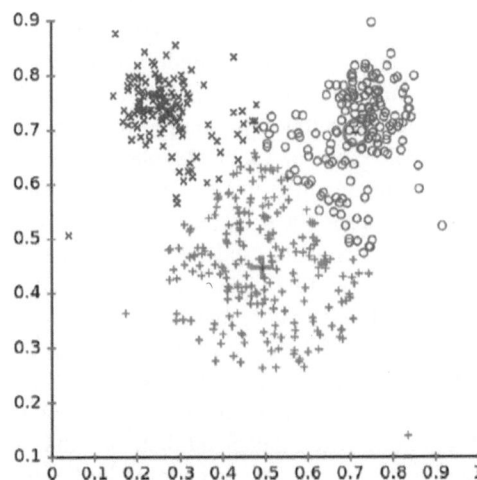

Figure 5.1 Identification of patterns within the data. Here, three main clusters are extracted via the K-means clustering method.

3. We repeat the previous step, revising our estimates for $C_1 = 4.75$ (rounded to the closest integer 5) and for $C_2 = 19.6$ (rounded to the closest integer 20) and our lists to $K_1 = (2, 3, 4, 10, 11, 12)$ and $K_2 = (20, 25, 30)$.
4. We again repeat the previous step to obtain new estimates for $C_1 = 7$ and for $C_2 = 25$ and lists of $K_1 = (2, 3, 4, 10, 11, 12)$ and $K_2 = (20, 25, 30)$.
5. We repeat the previous step and see no further change in the centroid values and lists of numbers, which concludes the k-means clustering process.

Therefore, this analysis allows us to identify two clusters within the data and classify the data into two classes. Furthermore, it provides a way to calculate the mean or centroids for the k clusters. The same working principle can be used to extract additional information from the data. For instance, rather than simply identifying where the k means are located, we can also determine how the data is distributed around each. Let's think of the data as a mixture of k Gaussian distributions. We can use the clustering algorithm to solve for the k μ values and solve iteratively for the variance, and covariance, of each of the k Gaussian distributions.

Another example of an unsupervised learning method is the Principle Component Analysis, or PCA, approach. It is often used to carry out what is known as dimensionality reduction or ordination. Dimensionality reduction allows us to determine which input and linear combination of information contributes the most to obtain a given output. It also provides a way to visualize the data better. This consists in converting high-dimensional data into data with fewer dimensions while minimizing the loss of information. Let us consider, for simplicity, a two-dimensional (2D) dataset. One way to reduce two-dimensional data, plotted on a 2D (x, y) graph, is to carry out a projection on, *e.g.*, the x-axis to obtain a 1D representation of the data. A dimensionality reduction method, such as PCA, aims to limit the loss of information while carrying out the projection. This can be achieved, for instance, by carefully choosing the axis on which the data is projected. If the axis is such that, *e.g.*, the distance of each data point from the axis is minimized, then the loss of information will be minimal. The new axis constitutes what is known as the first principle component axis. Generalizing the method to the case of, *e.g.*, 3D data, we find the second principle component axis by taking the axis perpendicular to the first component, and so on for multi-dimensional data.

5.1.3 LEARNING UNDER SUPERVISION

We have seen how unsupervised learning identifies k classes within the data via, for instance, the k-means clustering algorithm. On the other hand, if we have labeled data, that is data for which category labels have already been assigned, we can also think of training an ML algorithm to automatically give a category label to unseen data. This can be achieved by leveraging another learning style, supervised learning. In supervised learning, the model learns from input-output pairs to build a function that maps the input to the output. Going back to our examples, the inputs can be snapshots of the configurations of the molecular system, with each image associated with an output label that is either "vapor" or "solid". Alternatively, the inputs can be pairs of thermodynamic data, *e.g.*, (density, enthalpy), and the outputs either the "vapor" or "solid" category label. Through training on available labeled data, the model learns a function that, given a new image or a new (density, enthalpy) set, predicts the corresponding category label (here, the system's physical state). To illustrate how supervised learning methods work, we start with the example of the Support Vector Machine (SVM) algorithm. A graphical illustration of how SVM defines a boundary between two classes in the data is shown in Figure 5.2.

The SVM algorithm constructs what is known as the maximum margin separator. This decision boundary between the two classes (say, the "positive class" and the "negative class") exhibits the largest possible distance to the points provided in the dataset. An advantage of an SVM model is that it generalizes very well, meaning that its predictions are reliable when applied to unseen data. This is because SVM minimizes the generalization loss by choosing a separator that is farthest away from the examples seen so far in the data. As shown in Figure 5.2, this separator is surrounded by a region with no data points. This region is known as the margin and is bounded by dashed lines in Figure 5.2. Since no data point from the training dataset lies within the margin, it is unlikely that a new, unseen data point will fall within the margin. This is why SVM models minimize the expected generalization loss instead of focusing on the loss of the training data. The SVM algorithm thus finds the hyperplane (or line in 2D) equation that separates the data points into two distinct classes. More specifically, SVM identifies the vector **w** and intercept b that define the separator of equation

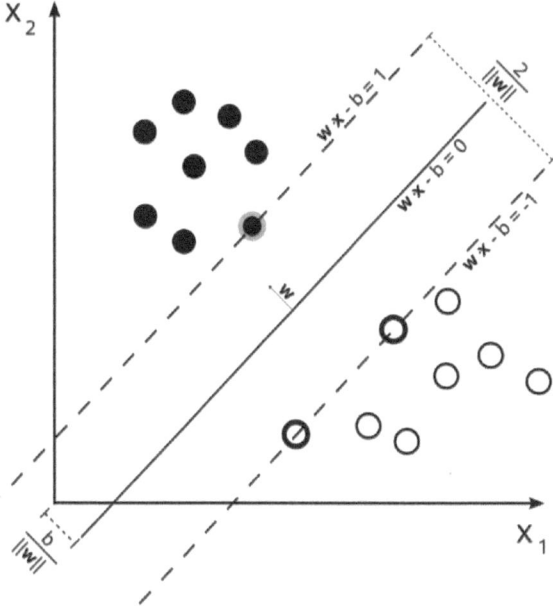

Figure 5.2 Classification via Support Vector Machine (SVM).

$\mathbf{w^T}.\mathbf{x} - b = 0$, and the two hyperplanes located at a distance of $\pm \frac{1}{||\mathbf{w}||}$ on either side of the separator (dashed lines on the plot).

Maximizing the margin involves minimizing $||\mathbf{w}||$, subject to the constraint that no data point falls within the margin. For each data point characterized by the index i, the constraint can be written as

$$\mathbf{w^T}.\mathbf{x_i} - b \geq 1 \tag{5.1}$$

if the corresponding output belongs to the "positive" class $y_i = 1$, and

$$\mathbf{w^T}.\mathbf{x_i} - b \leq 1 \tag{5.2}$$

if the corresponding output belongs to the "positive" class $y_i = -1$. The two conditions can be recast into a single constraint based on the sign of the product

$$y_i.(\mathbf{w^T}.\mathbf{x_i} - b) \geq 1 \tag{5.3}$$

This means that SVM can be understood as an optimization process, *i.e.*, the minimization of $||\mathbf{w}||$ subject to the constraint $y_i.\mathbf{w^T}.\mathbf{x_i} - b \geq 1$ for all data points i. The classifier C thus takes the form of $C(\mathbf{x} = \text{sign} (\mathbf{w^T}.\mathbf{x_i} - b)$. Interestingly, the weights in the $\mathbf{w}$ vector are completely determined by what the support vectors, defined by the data points that lie closest to the separator. SVM is just one of the many supervised learning methods that perform efficient classification tasks, and other examples include *e.g.*, decision trees, and random forest methods.

5.2 QUANTITATIVE MODELS FOR MOLECULAR GROUPS

5.2.1 TRAINING REGRESSION MODELS

What happens if we now want to build an ML model that makes *quantitative* predictions rather than classification tasks? The simplest case of such an ML model is a regression model with a univariate linear function. This method consists in fitting a line through the data (see Figure 5.3) or, in other words, finding the equation of the form $y = w_1 x + w_0$ in which y is the output, x is the input, and (w_0, w_1) are the coefficients, or weights, learned from the data. Linear regression finds the best (w_0, w_1) set that fits the data.

This set of weights is obtained by using a training set of n data points in the (x, y) plane and minimizing the empirical loss measured by the squared-error loss function summed over all the training examples:

$$\text{Loss}(w_0, w_1) = \sum_{i=1}^{n} (y_i - (w_1 x_i + w_0))^2 \tag{5.4}$$

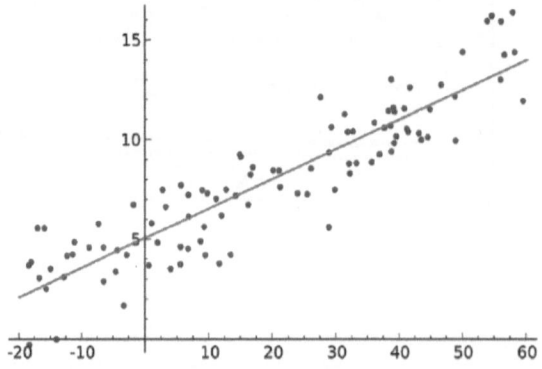

Figure 5.3 Linear regression model built from data points shown as filled circles.

To. minimize the loss, we need to write the corresponding mathematical condition, *i.e.*, that the partial derivatives of the loss function with respect to w_0 and w_1 are zero. This yields a system of two equations with two unknowns

$$\frac{\partial}{\partial w_0} \sum_{i=1}^{n} (y_i - (w_1 x_i + w_0))^2 = 0$$

$$\frac{\partial}{\partial w_1} \sum_{i=1}^{n} (y_i - (w_1 x_i + w_0))^2 = 0 \tag{5.5}$$

leading to the following solution

$$w_1 = \frac{n \left(\sum_i x_i y_i \right) - \left(\sum_i x_i \right) \left(\sum_i y_i \right)}{n \left(\sum_i x_i^2 \right) - \left(\sum_i x_i \right)^2}$$

$$w_0 = \frac{\sum_i y_i - w_1 \left(\sum_i x_i \right)}{n} \tag{5.6}$$

Alternatively, we could have also chosen to determine the weights through an iterative process that extends beyond the case of the univariate linear model for which the partial derivatives can be calculated analytically. Let us call the general function that maps the input-output relation as $h_\mathbf{w}$, where $\mathbf{w}$ denotes the weights in a vector form. The alternative solution consists of using gradient descent to minimize the loss and to update the value of each of the weights iteratively until the model is deemed to have converged. To this end, we introduce a parameter α, known as the learning rate, and update the weights according to

$$w_i \leftarrow w_i - \alpha \frac{\partial}{\partial w_i} \text{Loss}(\mathbf{w}) \tag{5.7}$$

To calculate the partial derivative, we invoke the chain rule to write that, for a training example (x, y), we have

$$\frac{\partial}{\partial w_i} \text{Loss}(\mathbf{w}) = \frac{\partial}{\partial w_i} (y - h_\mathbf{w}(x))^2$$

$$= 2 (y - h_\mathbf{w}(x)) \frac{\partial}{\partial w_i} (y - h_\mathbf{w}(x)) \tag{5.8}$$

This provides the following partial derivatives in the univariate case

$$\frac{\partial}{\partial w_0} \text{Loss}(\mathbf{w}) = -2 (y - h_\mathbf{w}(x))$$

$$\frac{\partial}{\partial w_1} \text{Loss}(\mathbf{w}) = -2 (y - h_\mathbf{w}(x)) \times x \tag{5.9}$$

and, in turn, to the following updating rules

$$w_0 \leftarrow w_0 + \alpha (y - h_\mathbf{w}(x))$$
$$w_1 \leftarrow w_1 + \alpha (y - h_\mathbf{w}(x)) \times x \tag{5.10}$$

Generalizing to the case of batch updating, when the updating is carried out on a set of n training examples rather than on a single example, we obtain

$$w_0 \leftarrow w_0 + \alpha \sum_{i=1}^{n} (y_i - h_\mathbf{w}(x_i))$$
$$w_1 \leftarrow w_1 + \alpha \sum_{i=1}^{n} (y_i - h_\mathbf{w}(x_i)) \times x_i \tag{5.11}$$

We finally add that the same reasoning applies, albeit in higher dimensions, to the case of multi-variable regression models in which the input $\mathbf{x}$ is now a vector with p elements and $\mathbf{w}$ is a $(p+1)$-element vector.

5.2.2 MAPPING NUMBERS: ARTIFICIAL NEURAL NETWORKS

An understanding of how rationality and intelligence emerge from the brain has remained elusive so far. Even though Aristotle noted in 335 BCE the uniqueness of the human brain ("Of all the animals, man has the largest brain in proportion to his size"), it was only much later, during the 19th century, that the brain became accepted universally as the undisputed seat of consciousness. This prompted many scientists to study and explore the role of neurons, also known as nerve cells, in the operation of the brain. A scheme of a neuron is shown in Figure 5.4. Neurons are at the center of all nerve tissues. They consist of a body (soma) containing the cell nucleus. Tree-like appendages known as dendrites branch out of the soma. The role of dendrites is to receive electrochemical signals and messages. There is also a long fiber, stretching out over centimeters, *i.e.*, two orders of magnitude longer than the cell body, the axon, that transmits messages from the cell body. The axon branches into strands that connect to the dendrites and somas of other neurons. These connections, also known as synapses, can involve hundreds of thousands of neurons. It was found that, as a result of stimulations, synaptic connections can undergo long-term changes and that neurons can form new connections with other neurons and sometimes migrate. It is thought that these changes are the basis of the learning process.

The human brain's neurons were the inspiration for the development of mathematical models for neurons, giving rise to the perceptron and, eventually, to the multi-layer neural network that is now used to learn complex mappings between inputs and outputs. Neural networks comprise nodes, also called units or neurons, connected by links, each associated with a numerical weight (see Figure 5.5). There is considerable freedom in defining a specific architecture for a neural network, so that we will focus here on the example of a feed-forward neural network. These weights allow for the storage of information in the network. Furthermore, learning takes place through the updating of these weights. Some neurons belonging to the input or output layers are directly connected to the exterior. In contrast, other neurons are only connected to other neurons from the network and, as such, belong to what is known as "hidden" layers. Every neuron receives "signals" from neurons belonging to the preceding layer, computes the weighted sum of these incoming signals, and calculates an activation level sent to neurons belonging to the next layer. The latter is performed by calculat-

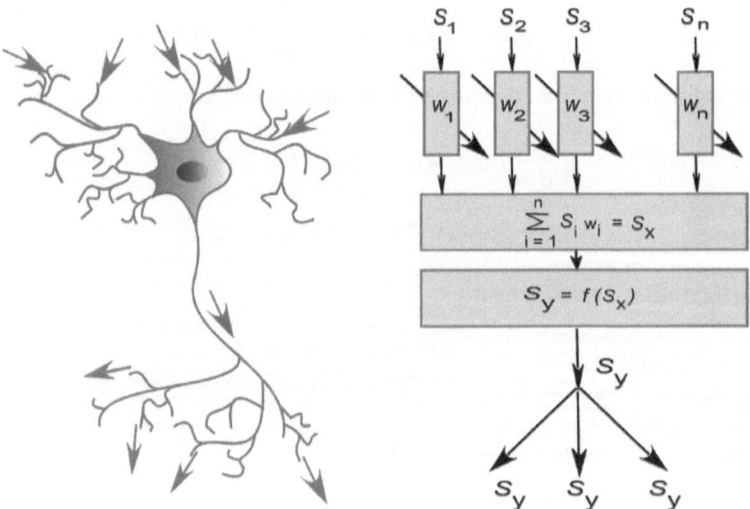

Figure 5.4 Neuron or the biological inspiration for an ANN. The neuron uses electrical impulses and chemical signals to transmit information. Just like dendrites collect signals in a neuron, the ANN collects input signals (S_i, $i = 1, 2, \ldots$), calculates a weighted sum of the input signals (weight factors w_i, $i = 1, 2, \ldots$, and applies the activation function f to the weighted sum to generate the output S_y.

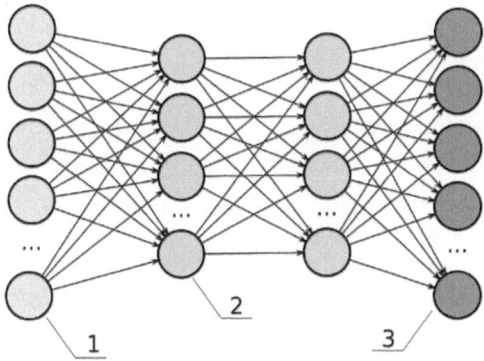

Figure 5.5 Neural network architecture with the input layer (labeled as 1), the hidden layers (labeled as 2), and the output layer (labeled as 3).

ing the value of an activation function $f(x)$, where x stands for the weighted sum of the incoming signals. Once the input "signals" have been through the entire network, we obtain a numerical value for the outputs as a function of the inputs. Taking the example of a 4-layer neural network, with an input layer, two hidden layers, and an output layer and calculating the value obtained for the first output neuron out_1, we get the following analytic expression.

$$out_1 = f_4[b_3 + \sum_{l=1}^{n_3} W(3,4,l,1)f_3(b_2 + \sum_{j=1}^{n_2} W(2,3,j,l)f_2[(b_1 + \sum_{i=1}^{n_1} W(1,2,i,j)G_i])] \qquad (5.12)$$

In this equation, $W(i-1,i,l,j)$ is the weight matrix connecting neuron l from layer $i-1$ to neuron j from layer i (the input layer, n_i is the number of neurons in layer i (with the convention that 1 denotes the input layer, 2 and 3 the hidden layers, and 4 the output layer), f_i are activation functions, b_i are bias nodes that act as adjustable offsets (*i.e.*, constant numbers that are added to shift the value of the incoming "signal"), and G_i are the input neurons. Once the weight values have been optimized, the output can be calculated quickly and efficiently for any set of input values through Eq. 5.12.

5.2.3 OPTIMIZATION THROUGH BACK-PROPAGATION

The next step consists of finding a rationale to update the weights and optimize the neural network's learning. While back-propagation is now widely recognized as an efficient learning method, the computational requirements of the algorithm slowed down its application to practical problems. They limited for several decades the performance of neural networks. While Bryson and Ho first proposed back-propagation in 1969 [182], it only came to fruition at the end of the 1980s [220] and was at the center of the success of deep learning since the beginning of the 2010s. We introduce this algorithm by considering what training a neural network entails. Training the neural network amounts to minimizing an error function. Let us consider the example of the mean-squared error as the error function Δ as

$$\Delta = \frac{1}{N_e} \sum_{i=1}^{N_e} \left[(out(x_i) - y_i)^2 \right] \qquad (5.13)$$

in which N_e is the number of examples (data points) used in the training dataset, x_i and y_i are the input and output pairs for a given example i, and $out(x_i)$ is the prediction made by the neural network for an input value x_i. In other words, Δ measures the departure of the neural network predictions from the actual outputs in the training dataset. In the back-propagation algorithm, the weights are updated according to an iterative process. Predictions are made based on the current weight values; in other words, the output is calculated after a forward pass of the neural network. Then, based on these predictions and the corresponding mean-squared error, all the weights are adjusted during a "backward" pass according to the following equation

$$W^{n+1}(i,j,k,l) = W^n(i,j,k,l) - \lambda \frac{\partial \Delta}{\partial W^n(i,j,k,l)} \tag{5.14}$$

in which W^n and W^{n+1} denotes the weights after n^{th} and $(n+1)^{th}$ iterations, respectively, and λ is a parameter known as the learning rate, which is a positive number, small enough to avoid numerical instabilities (typically of the order of $10^{-2} - 10^{-1}$). The partial derivatives of Δ with respect to $W^n(i,j,k,l)$ can be evaluated analytically since the neural network predictions are themselves an analytical function of the weights (Eq. 5.12). The entire process is then repeated until additional iterations, or epochs, do not lead to a change in the accuracy of the predictions Δ (Eq. 5.13).

How complex should the neural network be to yield highly accurate predictions? As with all models, Ockham's razor and the principle of parsimony should apply. This means we should balance complexity and simplicity to obtain the best neural network model. If the neural network contains too few neurons or is too shallow, then it will not be able to account for all the details and subtleties in the data. This will lead to poor accuracy and what is known as underfitting. On the other hand, if the networks contain too many layers and neurons and thus too many weight factors, the model will give rise to overfitting. This means that the accuracy of the training data of such a model will be satisfactory. Still, the model will exhibit poor generalization, and when applied to unseen data, its predictions will be inaccurate. One way to assess this is by splitting the available data into two datasets, a training dataset on which the optimization of the weights is carried out and a validation dataset on which the ability of the model to make accurate predictions is tested. A much larger error on the validation dataset than on the training dataset and increasing errors on the validation dataset as the number of epochs increases are signs that the model is overfitting the data.

5.3 BEYOND ARTIFICIAL NEURAL NETWORKS

5.3.1 LEARNING BY WATCHING: CONVOLUTIONAL NEURAL NETWORKS

What happens now if, instead of working with numerical data (density, enthalpy, free energy), we want to train an ML model on images such as *e.g.*, snapshots of configurations of the molecular system? This requires a different approach and using a specific type of neural network called a convolutional neural network (CNN). The idea here is to leverage the information in the image to train the model to predict its properties. How can an ML algorithm extract information from images? Color images are composed of pixels, and each pixel is associated with a triplet of integers corresponding to their RGB values. This stems from the fact that most humans have three types of color-sensitive receptors and that the principle of trichromacy applies. In essence, any color can be obtained by having a mixture of red (R), green (G), and blue (B), with the intensity of each color controlled by an integer between 0 and 255 (this range is imposed by the coding of information on 8 bits). Therefore, it is appropriate to take the RGB values from all the pixels in the image and treat the image as a vector. However, the location of pixels in the image and, more specifically, their adjacency also provide critical information that the model needs to learn to correctly identify shapes and their background. As discussed by Russell and Norvig [319], taking into account all this information without applying any filter would result in handling more than 9 trillion weights for a megapixel RGB image which, in turn, would require including a vast number of images in the training dataset and would lead to a substantial computational cost. CNNs are designed to address these issues by considering spatial local connections in the early layers of the network and by having patterns of weights, or kernels, replicated across the units of each layer. Applying a kernel to the pixels of an image is known as convolution.

We briefly discuss how a convolution layer works in the example of a 1D image. We start with a 1D image or, for simplicity, an input vector $\mathbf{x}$ of size n standing for the n pixels in the image. Let us consider, for example, an image with 9 pixels corresponding to

$$
\mathbf{x} = \begin{pmatrix} 8 \\ 7 \\ 8 \\ 2 \\ 9 \\ 6 \\ 7 \end{pmatrix} \tag{5.15}
$$

We decided to use a kernel to detect a dark pixel or, in other words, a pixel associated with a low number or low intensity. To this end, we apply a kernel vector $[+1, -1, +1]$ of size 3 with a stride of 2, meaning that we take snippets of 3 pixels with the centers of two successive snippets (or triplets of numbers) separated by a distance of two pixels. We can write this series of mathematical operations as the product of a matrix, based on this kernel, by the vector $\mathbf{x}$ to obtain

$$
\begin{pmatrix} +1 & -1 & +1 & 0 & 0 & 0 & 0 \\ 0 & 0 & +1 & -1 & +1 & 0 & 0 \\ 0 & 0 & 0 & 0 & +1 & -1 & +1 \end{pmatrix} \begin{pmatrix} 8 \\ 9 \\ 8 \\ 2 \\ 9 \\ 8 \\ 7 \end{pmatrix} = \begin{pmatrix} 7 \\ 15 \\ 8 \end{pmatrix} \tag{5.16}
$$

which highlights the presence of dark pixels in the middle region of the 1D image. Other types of layers in a CNN include pooling layers that replace adjacent units with a single value. This single value can either be the average of the average units (average-pooling) or the maximum value (max-pooling). This has the effect of coarsening the resolution and downsampling the image. The final architecture of the CNN will thus typically include convolution and pooling layers for the early layers, coupled with conventional layers in the outer layers that perform classification tasks or numerical predictions.

5.3.2   TIME SEQUENCES AND RECURRENT NEURAL NETWORKS

What happens now if we want to train a neural network to learn time-dependent information on the molecular system, such as, for instance, its temporal evolution in response to an external field? This can typically be achieved by a specific type of neural network known as a recurrent neural network (RNN). Such networks differ from conventional feed-forward networks because they allow cycles in the computation graph. In other words, the network units can take as input a value computed from their output at an earlier step in the computation. This means that RNNs have a memory since the inputs received earlier will impact the output of the RNN to the current information. As such, RNNs are exceptionally well suited to analyzing sequential data.

Through training, RNNs can identify a parametrized function $f_\mathbf{w}$ that models the dynamics of the system at all times. RNNs rely on what is known as a Markov assumption. Consider an RNN with an input layer $\mathbf{x}$, a hidden layer $\mathbf{z}$, and an output layer $\mathbf{y}$. The Markov assumption implies that the hidden state $\mathbf{z}_t$ of the network at time t encompasses the information from all previous inputs. For such an RNN, the update step will be carried out as follows.

$$
\mathbf{z}_t = f_\mathbf{w}(\mathbf{z}_{t-1}, \mathbf{x}_t) \tag{5.17}
$$

If we denote the activation function of the hidden layer by $\mathbf{g_z}$ and the activation function for the output layer by $\mathbf{g_y}$, we obtain the following equations for the model at time t

$$
\begin{aligned}
\mathbf{z}_t &= f_\mathbf{w}(\mathbf{z}_{t-1}, \mathbf{x}_t) \\
&= \mathbf{g}_z(\mathbf{w}_{z,z}\mathbf{z}_{t-1} + \mathbf{w}_{x,z}\mathbf{x}_t) \\
\hat{\mathbf{y}}_t &= \mathbf{g}_y(\mathbf{w}_{z,y}\mathbf{z}_t)
\end{aligned}
\tag{5.18}
$$

Considering a time sequence of finite length, for instance, with n steps, we can think of unrolling the RNN and turning this model into a feed-forward network with the weights $\mathbf{w}_{x,z}$, $\mathbf{w}_{z,z}$ and $\mathbf{w}_{z,y}$ shared across all time steps. The optimization of the weights can then take place similarly to "regular" neural networks, with the calculations of gradients in the usual way according to an algorithm known as back-propagation through time. For numerical reasons, an alternative architecture has been developed for RNNs known as long short-term memory RNNs or LSTM. Unlike an RNN that multiplies its memory by a weight matrix at every time step (and thus can eventually run into numerical issues), a memory cell $\mathbf{c}$ is copied from one time step to the next in an LSTM. New information is entered into the memory through updates and gating units. For instance, a forget gate $\mathbf{f}$ determines if each element of the memory cell is copied to the next time step (remembered) or reset to zero (forgotten). LSTMs have demonstrated excellent performance over a wide range of tasks.

5.3.3   UNDERSTANDING POLICIES: THE ADVENT OF REINFORCEMENT LEARNING

We have seen cases where we have extracted information from unlabeled data and identified functions that map input and output pairs when we work with labeled data. What happens now if we are provided only part of the information on the system rather than the complete label? Let us, for instance, consider a flow of particles and track how many particles successfully move from a starting point to an endpoint within a given time interval. In this case, being provided part of the information could be being only told of the outcome of the trajectory, *i.e.*, a positive result would mean that the particle has reached the endpoint. This means that we do not receive any feedback during the particle's motion on the progress of the particle toward the goal and the likelihood of the trajectory being successful.

Reinforcement learning (RL) is a framework that enables learning which sequence of actions is optimal to achieve success. This is done by passing on rewards, positive in case of success and negative otherwise, to the system. Rewards may be applied at the end of the trajectory, *i.e.*, for the terminal state, or at various stages during the trajectory. What is necessary to realize, however, is that the feedback does not give any insight into what would have been the best trajectory or strategy, to achieve a goal. It is indeed up to the system to alter its trajectory or actions to achieve the goal. In other words, a single unfortunate step can ruin the entire trajectory, and it will be up to the system to identify the wrong step since the only feedback provided is a negative outcome. RL has recently been used to find efficient navigation strategies and show how particles and microorganisms learn to move optimally by trial and error [61].

In RL, the system periodically receives rewards, also called reinforcements in psychology, that reflect how well it is doing. In other words, the system does not know the transition model (*i.e.*, what the outcome of an action in a given state is) or the reward function and has to make a move or take an action to learn more. RL is especially appealing since providing a reward signal to the system is generally much more straightforward than providing labeled examples of successful trajectories. Furthermore, specifying the reward function is generally direct when applied to the final state (has the system reached the end point?) or to intermediate steps (is the system closer to the endpoint?). There are many approaches to RL, including model-based RL, in which a transition model guiding the moves or actions, or "rules of the game", is provided to the system, and model-free RL, in which the system does not know or learn the transition model. Model-free RL can either be achieved through an action-utility learning approach, which determines a quality function, or Q-function

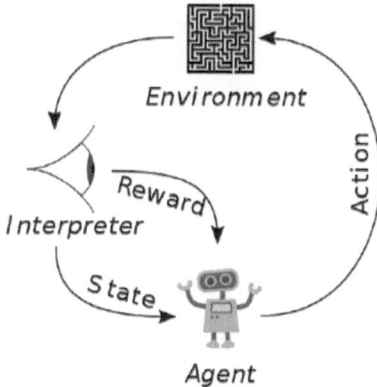

Figure 5.6 Schematic for the reinforcement learning approach.

$Q(s, a)$, that quantifies the sum of rewards from state s onward if action a is taken, or a policy-search approach, which finds a policy $\pi(s)$ that captures the mapping from states to actions (Figure 5.6).

Such decision problems are known as Markov decision processes. We illustrate how RL works on the example of the Q-learning method. Defining the utility of a given state as the reward and the expected utility of the successor states, it can be shown that the utility obeys the Bellman equation for an optimal policy

$$U(s) = \max_{a \in A(s)} \sum_{s'} P(s'|s, a) \left[R(s, a, s') + \gamma U(s') \right] \tag{5.19}$$

in which a the action, s and s' a set of states, $A(s)$ the set of actions in state s, U the utility function, R the reward function, P the transition model, and γ the discount rate for the rewards. The Bellman equation for $Q(s, a)$ is given by

$$Q(s, a) = \sum_{s'} P(s'|s, a) \left[R(s, a, s') + \gamma \max_{a'} Q(s', a') \right] \tag{5.20}$$

with the following Q-learning update rule

$$Q(s, a) \leftarrow Q(s, a) + \alpha \left[R(s, a, s') + \gamma \max_{a'} Q(s', a') - Q(s, a) \right] \tag{5.21}$$

in which α is the learning rate. This approach can be combined with deep learning to provide a more flexible and complex approximator for utility and Q functions.

Part II

Static Trends: Equilibrium Statistics

6 Polling a Molecular Population
Monte Carlo and Wang Landau Simulations

"The method of Monte Carlo integrations over configuration space seems to be a feasible approach to statistical mechanical problems which are as yet not analytically soluble."

Nicholas Metropolis, Arianna W. Rosenbluth, Marshall N. Rosenbluth,
Augusta H. Teller, and Edward Teller,
Equation of state calculations by fast computing machines [260]

We have laid the groundwork for a statistical, probabilistic approach to the study of groups and atoms, and molecules. We now need to build on these theoretical foundations to predict the properties of molecular groups. So the question we will address in this chapter is the following: how do we apply these tools to probe a molecular group's behavior and determine its collective response? As suggested by Metropolis and co-workers in the opening citation, a general answer will emerge through numerical and computational methods. In a social context, let us take a metaphor for statistical mechanics. The act of polling a group, or recording the opinion of this group, bears many similarities with the numerical evaluation of the properties of a molecular group. Indeed, polling relies on a statistical approach and hinges on carefully choosing a sample of individuals representative of the entire population when matters of broad interest, such as general elections or highly anticipated sporting contests, are the poll topics. Polling has been very successful over the years. It has now emerged as a very popular tool to gauge the opinion of social groups on matters as varied as, among others, politics, sports, philosophy, and preferences for competing commercial products. This is the same process that the Monte Carlo method, advocated by Metropolis and co-workers, follows but this time for a molecular group. The technique generates a sample of molecular configurations representative of the molecular group, obtains a molecular analog to an opinion (for instance, their energy), and averages the configurations' properties to determine thermodynamic properties and collective responses.

6.1 THE BIRTH OF THE MONTE CARLO METHOD

6.1.1 RANDOMNESS AND INTEGRATION

Although seemingly counter-intuitive, using random numbers or repeating a random set of actions can illuminate a well-defined, deterministic property or process. For instance [8], the French polymath Laplace suggested in 1812 that a series of random actions could constitute a practical method to determine the value of π! Laplace was referring to a practical realization of what was known as a Buffon experiment. Indeed, during the 18th century, the French naturalist Buffon showed that if a needle of length l is thrown randomly onto a set of equally spaced parallel lines, with a distance interval d between two successive lines ($d > l$), the probability that a needle falls across a line is $2l/\pi d$. This was later tested by the Italian mathematician Lazzarini, who claimed to have obtained the exact value of π up to its sixth decimal.

The method can be leveraged to evaluate any integral, which is precisely what we are trying to determine in statistical mechanics, as average values can be written as integrals of a property over the positions and momenta of all particles within the molecular group. To illustrate how the

DOI: 10.1201/9781003006411-8

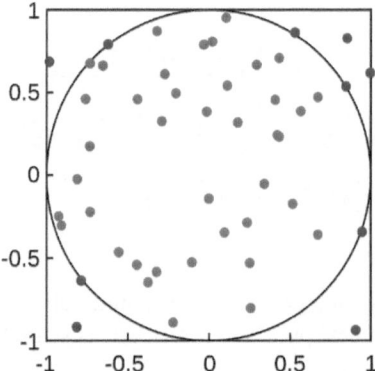

Figure 6.1 Example of a Monte Carlo integration process. Random coordinates are drawn and plotted in the square. The number of points with coordinates that fall within the unit circle is denoted by τ_{in} are shown in red, while the number of points with coordinates that fall outside the unit circle is denoted by τ_{out}. The ratio of τ_{in} to the number of points dawn randomly estimates π (see main text).

Monte Carlo method works, we look at a simple integral that, incidentally, allows us to determine a numerical estimate for π.

Figure 6.1 gives a graphical interpretation of the process. We start by considering the square shown in Figure 6.1 and the inscribed disk of unit radius. In terms of integrals, the areas can be written as

$$\mathcal{A}_{square} = \int\int_{square} dxdy = 4 \tag{6.1}$$

for the unit square, with the (square) being defined with both x and y coordinates ranging from -1 to 1, and

$$\mathcal{A}_{disk} = \int\int_{disk} dxdy = \pi \tag{6.2}$$

for the unit circle, with the (disk) being defined for x and y coordinates such that $(x^2 + y^2) \leq 1$.

To determine the value for π, we draw random coordinates, *i.e.*, two random numbers between -1 and $+1$ for the x-coordinate and the y-coordinate, respectively, for many points. We then plot these points in Figure 6.1. Points with coordinates that fall within the unit disk (τ_{in}) are shown in red, while points with coordinates that fall outside the unit circle (τ_{out}) are shown in blue. For a large number of random coordinates drawn, the ratio of τ_{in} to the total number of random attempts ($\tau_{in} + \tau_{out}$) becomes equal to

$$\frac{\tau_{in}}{\tau_{in} + \tau_{out}} \approx \frac{\mathcal{A}_{disk}}{\mathcal{A}_{square}} = \frac{\pi}{4} \tag{6.3}$$

giving us a direct access to π via the measure of τ_{in}.

6.1.2   SAMPLE MEAN APPROACH

We have just discussed how, by simply drawing random numbers for the coordinates, we were able to provide a numerical estimate for a 2D integral. To provide a more general and systematic method to calculate integrals, we now introduce the sample mean approach [3]. Let us first consider the calculation of the 1D-integral $\mathcal{I}$ defined by

$$\mathcal{I} = \int_a^b I(x)dx \tag{6.4}$$

Calculating the integral $\mathcal{I}$ requires the evaluation of the function $I(x)$ over the 1D-interval $[a,b]$. However, how do we select the values of x for which the function should be evaluated? A considerably wide choice exists in doing so (see Figure 6.2). A widely used example is the so-called

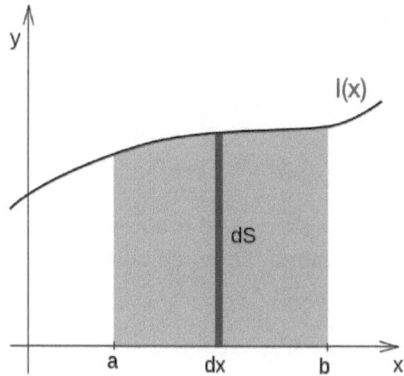

Figure 6.2 Evaluation of the integral $\mathscr{I}$, *i.e.*, area under the curve $I(x)$ over the interval $[a,b]$. This can be carried out by, among others, one of the following: (i) splitting the interval into very small intervals of length dx and evaluating the function at both ends of the intervals and at the midpoint (see Simpson's rule in the text), or (ii) drawing a large number τ of random points where the function $I(x)$ is evaluated (see, *e.g.*, the example of the uniform distribution in the text).

Simpson's rule which provides an accurate estimate via the following equation

$$\mathscr{I} \approx \frac{b-a}{6}\left[I(a) + 4I\left(\frac{a+b}{2}\right) + I(b)\right]\tag{6.5}$$

To improve accuracy, the interval $[a,b]$ may be split into smaller intervals, and Simpson's rule is applied to each interval.

Simpson's rule is very efficient in 1D, but it requires systematic and frequent function evaluations to retain a high accuracy. Thus, its performance decreases for high-dimensional integrals, *i.e.*, typically the type of integrals we need to evaluate in statistical mechanics where we have $6N$ phase space variables over which to integrate and $N \sim \mathcal{O}(10^2 - 10^3)$ particles. The idea is to use the sample mean approach and draw from a distribution $p(x)$ the values of x for which the function needs to be evaluated. Recasting the integral $\mathscr{I}$ under a form that explicitly depends on the distribution p, we have

$$\mathscr{I} = \int_a^b \left(\frac{I(x)}{p(x)}\right) p(x)dx\tag{6.6}$$

If we generate a random number η from the distribution p over the range $[a,b]$ and repeat this τ times, we can write for the integral

$$\mathscr{I} = \left\langle \frac{I(\eta)}{p(\eta)} \right\rangle_\tau\tag{6.7}$$

where $\langle ... \rangle$ indicates that the average is taken over all τ.

For instance, if $p(x)$ is an uniform distribution over the interval $[a,b]$, then

$$p(x) = \frac{1}{(b-a)} \quad a \leq x \leq b\tag{6.8}$$

which leads to the following estimate for $\mathscr{I}$

$$\mathscr{I} \approx \frac{(b-a)}{\tau}\sum_{i=1}^{\tau} I(\eta_i)\tag{6.9}$$

Of course, there are many more efficient choices than uniform distribution, as we will see in the next section with the concept of importance sampling.

6.1.3 THE CONCEPT OF IMPORTANCE SAMPLING

We have seen in the previous section how we could evaluate the integral $\mathscr{I}$ by drawing random numbers from a pre-defined distribution and examining the uniform distribution example. What happens now if we know that some regions of phase space or ranges of x in $p(x)$ have a greater weight in evaluating the integral? This is where importance sampling is incredibly crucial and efficient. Importance sampling involves drawing random numbers from a distribution that emphasizes the regions of phase space contributing the most to the integral. To illustrate the significance of importance sampling, we consider a molecular group containing N molecules and associated, for each configuration, to a phase space point Γ, *i.e.*, with a set of $6N$ phase space coordinates ($3N$ spatial coordinates for the position and $3N$ for the momenta). If the total energy, that is, the sum of the potential and kinetic energy, for the molecular group, is noted as $E(\Gamma)$, the canonical ensemble average $\langle B \rangle$ of any property B can be calculated through the following integral

$$\langle B \rangle = \frac{\int B(\Gamma) \exp\left(-\beta E(\Gamma)\right) d\Gamma}{Q_{NVT}} \tag{6.10}$$

in which Q_{NVT} denotes the canonical partition function for the system. In the above equation, the integrand is $B(\Gamma) \exp\left(-\beta E(\Gamma)\right)$. It can be quickly seen that whenever $\exp\left(-\beta E(\Gamma)\right)$ is close to 0 (for instance, when the total energy is tremendous, which happens when two molecules' coordinates are such that they are too close to one another and start repelling one another at short range), the whole integrand is close to 0. It does not contribute much to the ensemble average $\langle B \rangle$. Conversely, when the total energy is low, *i.e.*, when molecules are experiencing an overall favorable, attractive interaction with their neighbors, the weighting factor $\exp\left(-\beta E(\Gamma)\right)$ increases and the integrand becomes significant as a result.

Therefore, rather than drawing random coordinates from an arbitrary distribution $p(\Gamma)$, such as the uniform distribution, to generate a new configuration, it is extremely efficient to generate configurations according to their Boltzmann weight, *i.e.*, according to the probability distribution $p(\Gamma) = \exp\left(-\beta E(\Gamma)\right)$. Doing so decreases the number of times the function evaluation needs to be carried out (as well as the number of configurations that need to be generated) to evaluate an accurate value for the ensemble average. This implies that when the set of random configurations is generated according to such an importance sampling scheme, the ensemble average can be taken as a regular arithmetic average of the property B over this sample of configurations. The ensemble average then becomes

$$\langle B \rangle = \langle B \rangle_{\text{sample}} = \frac{1}{N_{\text{sample}}} \sum_{i=1}^{N_{\text{sample}}} B_i \tag{6.11}$$

where B_i is the evaluation of the property B for each configuration i in the sample of configurations generated according to the importance sampling scheme.

We must now define an importance sampling scheme that generates representative system configurations. This is one of the outstanding achievements of the Metropolis Monte Carlo method we discuss in the next section.

6.2 THE METROPOLIS METHOD

6.2.1 MARKOV CHAIN AND STOCHASTIC MATRIX

Metropolis and co-workers developed The Monte Carlo method towards the end of World War II. The name "Monte Carlo" comes from the extensive use of random numbers and random number generators to carry out the sampling of the configurations of the system. The development of the method was undoubtedly promoted by the emergence of high-performance computers, such as the MANIAC supercomputer of Los Alamos in the early 1950s, and the development of random number generators, as put forward by the RAND corporation in the mid-1940s and after [156].

The goal is thus to generate a series of configurations of the system, through random changes, such that, by the end of the process, we obtain a representative sample of the system in a given statistical ensemble, *i.e.*, a series of configurations in which each configuration appears several times proportional to its Boltzmann weight. To this end, we first have to introduce a concept invented by the Russian mathematician Andrey Markov (see Figure 6.3) in the early 1900s, who developed what is now known as a Markov chain and the underlying stochastic matrix, a work pursued, among others, by another Russian scientist Andrey Kolmogorov. The idea is to start with a representation of the "world" of possible configurations as a state space of dimension n. Generating a series of configurations involves transitioning from one state i to another state j with a probability $P_{\text{transition}}(j|i) = P_{i,j}$ leading to the following stochastic matrix

$$P_t = \begin{pmatrix} P_{1,1} & P_{1,2} & P_{1,j} & \cdots & P_{1,n} \\ P_{2,1} & P_{2,2} & P_{2,j} & \cdots & P_{2,n} \\ \vdots & \vdots & \vdots & \ddots & \vdots \\ P_{i,1} & P_{i,2} & P_{i,j} & \cdots & P_{i,n} \\ \vdots & \vdots & \vdots & \ddots & \vdots \\ P_{n,1} & P_{n,2} & P_{n,j} & \cdots & P_{n,n} \end{pmatrix}$$

The matrix is termed a right stochastic matrix since the sum over an entire row equals 1. Indeed, the sum of the transition probabilities from any state i to all other states is equal to 1

$$\sum_{j=1}^{n} P_{i,j} = 1 \tag{6.12}$$

Another essential feature of a Markov chain is that the outcome of a trial move (transition) from a state i to a state j depends on the state i but not on its prior history, *i.e.*, on the sequence of steps that brought the system to state i.

The Markov chain composed of the successive states of the system can be written mathematically in a matrix-vector form thanks to the stochastic matrix. If we represent the starting state, or initial probability, with a row vector of dimension n and note this vector as $\rho^{(1)}$, we can then use the

Figure 6.3 Andrey Markov (1856–1922).

stochastic matrix to connect the starting state to the next state (2) in the Markov chain according to

$$\rho^{(2)} = \rho^{(1)} P_t \tag{6.13}$$

Repeating this operation to obtain state (3) in the chain, we have

$$\rho^{(3)} = \rho^{(2)} P_t = \rho^{(1)} P_t P_t = \rho^{(1)} P_t^2 \tag{6.14}$$

and, by the same token, after m steps, we obtain

$$\rho^{(m)} = \rho^{(1)} P_t^m \tag{6.15}$$

This means that the limiting distribution ρ, $e.g.$, in our case, the canonical distribution, must satisfy the following eigenvalue equation

$$\rho = \rho P_t \tag{6.16}$$

In other words, the distribution must be stationary when the process has converged. The next step is finding a solution by generating a sequence of states following a limiting canonical distribution. We look over the solution proposed by Metropolis in the next section.

6.2.2 RANDOMNESS AND ACCEPTANCE

We now focus on finding a solution to Eq. 6.16 and identifying a transition matrix yielding the canonical distribution ρ. To do this, we adopt the two following sufficient conditions, $i.e.$, a condition that guarantees that the canonical distribution is recovered. The first of these two conditions is known as "microscopic reversibility". This condition states that the probability of a "forward" transition in the Markov chain ($e.g.$, from state i to state j) is equal to the probability of a "reverse" transition (from state j to state i) and is given by

$$\rho_i P_{i,j} = \rho_j P_{j,i} \tag{6.17}$$

Summing over all states i, we obtain

$$\sum_i \rho_i P_{i,j} = \sum_i \rho_j P_{j,i} = \rho_j \sum_i P_{j,i} = \rho_j \tag{6.18}$$

since P_t is a right stochastic matrix (see Eq. 6.12).

Metropolis and co-workers [260] then suggested adopting the following choice for the transition matrix P_t:

$$P_{i,j} = \alpha_{i,j} \quad \rho_j \geq \rho_i \quad i \neq j \tag{6.19}$$

$$P_{i,j} = \alpha_{i,j}(\rho_j/\rho_i) \quad \rho_j < \rho_i \quad i \neq j \tag{6.20}$$

in which α is a symmetrical stochastic matrix (($\alpha_{i,j} = \alpha_{j,i}$). We are now left with specifying the elements of matrix α.

To generate a new configuration, we carry out a trial random move of, $e.g.$, an atom of the system. To this end, we start from the initial location of the atom and draw a random displacement of $\pm d$ in all three directions of space. This means that the new location of the atom (see Figure 6.4) will be one of the positions within a cube, centered on the initial position of the atom and edge $2d$. In other words, we have for the matrix elements $\alpha_{i,j} = \frac{1}{N_{cube}}$ where N_{cube} denotes all accessible positions in the cube. This means that to generate a new trial position for this atom, we can pick one of the positions, and all positions are equiprobable (hence the same value for $\alpha_{i,j}$ for all positions). We add that, since a computer only stores information with limited precision, $e.g.$, double precision, there is a finite number of accessible positions N_{cube}, rather than an infinite number of positions if the computer were able to handle actual real numbers with infinite precision. Since, regardless of

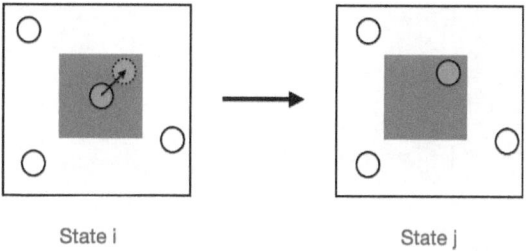

State i   State j

Figure 6.4   Transition from state i to state j. The generation of state j from state i is achieved by randomly drawing a displacement vector (shown with an arrow) for the central atom (solid circle) and displacing it to a new position (dashed circle).

the position selected, we can set the probability of generating the new trial position and the $\alpha_{i,j}$ matrix element to 1. We have just seen how we already use a significant number of times a random number to select a random atom (the central atom in our example) and to decide how much of a random displacement we apply in the three directions of space (3 more random numbers, one for each direction) to justify the "Monte Carlo" name for the method. However, we need to draw one more random number to decide if we accept the $i \to j$ move and include the randomly generated j state in our Monte Carlo Markov chain.

Now that we have specified how the new state could be generated, together with its underlying stochastic matrix, we have to get back to the equation for the elements of the transition matrix (Eqs. 6.19 and 6.20) and input the value of the Boltzmann weights (or limiting probability) for each of the states i and j. These two equations lead us to obtain the probability of acceptance for a trial (Monte Carlo) move as

$$\mathrm{acc}(i \to j) = \min\left(1, \frac{\rho_j}{\rho_i}\right) \tag{6.21}$$

in which "min" denotes the minimum value of the two arguments. This equation summarizes the Metropolis acceptance criterion for a Monte Carlo simulation. This means that if j is associated with a greater Boltzmann weight and is thus more probable than state i, the random move is automatically accepted (hence the "1" in the acceptance rule). If not, the key is to pick a random number η between 0 and 1 and accept the move if η is less than the $\frac{\rho_j}{\rho_i}$ ratio. In other words, the algorithm will allow the Markov chain to include a slightly less probable state j, i.e., less than 1 but still very likely close to 1 since η acts as a lower bound. This second possibility $\left(\frac{\rho_j}{\rho_i}¡1\right)$ means that all random moves are not accepted and that the probability of acceptance is thus, at most, equal to 1. In practice, if the random move from state i to state j is not accepted, the system will remain in state i, and state i will be repeated in the Markov chain until a trial move gets accepted.

The first practical implementation of the Monte Carlo algorithm was coded by Arianna W. Rosenbluth in 1953 and tested on the *MANIAC I* computer at Los Alamos. This was a significant achievement that illustrated how these early technological advances, like the MANIAC I built under the direction of Metropolis according to the Von Neumann architecture of the IAS machine (see Figure 6.5), set the stage for the computer simulations of molecular systems.

6.2.3   IMPLEMENTATION AND TESTING

To provide a first example of a Monte Carlo simulation, we examine the case of an atomic system and leverage the Monte Carlo method to determine the properties of the system in the canonical ensemble, i.e., for a constant number of atoms N, constant volume V, and constant temperature T. In this case, we can only think of a single type of Monte Carlo moves corresponding to a random displacement of a randomly selected atom. Let us write explicitly the Metropolis criterion in this case by stating explicitly the value taken by the two Boltzmann weighting factor ρ_i and ρ_j. As our

Figure 6.5 An Institute of Advanced Study (IAS) computer from the late 1940s to early 1950s, currently located at the Smithsonian Institute. This machine was built under the direction of John von Neumann (1903–1957).

discussion of the Monte Carlo has led on in the previous section, we essentially explore the configurational part of the phase space since we are generating new configurations through random changes in the positions of the atoms. This means the momenta can be "integrated out" in determining the Boltzmann weights. For instance, for ρ_i, we can write

$$Q_{N,V,T} = \frac{V^N}{N!\Lambda^{3N}} \int \exp\left(-\beta V_i\right) d\mathbf{r}^N \tag{6.22}$$

in which V_i denotes the potential energy of configuration i, which solely depends on the positions of the atoms in the system.

This, in turn, leads to the following ratio of Boltzmann factors

$$\frac{\rho_i}{\rho_j} = \frac{\exp\left(-\beta V_i\right)}{\exp\left(-\beta V_j\right)} = \exp\left(-\beta \delta V\right) \tag{6.23}$$

in which $\delta V = V_i - V_j$ denotes the change in potential energy between the two configurations i and j.

The Metropolis criterion then becomes

$$\mathrm{acc}(i \to j) = \min\left(1, \exp\left(-\beta \delta V\right)\right) \tag{6.24}$$

If $\delta V < 0$ and thus $\exp\left(-\beta \delta V\right) > 1$, the trial move is automatically accepted, and the new configuration j becomes part of the Markov chain. Alternatively, if If $\delta V \geq 0$ and thus $\exp\left(-\beta \delta V\right) \leq 1$, a random number η ($0 \leq \eta < 1$) is drawn and the move is accepted if $\eta < \exp\left(-\beta \delta V\right)$. As we have seen above, this means that a less probable configuration j can be incorporated into the Markov chain provided that it is still very likely or, in other words, that the increase in potential energy δV is moderate.

We summarize in the pseudo-code shown in Figure 6.6 the main steps of a Monte Carlo code. While random translations are the only trial moves one can apply to atoms, many different trial moves can be tailored for molecules, including the rotation by a random angle and around a randomly chosen axis for a polyatomic molecule or a random change in conformation of a flexible polyatomic molecule.

Beyond the implementation of the Metropolis rules and the computation of energies, there are also a couple of essential tricks to mention at this stage. First, simulations are often run with rules known as periodic boundary conditions. From a practical standpoint, the system's atoms are placed within a simulation cell of volume V. The simulation cell is often chosen to be cubic with a side

Algorithm   Monte Carlo

1: Start from a configuration Γ of energy $E = E(\Gamma)$

2: **while** i_{move} is less than the number of attempted MC moves N_{move} **do**

3:    Generate a new configuration Γ' through a random move. For example, choose a molecule

    at random and translate by a random vector, or rotate it by a random angle...

4:    Compute the energy for the new configuration $E' = E(\Gamma')$

5:    Compute the change in energy of the system $\Delta E = E' - E$ as a result of the random move

6:    **if** $\Delta E \leq 0$ **then**

7:        Accept move and update current configuration state, *i.e.*, $\Gamma \leftarrow \Gamma'$

8:        Update the current value for the energy, *i.e.*, $E \leftarrow E'$

9:    **else**

10:        Compute the transition probability $W = e^{-\beta \Delta E}$

11:        Generate a random number r in the interval $[0, 1]$

12:        **if** $r \leq W$ **then**

13:            Accept the new configuration

14:            Update current configuration state, *i.e.*, $\Gamma \leftarrow \Gamma'$

15:            Update the current value for the energy, *i.e.*, $E \leftarrow E'$

16:        **else**

17:            Reject move and keep Γ as current configuration state

18:            Keep E as current value for the energy

19:        **end if**

20:    **end if**

21:    Accumulate the current value for the energy in the running total for the energy

22:    Accumulate running total for other quantities of interest

23: **end while**

24: Compute averages over all configurations, *i.e.*, divide running totals by N_{move}

Figure 6.6   Pseudo-code for a (N, V, T) Monte Carlo simulation.

L such that $L = V^{1/3}$. If one wants to simulate a bulk (be it solid, liquid, or vapor), there are no explicit boundaries or physical walls, and one needs to employ periodic boundary conditions. In essence, this means that an atom i which, after a random move is accepted, sees one of its position coordinates exceed the dimension of the simulation cell (for instance, $x'_i > L$), the atom needs to be placed back inside the simulation cell ($x'_i \leftarrow x'_i - L$). In other words, the position of atom i is defined as $(\mathbf{r}_i)_{modL}$ in the modulo notation. The same reasoning is also applied to the computation of interparticle energies. This implies that an atom can interact with a periodic image of another atom when this image is closer than the other atom itself. We illustrate the way periodic boundary conditions operate in a simulation in Figure 6.7.

    The second trick pertains to what is known as tail corrections. Interatomic interactions often have a short range. This is, for instance, the case of the two terms involved in the Lennard-Jones

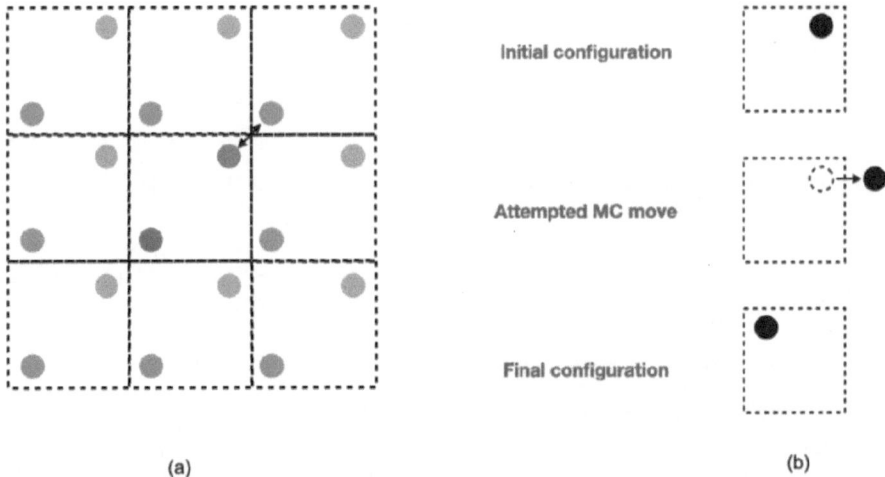

(a)   (b)

Figure 6.7 Periodic boundary conditions in simulations. (a) The central simulation cell is highlighted and is the unit cell replicated in the three spatial directions. As shown by the line with two arrowheads, the atom in the top right corner of the central cell interacts with the periodic image of the other atom. (b) When an accepted random displacement moves an atom outside the central simulation cell, it is placed back inside the cell at a new position $x'_i \leftarrow x'_i - L$, *i.e.*, on the other side of the simulation cell.

potential, which vary with r as r^{-12} for the repulsive interactions and as r^{-6} for the attractive interactions. It is, therefore, customary to calculate explicitly the interaction energy of pairs of atoms separated by a distance smaller than a cutoff distance, integrate these interactions beyond the cutoff distance, and add the result (tail correction) to the overall potential energy. Assuming a uniform spatial distribution, the overall potential energy becomes

$$U = U_c + 2\pi\rho N \int_{r_c}^{\infty} r^2 v(r) dr \tag{6.25}$$

where U_c is the sum of interaction energies over pairs of atoms separated by a distance smaller than a cutoff distance r_c, N the number of atoms in the system, and $\rho = N/V$ the density of atoms in the volume V, we add that alternative schemes, based for instance, on Ewald sums, are employed for slowly (conditionally) converging interactions, such as Coulombic interactions, which vary as r^{-1}.

6.3 WANG-LANDAU SAMPLING

6.3.1 A PARADIGM SHIFT: EVALUATING THE DENSITY OF STATES

The Metropolis importance sampling method has the tremendous advantage of obtaining the correct limiting distribution without explicitly calculating the partition function. However, simulations can only be carried out for a limited amount of time, so only a limited number of configurations can be included in the Monte Carlo Markov chain. Since configurations appear several times that are proportional to their Boltzmann weight, this implies that the configurations with a very low probability or, in other words, configurations that belong to the tail of the distributions will very likely be absent from the finite Markov chain generated during simulations. This is a direct consequence of the Metropolis criterion in Eq. 6.21. Since moves from a configuration $i \to j$ are accepted based on the probability ratio $\frac{\rho_j}{\rho_i}$, configurations j that are associated with a low probability ρ_j will rarely be, if at all accepted and sampled during a Monte Carlo simulation. However, this is crucial if one wants, for instance, to study a phase transition process since the configurations that span the transition pathway and contain, *e.g.*, incipient nuclei of the new phase have a very low Boltzmann weight. This requires then the development of methods that can sample evenly all possible system configurations. In the early 2000s, Fugao Wang and David P. Landau introduced a new type of Monte Carlo method that achieves a uniform sampling of all configurations of a system [217,390]. The Wang-Landau algorithm is a general method and can be applied to any system characterized by a cost function such as an energy function, *e.g.*. It has been applied to a wide range of model systems [151,390], polymers [286,325,353,354], protein models [131,300,347,394], as well as molecules [90,91,97].

The goal of the Wang Landau algorithm is thus dramatically different from that of conventional Monte Carlo simulations. As discussed by Wang and Landau in their original paper [390], "Unlike conventional Monte Carlo methods that directly generate a canonical distribution at a given temperature $g(E)\exp(-E/(k_B T))$, our approach is to estimate the density of states $g(E)$ accurately via a random walk which produces a flat histogram in energy space" (note that, in the set of notations used so far, the limiting probability ρ corresponds to $g(E)\exp(-E/(k_B T))$ in the Wang and Landau paper). In the conventional Monte Carlo or Metropolis approach, we recall that only a ratio of probabilities appears in the acceptance rule, thereby bypassing the need to explicitly know the value of ρ for each energy value and configuration. However, as discussed above, this, in turn, limits the sampling to only low energy values and high probability configurations. On the other hand, the Wang Landau algorithm calculates the density of states $g(E)$ and proceeds to sample all configurations evenly. This is because, as stated by Wang and Landau [390], "our algorithm is based on the observation that if we perform a random walk in energy space with a probability proportional to the reciprocal of the density of states $\frac{1}{g(E)}$, then a flat histogram is generated for the energy distribution". Here, Wang and Landau use the term "flat histogram" to indicate that all configurations are sampled evenly and thus that all energy values are observed the same number of times in the Markov chain. In other words, if one were to draw a histogram that keeps track of the number of times a given energy value is generated in the Markov chain, one would obtain a flat histogram for the energy distribution. We will see that such a histogram is collected during simulations following the Wang Landau algorithm. This leads then to the use of an acceptance criterion that reads, for a transition from configuration $i \to j$, as

$$\text{acc}(i \to j) = \min\left(1, \frac{g(E_i)}{g(E_j)}\right) \qquad (6.26)$$

In the following two sections, we discuss why and how the Wang Landau algorithm works.

6.3.2 THE BIASED DISTRIBUTION

We now examine how the Wang Landau sampling can lead to determining the density of states and to the partition function, as well as how the acceptance criterion of Eq. 6.26 is obtained. We start by examining a system of N molecules in a volume V and with an energy E. If we denote by Γ the set of coordinates for the atoms of the system in a specific configuration, we can define by $p(\Gamma, E)$ the joint probability of Γ and E, leading to the probability for the energy E of

$$p(E) = \int p(\Gamma, E) d\Gamma \qquad (6.27)$$

To perform a uniform sampling of the energy E, the idea of the Wang Landau sampling is to use a joint biased distribution. If we define the joint biased distribution as

$$p_{\text{bias}}(\Gamma, E) = \frac{p(\Gamma, E)}{p(E)} \qquad (6.28)$$

then the biased distribution for E, $p_{\text{bias}}(E)$, is such that it is the same for all energy values since, for all E, we have

$$p_{\text{bias}}(E) = \int p_{\text{bias}}(\Gamma, E) = \frac{\int p(\Gamma, E) d\Gamma}{p(E)} = 1 \qquad (6.29)$$

By applying the principle of "microreversibility" (also known as detailed balance equation), we can obtain the Metropolis criterion for a simulation performed according to a biased (Wang-Landau) sampling as

$$\text{acc}(i \to j) = \min\left(1, \frac{p_{\text{bias}}(\Gamma_j, E_j)}{p_{\text{bias}}(\Gamma_i, E_i)}\right) \tag{6.30}$$

Considering a Monte Carlo simulation that samples the energy E uniformly for a given number of molecules N and volume V, we have the following probability for a specific configuration Γ

$$p(\Gamma) = \frac{V^N \exp[-\beta U(\Gamma)]}{N! \Lambda^{3N} Q(N, V, T)} \tag{6.31}$$

in which $Q(N, V, T)$ is the canonical partition function. The probability distribution for the energy $E = U\Gamma$ is

$$p(E) = \int p(\Gamma) \delta(U(\Gamma) - E) d\Gamma = \frac{V^N g(E) \exp(-\beta E)}{N! \Lambda^{3N} Q(N, V, T))} \tag{6.32}$$

If we now define the biased distribution $p_{\text{bias}}(\Gamma, E) = \frac{p(\Gamma)}{p(E)}$, we obtain

$$p_{\text{bias}}(\Gamma, E) = \frac{1}{g(E)} \tag{6.33}$$

leading to the Metropolis criterion listed in Eq. 6.26.

Once the density of states is determined numerically, which is the topic of the next section, we can calculate the canonical partition function through a Laplace transform of the biased distribution. Specifically, for the canonical partition function, we have

$$Q(N, V, T) = \frac{V^N}{N! \Lambda^{3N}} \int \exp(-\beta U(\Gamma) d\Gamma) = \frac{V^N}{N! \Lambda^{3N}} \int g(E) \exp(-\beta E) dE \tag{6.34}$$

This means that a simple (numerical) integration of the function $g(E)$ provides the value of the partition function and, in turn, all the system's thermodynamic properties via the formalism of statistical mechanics.

6.3.3 A TWIST IN THE MONTE CARLO PLOT

The last thing we are left to specify is how the Wang-Landau sampling scheme works in practice. This algorithm leads to a self-consistent determination of the density of states $g(E)$ and thus of the partition function $Q(N, V, T)$. First, we need to split the range of energies taken by the system into N_{int} intervals of width δE such that

$$\delta E = \frac{E_{\text{max}} - E_{\text{min}}}{\delta E} \tag{6.35}$$

in which E_{max} and E_{min} stand for the maximum and minimum energy, providing the upper and lower bound, respectively, for the range of energies. Second, we need to define two histograms that will be constantly updated during the simulation: (i) the histogram $H(E_{\text{int}})$ for the number of visits of an energy interval, centered around E_{int}, and (ii) the histogram for the density of states $g(E_{\text{int}})$. The determination of $g(E)$, and thus of the partition function $Q(N, V, T)$ relies on collecting these two histograms. At the beginning of the simulation, we initialize the histogram for the density of states to $g(E_{\text{int}}) = 1$ for all energy intervals. We also initialize the histogram for the number of visits to 0, meaning that, for all E_{int}, we have $H(E_{\text{int}}) = 0$. During the simulation, whenever the system takes an energy value E_n due to an MC move, we update the value of the histogram for the number of visits and the histogram for the density of states. Specifically, if E_n is such that $E_{\text{int}} - \delta E/2 < E_n < E_{\text{int}} + \delta E/2$, then we apply the following updates

$$H(E_{\text{int}}) \leftarrow H(E_{\text{int}}) + 1$$
$$g(E_{\text{int}}) \leftarrow g(E_{\text{int}}) \times f \qquad\qquad (6.36)$$

where f is a convergence factor greater than 1. Once the histogram for the number of visits is close to being flat (for instance, if the lowest number of visits is greater than 80% the histogram value for the most visited energy interval), the convergence factor is reduced ($f \leftarrow \sqrt{f}$), the histogram for the number of visit is initialized back to 0 for all energy intervals. The simulation is rerun with this new convergence factor. This process is repeated until f reaches a minimum threshold value (typically, $10^{-8} < \ln f < 10^{-5}$), leading to an accurate numerical estimate for $g(E)$ for all E_{int}. We add that the convergence of the Wang-Landau algorithm has been proved formally, and there are many possible choices for the criterion assessing the "flatness' of the histogram for the number of visits.

In the pseudo-code shown in Figure 6.8, we summarize the main steps in a Monte Carlo simulation according to a Wang Landau sampling. The great advantage of this type of simulation is that the histogram for the density of states is obtained up to a multiplicative constant. We will discuss in the following chapters how the exact density of states can be retrieved, allowing for the determination of the absolute thermodynamic properties of a system.

Algorithm   Wang-Landau

1: Initialize the histogram $g(E_i) = 1$ for all $i = 1, .., n$

2: Initialize the histogram for the number of visits $H(E_i) = 0$ for all $i = 1, ..., n$

3: Start with a modification factor, *e.g.*, $f = 2.71$

4: Start from a configuration Γ of energy $E = E(\Gamma)$

5: **while** $\ln f$ is greater than, *e.g.*, 10^{-8} **do**

6:    **while** H is not flat enough (until all $H(E_i)$ are at least 80% of the maximum value) **do**

7:       Generate a new configuration Γ' via a random translation, rotation,...

8:       Calculate the energy $E' = E(\Gamma')$ for the new configuration

9:       Accept move from configuration Γ to Γ' according to $acc(\Gamma \rightarrow \Gamma') = \min\left\{1, \frac{g(E)}{g(E')}\right\}$

10:      Update current configuration state, *i.e.*, either $\Gamma \leftarrow \Gamma'$ if move accepted or keep Γ as is if move rejected

11:      Update the current value for the energy, *i.e.*, either $E \leftarrow E'$ if move accepted or keep E as is if move rejected

12:      Determine in which interval centered around E_i the current energy E falls

13:      Adjust the density $g(E_i) \leftarrow g(E_i) \times f$

14:      Update visit histogram $H(E_i) \leftarrow H(E_i) + 1$

15:   **end while**

16:   Reduce f to $f \leftarrow \sqrt{f}$

17:   Reset the visit histogram for the number of visits $H(E_i)$ for all $i = 1, .., n$

18: **end while**

19: return $g(E_i)$ for all i

Figure 6.8 Pseudo-code for a (N, V, T) Monte Carlo simulation following the Wang-Landau sampling scheme.

7 Molecular Networking in Insulation

Adiabatic Ensembles

"The equilibrium properties of a pure substance can be described in at least eight distinct ways in thermodynamics and statistical mechanics. These eight different descriptions are interrelated through a Laplace-Legendre transformation scheme."

H. Will Graben and John R. Ray,
Eight physical systems of thermodynamics, statistical mechanics,
and computer simulations [169]

We have seen two main frameworks for evaluating the properties of groups of molecules. The first framework we came across is the microcanonical ensemble or (N,V,E) ensemble. This is the statistical ensemble typically used when the group under study is an isolated system, *i.e.*, contains a fixed number of molecules N (no exchange of matter allowed), has a fixed volume V (no work done, *e.g.*, on the system by the surroundings through the contraction/expansion of the system) and has fixed total energy E (no exchange of energy with the surroundings, including heat transfer through the boundaries of the system). The second framework we have discussed throughout the first chapters is the canonical ensemble or (N,V,T) ensemble. This is the statistical ensemble that corresponds to a system that contains a fixed number of molecules N (no exchange of matter allowed), has a fixed volume V (no work done, *e.g.*, on the system by the surroundings through the contraction/expansion of the system), and is maintained at a constant temperature T. This means that, in the case of the canonical ensemble, the exchange of energy with the surroundings is allowed and most often results from the transfer of heat through the boundaries. These two frameworks only reflect a few scenarios that may arise from a practical standpoint. For instance, these frameworks do not account for systems under constant pressure. Another example that these two frameworks cannot model is the case of open systems that allow for the number of molecules N to change due to an exchange of matter with the surroundings when the boundary is a permeable membrane. As we will see in this chapter, these scenarios require the definition of alternative frameworks and lead to at least eight statistical ensembles as indicated by H. W. Graben and J. R. Ray [169].

7.1 ADIABATIC PROCESSES AND ENSEMBLES

7.1.1 ADIABATIC VS. ISOTHERMAL

Before we discuss the statistical mechanics of adiabatic and isothermal ensembles, we start by recalling the difference between an adiabatic and an isothermal process. First, we define an isothermal thermodynamic process as occurring under a constant temperature T. This has important consequences on how energy is distributed across the system and how it flows in and out of the system. For instance, let us consider a system composed of n moles of an ideal gas at a temperature T_1 undergoing an isothermal process, *e.g.*, an isothermal expansion from a volume V to a volume V'. Therefore, the system can exchange energy with its surroundings to maintain its temperature constant, for instance, through heat transfer. The first law of thermodynamics we discussed in Chapter 1 states that the change in internal energy is given by

$$dE = \delta W + \delta Q = -PdV + \delta Q \qquad (7.1)$$

DOI: 10.1201/9781003006411-9

Since we have an ideal gas, its internal energy only depends on temperature. Since temperature is constant, the internal energy is also constant during an isothermal process and $dE = 0$. In other words, the heat transferred during the isothermal expansion is given by the integral of PdV over the volume change $V \rightarrow V'$ as

$$Q = \int_V^{V'} PdV \tag{7.2}$$

and the dependence of pressure upon volume is simply given by the ideal gas law

$$P = \frac{nRT}{V} \tag{7.3}$$

in which the numerator is constant. We plot the resulting hyperbolic function, also known as isotherm, in Figure 7.1 and outline the portion that corresponds to the isothermal compression $(P,V) \rightarrow (P',V')$.

Let us now turn to an adiabatic process that starts from the same conditions (T,P,V) as the isothermal process we discussed above and corresponds to the adiabatic compression of the system to the new volume V'. In this case, the system is thermally insulated from its surroundings. This implies that the amount of heat transferred is zero, leading to $\delta Q = 0$, and thus to an overall heat exchange $Q = 0$ over the adiabatic process. As a result, the first law of thermodynamics gives the following change in internal energy

$$dE = \delta W + \delta Q = -PdV \tag{7.4}$$

Since there is no heat transfer, the system's temperature changes as it undergoes adiabatic compression. We now replace dE by its expression as a function of temperature, $dE = nC_v dT$, in which C_v is the heat capacity at constant volume. Using the ideal gas law, Eq. 7.4 can then be recast into

$$nC_v dT + \frac{nRT}{V} dV = 0 \tag{7.5}$$

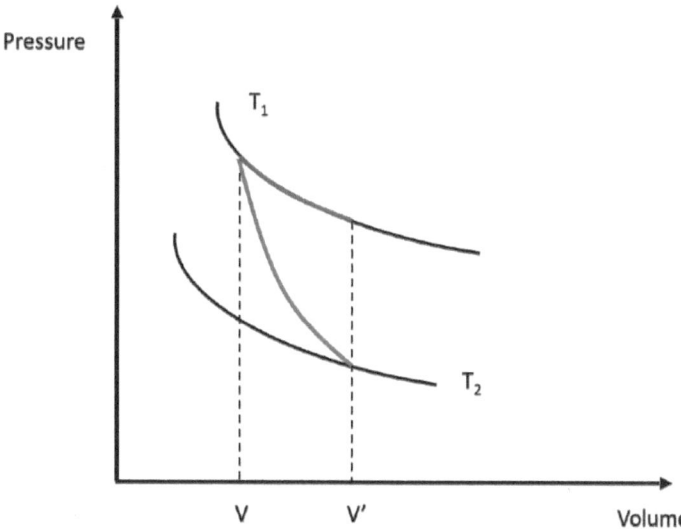

Figure 7.1 Comparison between the thermodynamic paths followed during an isothermal expansion from V to V' (corresponding to a portion of the top isotherm at $T = T_1$ and an adiabatic expansion from V to V' (line that jumps from the top isotherm at $T = T_1$ to the bottom isotherm at $T = T_2$.

and, after rearranging the equation, into

$$\frac{1}{T}dT + \frac{R}{C_v V}dV = 0 \tag{7.6}$$

For an ideal gas, we have $C_p - C_v = R$, according to the Mayer relation between the heat capacity at constant pressure C_p and C_v. We then introduce the ratio $\gamma = \frac{C_p}{C_v}$ and obtain the following relation by integrating Eq. 7.6

$$TV^{\gamma-1} = \text{Constant} \tag{7.7}$$

according to which the product of the temperature by the volume, raised to the power $(\gamma - 1)$, is constant throughout the adiabatic compression. Alternatively, using the ideal gas law, we obtain the following relation involving pressure and volume

$$PV^{\gamma} = \text{Constant} \tag{7.8}$$

This power law for pressure notably departs from the hyperbolic law followed by pressure during the isothermal process. We plot the resulting power law in Figure 7.1. This highlights how the adiabatic process follows a thermodynamic path that differs significantly from the isothermal path, essentially switching from one isotherm at $T = \frac{PV}{nR}$ to a second isotherm at $T = \frac{P'V'}{nR}$.

7.1.2   THE CONCEPT OF HEAT FUNCTION

The comparison between an isothermal process and an adiabatic process provides excellent insight into what distinguishes an adiabatic ensemble from an isothermal ensemble. Isothermal ensembles are in thermal contact with a reservoir, while adiabatic ensembles are thermally insulated. This means that, in Gibbs' picture, isothermal ensembles will contain replicas of systems at the same temperature T. Indeed, the heat exchanges between the system and its surroundings, which play the role of a thermal reservoir, allow for the system's temperature to be constant and the same as that of the thermal reservoir. On the other hand, since adiabatic ensembles are thermally insulated from their surroundings, heat exchanges will not occur with the surroundings. This does not preclude different types of energy transfers from occurring, including, *e.g.*, work from being done on the systems. As a result, the temperature of the system will not be constant. Yet, since no heat transfer takes place, we can identify another quantity, akin to energy, that will remain constant for each of the adiabatic ensembles we will be able to define. This quantity is called heat function, and, as we will see shortly, its exact expression depends on the nature of the other thermodynamic variables held fixed in the definition of the ensemble.

We look into the connection between the canonical ensemble (N,V,T) and the microcanonical ensemble (N,V,E) to see how heat functions can be obtained. At this point, we use a mathematical tool defined in the book's first chapter, *i.e.*, the Legendre transform. In thermodynamics, Legendre transforms can connect different state functions, *e.g.*, the internal energy E, and the enthalpy H through the relation.

$$H = E + PV \tag{7.9}$$

Let us see how Legendre transforms can be leveraged to connect the thermodynamic potential of an isothermal and adiabatic ensemble. We recall that a Legendre transform of a thermodynamic function $w(x,y,z)$ with respect to x defines a new function v of the variables $u = \partial w/\partial x$, y, and z by the relation

$$v(u,y,z) = w(x,y,z) - ux \tag{7.10}$$

By doing so, the variables u, y, and z have become the natural variables of the new function v. Let us apply this approach here to the entropy function $S(N,V,E)$, which is the thermodynamic potential of the microcanonical ensemble $S(N,V,E)$. We start the differential equation for the internal energy E

$$dE = TdS - pdV + \mu dN \tag{7.11}$$

and rearranging this equation to focus on entropy

$$dS = \left(\frac{1}{T}\right)dE + \left(\frac{P}{T}\right)dV - \left(\frac{\mu}{T}\right)dN \qquad (7.12)$$

If we use Legendre transforms of the entropy $S(N,V,E)$ to define new thermodynamic functions, the new natural variables playing the role of u in Eq. 7.10 will be either $\frac{1}{T} = \frac{\partial S}{\partial E}$, or $\frac{-\mu}{T} = \frac{\partial S}{\partial N}$, or $\frac{P}{T} = \frac{\partial S}{\partial V}$. Here we choose to work with $\frac{1}{T}$ or, more precisely, $\beta = \frac{1}{k_B T}$. Doing so allows us to obtain a new thermodynamic function, the Helmholtz free energy $A(N,V,\beta)$ or, equivalently, $A(N,V,T)$ through

$$-\beta A(N,V,T) = k_B^{-1} S(N,V,E) - \beta E \qquad (7.13)$$

which gives the well-known thermodynamic equation

$$A(N,V,T) = E - TS(N,V,E) \qquad (7.14)$$

Through this Legendre transform, we have therefore connected $A(N,V,T)$, *i.e.*, the thermodynamic potential of the canonical ensemble, to $S(N,V,E)$, *i.e.*, the thermodynamic potential of the microcanonical ensemble. The variables change from the isothermal to the adiabatic ensemble can be summarized as $(N,V,T) \rightarrow (N,V,E)$. In other words, when N and V are fixed, we can replace the temperature T by the heat function, here the internal energy E, to obtain an adiabatic ensemble from an isothermal ensemble, and vice-versa for the inverse switch. The internal energy E is one of several heat functions that can be obtained. For instance, holding N and P constant rather than N and V will yield another heat function. This is what we explore in the next section.

7.1.3 EIGHT STATISTICAL ENSEMBLES

Before we define these eight statistical ensembles, let us travel back in time to the early stages of the development of statistical mechanics. Four statistical ensembles were identified early on as statistical mechanics developed in the first part of the 20th century. Those were the first adiabatic ensemble, *i.e.*, the microcanonical ensemble (N,V,E), and three isothermal ensembles that included the canonical ensemble (N,V,T), the grand-canonical ensemble in which the three fixed thermodynamic quantities are (μ,V,T), and the isothermal-isobaric ensemble for constant (N,P,T). In 1939, Guggenheim published a paper that discusses two important results [171]. First, Guggenheim highlights the significance of the partition function of a statistical ensemble to supplement the concept of "thermodynamic probability". To this end, he generalizes the relation between thermodynamics and statistical mechanics for all existing ensembles. Second, Guggenheim introduces a fifth ensemble with constant (μ,P,T), now known as the generalized or Guggenheim ensemble. This ensemble is the only ensemble for which the three fixed thermodynamic quantities are intensive properties. As a result, the generalized ensemble is associated with a partition function equal to zero. In 1939, together with Fowler, Guggenheim also published the first textbook on Statistical Thermodynamics [139] and cemented the emergence of this new field. Following Guggenheim's ideas, several scientists discovered three additional statistical ensembles over the next few decades. Indeed, using the Legendre-Laplace mapping procedure, Brown, Hill, and Ray develop three other ensembles, known as the isoenthalpic-isobaric ensemble (N,P,H) [38,174], the grand-isochoric adiabatic ensemble (μ,V,L) [168,304] and the grand-isobaric adiabatic ensemble [168,303,305] (μ,P,R). Together with the microcanonical ensemble, these four ensembles form a set of adiabatic ensembles for which the value taken by a heat function is constant. This heat function is either the internal energy E (microcanonical ensemble), the enthalpy H (isoenthalpic-isobaric ensemble), or the Hill energy L (grand-isochoric adiabatic ensemble), or the Ray energy (grand-isobaric adiabatic ensemble). These three heat functions are defined through Legendre transforms as follows:

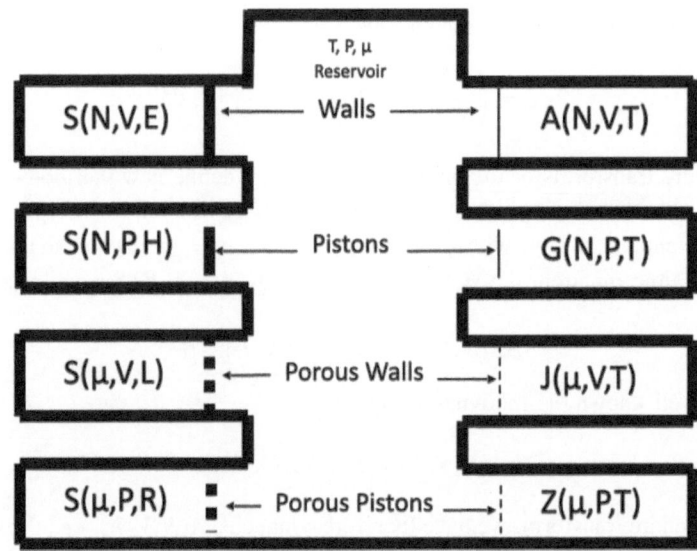

Figure 7.2 Diagram summarizing the eight statistical ensembles, including the four isothermal ensembles on the right and the four adiabatic ensembles on the left.

$$H = E + PV$$
$$L = E - \mu N \qquad (7.15)$$
$$R = E + PV - \mu N$$

We discuss below in greater detail the statistical mechanics of the corresponding statistical ensembles. A graphical summary of the eight statistical ensembles is also presented in Figure 7.2.

7.2 MECHANICS OF ADIABATIC ENSEMBLES

7.2.1 MICROCANONICAL DISTRIBUTION AND THERMODYNAMIC EQUATIONS

Legendre transforms provide relations between thermodynamic state functions. For instance, Eq. 7.13 connects the Helmholtz free energy A and the entropy S. Taking now a viewpoint from an ensemble perspective, Eq. 7.13 relates two thermodynamic potentials since A and S dictate how a system evolves and reaches equilibrium. Indeed, a system at fixed N, V, and T will evolve to minimize its Helmholtz free energy $A(N,V,T)$ and reach equilibrium for a minimum of $A(N,V,T)$. Similarly, a system at constant N, V, and E will evolve to maximize its entropy $S(N,V,E)$ and will be at a maximum of $S(N,V,E)$ when it reaches equilibrium. This means that the Legendre transform of Eq. 7.13 provides a connection between two different statistical ensembles, the canonical ensemble (N,V,T) and the microcanonical ensemble (N,V,E). Of course, this relation between thermodynamic properties can be extended to statistical mechanical functions. Regarding mechanics, Eq. 7.13 relates the partition function of the isothermal ensemble to the phase space volume, or number of microstates, of the microcanonical ensemble. Looking at the statistical mechanical definition for the Helmholtz free energy, we see a direct link emerge between $A(N,V,T)$ and the canonical partition through $A(N,V,T) = -k_B T \ln Q(N,V,T)$. Similarly, considering the statistical mechanical definition for entropy, we have $S(N,V,E) = k_B \ln \Omega(N,V,E)$ through the well-known Boltzmann formula, with $\Omega(N,V,E)$ denoting the number of microstates in the microcanonical ensembles.

7.2.2 THE (μ, P, R) ENSEMBLE

A similar reasoning applies to the grand-isobaric adiabatic (μ, P, R). Here we focus on identifying the connection between the grand-isobaric adiabatic (μ, P, R) ensemble and its isothermal counterpart, the generalized (μ, P, T) ensemble introduced by Guggenheim. The two thermodynamic potentials are, in this case, the entropy $S(\mu, P, R)$ for the grand-isobaric adiabatic ensemble and the Guggenheim thermodynamic potential $Z(\mu, P, T)$ for the (μ, P, T) ensemble. Applying a Legendre transform to the entropy $S(\mu, P, R)$, we obtain

$$-\beta Z(\mu, P, T) = k_B^{-1} S(\mu, P, R) - \beta R \tag{7.16}$$

The internal energy E characterizes a macrostate associated with many microstates in the microcanonical ensemble. Similarly, in the (μ, P, R) ensemble, the specification of R depicts a macrostate consisting of many microstates. Precisely as the determination of the number of microstates $\Omega(N, V, E)$ gives the entropy $S(N, V, E) = k_B \ln \Omega(N, V, E)$ in the microcanonical ensemble, the computation of the number of such microstates $Q(\mu, P, R)$ yields the entropy of the system through $S(\mu, P, R) = k_B \ln Q(\mu, P, R)$ in the grand-isobaric adiabatic ensemble.

The grand-isobaric adiabatic ensemble has a distinct advantage over the microcanonical ensemble. Indeed, unlike the microcanonical ensemble, the (μ, P, R) ensemble enables a straightforward evaluation of the system's entropy. This stems from the precise nature of the Guggenheim potential $Z(\mu, P, T)$. $Z(\mu, P, T)$ is the only thermodynamic potential for a system with three fixed intensive variables μ, P, and T. In contrast, the other ensembles only impose at most two intensive variables (*e.g.*, μ, P, and/or T) and at least one extensive variable (*e.g.*, N, V, or R). As discussed by Guggenheim [171], the thermodynamic potential $Z(\mu, P, T)$ is exactly 0 for this very reason. Indeed, when the independent variables μ, P, and T are fixed, the Gibbs-Duhem equation states that $Nd\mu + VdP - SdT = 0$ (Gibbs-Duhem relation). Plugging $Z(\mu, P, T) = 0$ into Eq. 7.16, we obtain the following relation between the entropy $S(\mu, P, R)$ and the heat function (Ray energy) R

$$k_B^{-1} S(\mu, P, R) = \beta R \tag{7.17}$$

The entropy $S(\mu, P, R)$ is thus a simple ratio of the Ray energy R and of the temperature T through

$$S(\mu, P, R) = \frac{R}{T} \tag{7.18}$$

R is a constant specified in the definition of the macrostate and ensemble or, in other words, a fixed input parameter during a simulation of a system in the grand-isobaric adiabatic ensemble. This implies that to calculate the entropy during a simulation in the (μ, P, R) ensemble, we only need to evaluate the average temperature T of the system through a running average of the following quantity

$$\langle T \rangle = \frac{\langle 2K \rangle}{3k_B} = \frac{\langle 2(R - U + \mu N - PV) \rangle}{3k_B} \tag{7.19}$$

We will discuss in Section 7.3.3 a practical implementation of Monte Carlo (MC) simulations in the (μ, P, R) ensemble and its application to determining entropy.

7.2.3 A FULL PICTURE FOR THE FOUR ADIABATIC ENSEMBLES

The same approach can be extended to the remaining two adiabatic ensembles, thereby providing a relation between each adiabatic ensemble and its isothermal counterpart [139,169,184]. First, we examine the pair of fixed thermodynamic variables (N, P, T) and (N, P, H). These two sets are associated with the isothermal-isobaric ensemble and the isoenthalpic-isobaric ensemble, respectively. The two thermodynamic potentials are then the Gibbs free energy G, given by $-\beta G(N, P, T) = \ln Q(N, P, T)$ in which $Q(N, P, T)$ is the isothermal-isobaric partition function, and the entropy, given by $S(N,P,H) = k_B \ln Q(N,P,H)$ in which $Q(N,P,H)$ is the number of microstates in the isoenthalpic-isobaric ensemble. Through a Legendre transform, we obtain the following relation between $G(N,P,T)$ and $S(N,P,H)$

$$-\beta G(N,P,T) = k_B^{-1} S(N,P,H) - \beta H \tag{7.20}$$

Similarly, let us now consider the pair of fixed thermodynamic variables (μ,V,T) and (μ,V,L). These two sets are associated with the grand-canonical ensemble and the grand-isochoric ensemble, respectively. The corresponding thermodynamic potentials are thus the Landau free energy, or grand potential J, defined as $-\beta J(\mu,V,T) = \ln Q(\mu,V,T)$, in which $Q(\mu,V,T)$ is the grand partition function, and the entropy, given by $S(\mu,V,L) = k_B \ln Q(\mu,V,L)$ in which $Q(\mu,V,L)$ is the number of microstates in the grand-isochoric ensemble. A Legendre transform of $S(\mu,V,L)$ provides a relation between $J(\mu,V,T)$ and $S(\mu,V,L)$

$$-\beta J(\mu,V,T) = k_B^{-1} S(\mu,V,L) - \beta L \tag{7.21}$$

As previously discussed, this series of relations between thermodynamic potentials, obtained through Legendre transforms, also implies relations between partition functions for isothermal ensembles and numbers of microstates for adiabatic ensembles. Interestingly, the second series of relations is obtained through a second type of transforms, known as Laplace transforms. We provide below the equations summarizing these adiabatic-isothermal connections (see also Figure 7.3) and complete the statistical mechanical picture for adiabatic ensembles

$$Q(N,V,T) = \beta \int_0^\infty \exp(-\beta E)\Omega(N,V,E)dE \tag{7.22}$$

$$Q(N,P,T) = \beta \int_0^\infty \exp(-\beta H)Q(N,P,H)dH \tag{7.23}$$

$$Q(\mu,V,T) = \beta \int_0^\infty \exp(-\beta L)Q(\mu,V,T)dL \tag{7.24}$$

$$Q(\mu,P,T) = \beta \int_0^\infty \exp(-\beta R)Q(\mu,P,R)dR \tag{7.25}$$

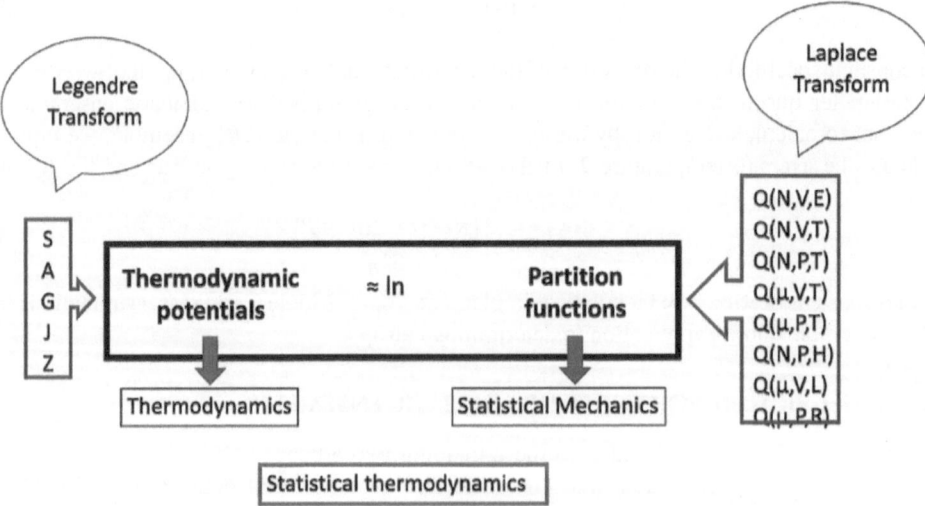

Figure 7.3 Summary of the eight thermodynamic potentials and partition functions, together with the two types of transforms involved in the relations between thermodynamic potentials (Legendre transforms) and partition functions (Laplace transforms).

7.3 MONTE CARLO EXPLORATION OF ADIABATIC ENSEMBLES

7.3.1 EXPLORING THE MICROCANONICAL ENSEMBLE

We introduced in the previous chapter how Monte Carlo simulations could be carried out in an isothermal ensemble and discussed the specific example of the canonical ensemble. We now examine how the Monte Carlo simulation framework can be extended to an adiabatic ensemble, starting with the example of the microcanonical ensemble. As we discussed, the critical step involves determining the Metropolis acceptance rules for each type of random move. There is, however, a fundamental difference between isothermal and adiabatic ensembles. In isothermal ensembles, random moves are accepted based on their Boltzmann probability, which depends on the value of the change in potential energy due to the moves. Indeed, accepting the move is based on comparing a random number and the quantity $\exp(-\beta \Delta U)$, in which ΔU denotes the change in potential energy following the move. On the other hand, in adiabatic ensembles, the Metropolis acceptance rules hinge on the change in kinetic energy resulting from the random moves. To better understand how the system's kinetic energy plays such a significant role in the Metropolis acceptance rules, we examine how ensemble averages are calculated in the microcanonical ensemble. To this end, let us consider a phase space function B. Its ensemble average is given by

$$<B>= \frac{1}{\omega} \left(\frac{2\pi m}{h^2} \right)^{3N/2} \frac{1}{N!\Gamma(\frac{3N}{2})} \int B(q)[E - U(q)]^{3N/2-1}\Theta[E - U(q)]dq^{3N} \tag{7.26}$$

As shown in the equation above, the quantity $E - U(q)$ is central to the definition of a microcanonical ensemble average and plays a similar role to the $\exp(-\beta U)$ factor in a canonical ensemble average. The probability density, and thus the acceptance rule, can be inferred from the ensemble average by realizing that the ensemble average $<B>$ can be recast as

$$<B>= \int B(\mathbf{q})\rho(\mathbf{q})dq^{3N} \tag{7.27}$$

in which $\rho(\Gamma)$ is the probability density.

By identification, we have the probability density defined as

$$W_E(\mathbf{q}) = C[E - U(\mathbf{q})]^{3N/2-1}\Theta[E - U(\mathbf{q})] \tag{7.28}$$

in which C is a constant, and the Metropolis acceptance probability associated with a random move for a molecule is given by

$$\text{acc}(\mathbf{q} \to \mathbf{q}') = \min\left(1, \frac{W_E(\mathbf{q}')}{W_E(\mathbf{q})}\right) = \min\left(1, \frac{[E - U(\mathbf{q}')]^{3N/2-1}}{[E - U(\mathbf{q})]^{3N/2-1}}\right) \tag{7.29}$$

in which $\mathbf{q}$ and $\mathbf{q}'$ denote the set of positions before the random move and the set of positions after the random move, respectively. Using kinetic energy instead of the potential energy in the acceptance rule is the main difference between Monte Carlo simulations in the microcanonical and canonical ensembles, with the rest of the algorithm following the protocol outlined in the previous chapter. We provide in Figure 7.4 a pseudo-code for a Monte Carlo simulation in the microcanonical ensemble.

7.3.2 MUSING IN THE (N,P,H) ENSEMBLE

Exploring the other adiabatic ensembles can be implemented in Monte Carlo simulations using the same reasoning for the (N,V,E) ensemble. There are, however, a few changes due, most notably,

---

Algorithm 3 Monte Carlo - Adiabatic NVE

1: Start from a configuration Γ of potential energy $U = U(\Gamma)$ and kinetic energy $K = K(\Gamma) =$
$E - U(\Gamma)$

2: **while** i_{move} is less than the number of attempted MC moves N_{move} **do**

3:      Generate a new configuration Γ' through a random move

4:      Compute the kinetic energy for the new configuration $K' = K(\Gamma')$

5:      Compute the ratio of the kinetic energies $f = \frac{K'}{K}$

6:      **if** $K' > 0$ **then**

7:          Compute the logarithm of the probability of acceptance as $\ln p = (\frac{3N}{2} - 1) \ln f$

8:          Generate a random number r in the interval $[0, 1]$

9:          **if** $r < p$ **then**

10:              Accept move and update current configuration state, *i.e.*, $\Gamma \leftarrow \Gamma'$

11:              Update the current value for the potential energy, *i.e.*, $U \leftarrow U'$

12:          **end if**

13:      **else**

14:          Reject move and keep Γ as current configuration state

15:          Keep K and U as current values for the kinetic and potential energy, respectively

16:      **end if**

17:      Accumulate the current value for the kinetic and potential energies in the running totals

18:      Accumulate running total for other quantities of interest

19: **end while**

20: Compute averages over all configurations, *i.e.*, divide running totals by N_{move}

Figure 7.4 Pseudo-code for a Monte Carlo simulation in the microcanonical (N, V, E) ensemble.

to the external constraint applied to the system, which, in turn, requires different types of random moves to achieve the sampling of the ensemble. For instance, if we consider the isoenthalpic-isobaric (N, P, H) ensemble, the external constraint differs from the (N, V, E) ensemble since pressure, rather than volume, is constant. This means that to ensure constant pressure, the system's volume will change during the simulations and fluctuate around its average value once equilibrium has been reached. A new type of random move, based on a random volume change and rescaling of the position coordinates of the molecules, will need to be implemented to perform the sampling of the (N, P, H) ensemble.

To derive the corresponding Metropolis rules, we start with the ensemble average of a phase-space function B in the isoenthalpic-isobaric ensemble. The average ensemble $< B >$ is given by

$$< B >= \frac{(2\pi m/h^2)^{3N/2}}{\omega N! \Gamma(\frac{3N}{2})} \int B(\mathbf{s}, V) V^N [H - PV - U(\mathbf{s})]^{3N/2-1} \Theta[H - PV - U(\mathbf{s})] ds^{3N} dV \quad (7.30)$$

In this equation, we have scaled the positions $\mathbf{q}$ by the edge L of the simulation cubic cell. This leads to the $3N$ scaled coordinates $s = L^{-1}q$ and a factor of $L^{3N} = V^N$ in the equation. The probability density is then given by

$$W_H(\mathbf{s}, V) = CV^N [H - PV - U(\mathbf{s})]^{3N/2-1} \Theta[H - PV - U(\mathbf{s})] \quad (7.31)$$

in which C is a constant and $\mathbf{s}$ is the set of scaled positions defining the system's configuration in the volume V. We emphasize that, once again, the kinetic energy $K = H - PV - U(\mathbf{s})$ is central to the probability density and thus to the Metropolis acceptance rules, as we will see below.

This gives the Metropolis acceptance rule for a random translation move $\mathbf{s} \to \mathbf{s}'$

$$\mathrm{acc}(\mathbf{s} \to \mathbf{s}') = \min\left(1, \frac{W_H(\mathbf{s}',V)}{W_H(\mathbf{s},V)}\right) = \min\left(1, \frac{[H - PV - U(\mathbf{s}',V)]^{3N/2-1}}{[H - PV - U(\mathbf{s},V)]^{3N/2-1}}\right) \quad (7.32)$$

For a random volume change, the Metropolis acceptance rule for $(\mathbf{s},V) \to (\mathbf{s}',V')$ is

$$\mathrm{acc}((\mathbf{s},V) \to (\mathbf{s}',V')) = \min\left(1, \left(\frac{V'}{V}\right)^N \frac{[H - PV' - U(\mathbf{s}')]^{3N/2-1}}{[H - PV - U(\mathbf{s})]^{3N/2-1}}\right) \quad (7.33)$$

In Figure 7.5, we provide a pseudo-code showing the main steps involved in a Monte Carlo simulation in the (N,P,H) ensemble. We add that simulations in the isoenthalpic-isobaric ensemble allow for the evaluation of several thermodynamic quantities, including the isobaric heat capacity C_p, the adiabatic compressibility κ_S, and the isobaric volume expansivity α_p, according to

$$\frac{k_B}{C_p} = 1 - \left(1 - \frac{2}{3N}\right) <K><K^{-1}> \quad (7.34)$$

$$<V>\kappa_s = \left(\frac{3N}{2} - 1\right)[<V^2 K^{-1}> - 2 <V><VK^{-1}> + <V>^2<K^{-1}>] \quad (7.35)$$

$$\frac{<V>\alpha_p}{C_p} = \left(\frac{3N}{2}\right)[<VK^{-1}> - <V><K^{-1}>] \quad (7.36)$$

7.3.3   DIRECT ENTROPY EVALUATIONS IN THE (μ,P,R) ENSEMBLE

We finally turn to the (μ,P,R) ensemble, in which the chemical potential μ, pressure P, and heat function R are held fixed. This implies that the temperature T, number of molecules N, and volume V are allowed to vary. We define four types of random moves for simulations in this ensemble to sample the grand-isobaric adiabatic ensemble. These consist of random displacements of a single molecule, deletion of a randomly selected molecule, insertion of an additional molecule at a random position in the system, and random volume changes. We give in Figure 7.6 a pseudo-code for a Monte Carlo simulation in the grand-isobaric ensemble and detail below how the acceptance rules are obtained for each type of random move. The probability associated with a configuration Γ containing N particles in a volume V as

$$P(\Gamma,N,V) = \frac{(bV)^N}{\Gamma(3N/2)Q(\mu,P,R)}[R - PV + \mu N - U(\Gamma)]^{3N/2-1} \quad (7.37)$$

in which $b = (2\pi m/h^2)^{3/2}$.

Let us consider first a Monte Carlo move corresponding to the random displacement of a molecule. If we define the start configuration as Γ and the end configuration Γ', the Metropolis acceptance rule is

$$\mathrm{acc} = \min\left[1, \frac{P(\Gamma',N,V)}{P(\Gamma,N,V)}\right] = \min\left[1, \frac{[R - PV + \mu N - U(\Gamma')]^{3N/2-1}}{[R - PV + \mu N - U(\Gamma)]^{3N/2-1}}\right] \quad (7.38)$$

Algorithm 4 Monte Carlo - Adiabatic NPH

1: Start from a configuration Γ of potential energy $U = U(\Gamma)$ and kinetic energy $K = K(\Gamma) = H - PV - U(\Gamma)$

2: **while** i_{move} is less than the number of attempted MC moves N_{move} **do**

3:     Generate a new configuration Γ' through a random move

4:     Store the volume for the new configuration as $V' = V(\Gamma')$

5:     Compute the kinetic energy for the new configuration $K' = K(\Gamma')$

6:     Compute the ratio of the kinetic energies $f = \frac{K'}{K}$

7:     **if** $K' > 0$ and $V' > 0$ **then**

8:         **if** random volume change **then**

9:             Compute the logarithm of the probability of acceptance as $\ln p = (\frac{3N}{2} - 1)\ln f + N\ln\frac{V'}{V}$

10:         **else**

11:             Compute the logarithm of the probability of acceptance as $\ln p = (\frac{3N}{2} - 1)\ln f$

12:         **end if**

13:         Generate a random number r in the interval $[0, 1]$

14:         **if** $r < p$ **then**

15:             Accept move and update current configuration state, *i.e.*, $\Gamma \leftarrow \Gamma'$

16:             Update the current value for the volume, *i.e.*, $V \leftarrow V'$

17:             Update the current value for the potential energy, *i.e.*, $U \leftarrow U'$

18:         **end if**

19:     **else**

20:         Reject move and keep Γ as current configuration state

21:         Keep K, U, and V as current values for the kinetic energy, potential energy, and volume, respectively

22:     **end if**

23:     Accumulate the current value for the kinetic and potential energies, as well as volume, in the running totals

24:     Accumulate running total for other quantities of interest

25: **end while**

26: Compute averages over all configurations, *i.e.*, divide running totals by N_{move}

Figure 7.5 Pseudo-code for a Monte Carlo simulation in the isoenthalpic-isobaric (N, P, H) ensemble.

The acceptance rule for the deletion of a randomly selected molecule that is, from the set $(\Gamma, N) \to (\Gamma', (N-1))$, follows as

$$\text{acc} = \min\left[1, \frac{N\Gamma(3N/2)}{bV\Gamma(3(N-1)/2)} \times \frac{[R - PV + \mu(N-1) - U(\Gamma')]^{3(N-1)/2-1}}{[R - PV + \mu N - U(\Gamma)]^{3N/2-1}}\right] \tag{7.39}$$

The acceptance rule for the insertion of an additional molecule *i.e.* from $(\Gamma, N) \to (\Gamma', (N+1))$, is given by

---

Algorithm 5 Monte Carlo - Adiabatic μPR

1: Start from a configuration $\mathbf{\Gamma}$ of potential energy $U = U(\mathbf{\Gamma})$ and kinetic energy $K = K(\mathbf{\Gamma}) = R - PV + N\mu - U(\mathbf{\Gamma})$

2: **while** i_{move} is less than the number of attempted MC moves N_{move} **do**

3:     Generate a new configuration $\mathbf{\Gamma'}$ through a random move

4:     Store the volume for the new configuration as $V' = V(\mathbf{\Gamma'})$

5:     Store the number of particles for the new configuration as $N' = N(\mathbf{\Gamma'})$

6:     Compute the kinetic energy for the new configuration $K' = K(\mathbf{\Gamma'})$

7:     Compute the ratio of the kinetic energies $f = \frac{K'}{K}$

8:     **if** $K' > 0$ and $V' > 0$ and $N' > 0$ **then**

9:         **if** random volume change **then**

10:             Compute the logarithm of the probability of acceptance as $\ln p = (\frac{3N}{2} - 1)\ln f + N\ln\frac{V'}{V}$ random insertion

11:             Compute the logarithm of the probability of acceptance as $\ln p = (\frac{3N}{2} - 1)\ln f$ random deletion

12:             Compute the logarithm of the probability of acceptance as $\ln p = (\frac{3N}{2} - 1)\ln f$

13:         **else**

14:             Compute the logarithm of the probability of acceptance as $\ln p = (\frac{3N}{2} - 1)\ln f$

15:         **end if**

16:         Generate a random number r in the interval $[0, 1]$

17:         **if** $r < p$ **then**

18:             Accept move and update current configuration state, *i.e.*, $\mathbf{\Gamma} \leftarrow \mathbf{\Gamma'}$

19:             Update the current value for the volume, *i.e.*, $V \leftarrow V'$

20:             Update the current value for the potential energy, *i.e.*, $U \leftarrow U'$

21:         **end if**

22:     **else**

23:         Reject move and keep $\mathbf{\Gamma}$ as current configuration state

24:         Keep $K, U,$ and V as current values for the kinetic energy, potential energy, and volume, respectively

25:     **end if**

26:     Accumulate the current value for the kinetic and potential energies, as well as volume, in the running totals

27:     Accumulate running total for other quantities of interest

28: **end while**

29: Compute averages over all configurations, *i.e.*, divide running totals by N_{move}

Figure 7.6   Pseudo-code for a Monte Carlo simulation in the grand-isobaric adiabatic (μ, P, R) ensemble.

$$\text{acc} = \min\left[1, \frac{bV\Gamma(3N/2)}{(N+1)\Gamma(3(N+1)/2)} \times \frac{[R-PV+\mu(N+1)-U(\Gamma')]^{3(N+1)/2-1}}{[R-PV+\mu N-U\Gamma)]^{3N/2-1}}\right] \quad (7.40)$$

Finally, for a random volume change from $(\Gamma, V) \rightarrow (\Gamma', V')$, the Metropolis criterion is

$$\text{acc} = \min\left[1, \frac{V'^N[R-PV'+\mu N-U(\Gamma')]^{3N/2-1}}{V^N[R-PV+\mu N-U(\Gamma)]^{3N/2-1}}\right] \quad (7.41)$$

Below is an example of properties calculated during simulations in the (μ, P, R) ensemble [104]. To this end, we consider a system composed of Copper atoms, modeled with an embedded-atom potential (EAM) [68,138,256,346] known as the quantum-corrected Sutton-Chen [231] embedded atoms (qSC-EAM) potential. The qSC-EAM potential is a density-dependent force field in which the potential energy U of a system containing N atoms is calculated as the sum of a two-body term and of a many-body term.

$$U = \frac{1}{2}\sum_{i=1}^{N}\sum_{j\neq i}^{N}\varepsilon\left(\frac{a}{r_{ij}}\right)^n - \varepsilon C\sum_{i=1}^{N}\sqrt{\rho_i} \quad (7.42)$$

in which r_{ij} is the distance between two atoms i and j and the density term ρ_i is given by

$$\rho_i = \sum_{j\neq i}\left(\frac{a}{r_{ij}}\right)^m \quad (7.43)$$

In the example below, the following set of parameters [231], with $\varepsilon = 0.57921 \times 10^{-2}$ eV, $C = 84.843$, $a = 3.603$, $n = 10$ and $m = 5$, and a cutoff for the calculation of interactions set to $2a$ were used. During the simulations, 33% of the attempted Monte Carlo moves are random translations of atoms, 33% are random insertions of an additional atom, 33% are deletions of a randomly selected atom, and the remaining 1% are random volume changes.

Focusing here on the vapor-liquid transition, (μ, P, R) simulations are performed along isobars (see Figure 7.7). For each isobar, μ is gradually increased from a low value (corresponding to a

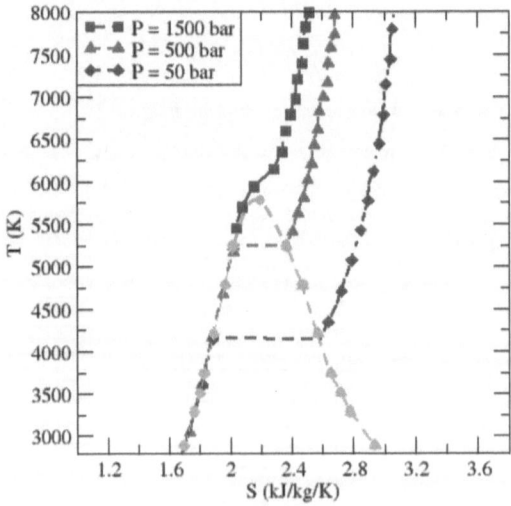

Figure 7.7 (μ, P, R) simulations of copper. The plot shows simulation results (filled symbols). Symbols along the bell-shaped curve outline the conditions identified from the simulations for the onset of the vapor-liquid coexistence for copper.

high temperature-low density fluid) to a high value of μ (corresponding to a low temperature-high density fluid). Once a (μ, P) has been selected, the value of R only controls the system size, *i.e.*, how many atoms compose the system. R is set so that there are enough particles in the system (typically, a few hundred) to ensure statistically reliable results. Increasing μ along the isobar results in two drastically different behaviors. For the isobars such that $P < 1500$ bars, entropy undergoes a steady decrease as μ increases along the isobar until the system reaches the vapor-liquid coexistence curve, or binodal curve. At this point, there is a discontinuity in the entropy as the system experiences a steep change in density. Indeed, the system, initially in the vapor phase (high entropy branch of the binodal curve), becomes a liquid (low entropy branch of the binodal). Copper undergoes a first-order phase transition, with entropy and molar volume discontinuities at the transition. There is, however, no discontinuity along the $P = 1500$ bars isobar, establishing that Copper is a supercritical fluid under these conditions. This example shows how entropy can be evaluated rapidly and how locating the onset of phase transitions using simulation in an adiabatic ensemble is straightforward.

8 Networking under One (or More) Cues

Isothermal Ensembles

"It is shown, in accordance with the accepted principles of statistical mechanics, that the establishment of a grand canonical distribution is to be expected in an ensemble which represents the attainment of equilibrium in an "open" system."

Richard C. Tolman,
On the establishment of grand canonical distributions [363]

We have seen in the previous chapter the policies and probabilities that govern the behavior of adiabatic systems for which a heat function, *i.e.*, the value of some energy, is conserved. The microcanonical (N,V,E) ensemble requires the system to be isolated so that no energy flows in or out of the system. The difference with the corresponding isothermal ensemble (canonical (N,V,T) ensemble) is thus that, in the case of the microcanonical ensemble, a thermally insulating boundary needs to surround the system to avoid any heat flow and allow the system to be in thermal equilibrium with the outside world and thus be at the same temperature. What happens now if the boundary between the system and its surroundings allows for the onset of chemical equilibrium? Or of a mechanical equilibrium? The first scenario will occur when the boundary is porous, allowing for the passage of molecules and the onset of chemical equilibrium characterized by equal chemical potentials inside and outside the system. The second scenario will occur when the vessel containing the system is deformable, allowing for volume fluctuations and the onset of mechanical equilibrium characterized by equal internal and external pressures. In other words, the surroundings can exert one or several constraints on the system by imposing its temperature, chemical potential, or pressure. As we have seen through the comparison between the (N,V,E) and (N,V,T) ensembles, this dramatically impacts the rules dictating the evolution of the system and the mathematical functions characterizing its states. We examine here two cases where external "reservoirs" ensure that the system reaches either chemical equilibrium, *i.e.*, the case of open systems and the grand-canonical (μ,V,T) ensemble, or mechanical equilibrium, *i.e.*, the case of the isothermal-isobaric (N,P,T) ensemble.

8.1 THERMAL AND CHEMICAL CUES

8.1.1 THE GRAND-CANONICAL ENSEMBLE

What happens if the only information we have on a system is that this system is in equilibrium with another phase? How should we proceed if we would like to determine how much hydrogen gas we can store in a porous material for energy applications or how much of a pollutant we can sequestrate in another porous material for environmental applications? In this case, the only information we have is about the "other phase", *e.g.*, the density of the hydrogen gas that is produced or the concentration of the pollutant in the exhaust that is released. While we have information on this "other phase, we want to determine how much of these compounds will become adsorbed in the porous materials to form the "adsorbed phase" (see Figure 8.1). This can be determined by considering a new ensemble, the grand-canonical ensemble.

To build this ensemble, we first go back to the definition of the canonical ensemble. The canonical ensemble consists of many replicas of a system of N molecules, enclosed in a rigid vessel of volume

DOI: 10.1201/9781003006411-10

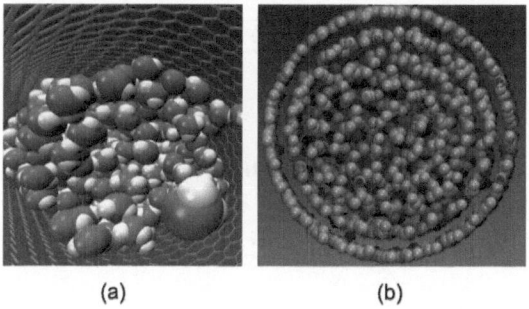

(a)   (b)

Figure 8.1 Adsorption of compounds in porous materials. (a) Water adsorbed in a carbon naotube. (b) Structure of confined Argon in the porous material MCM-41, in which the snapshot shows the layering in confined Argon, and the container is omitted for clarity. In both cases, simulations in the grand-canonical ensemble yield the properties of the adsorbed phase, such as, *e.g.*, how many molecules are held within the porous material.

V and held at constant temperature T. The temperature is maintained through heat exchanges with a heat reservoir or heat bath. This reservoir provides heat to the system or absorbs heat as needed to ensure that T remains fixed. The grand-canonical ensemble derives from the canonical ensemble by imagining that the vessel is a porous membrane in addition to allowing for heat exchange. This porous boundary enables the free passage of molecules in and out of the system to ensure that, in addition to T, the chemical potential μ for the molecular species contained in the system remains constant too. This accounts for a chemical equilibrium between the system and its surroundings. Chemical equilibrium arises thanks to the ability of the system to have its density fluctuate through variations in N. In other words, the replica that constitutes a grand-canonical ensemble all have the same temperature and chemical potential but contain different N and have thus different densities.

To obtain the grand partition function $\Theta(\mu, V, T)$ for the grand-canonical ensemble, we can build on the approach we followed in Chapter 3 for the canonical ensemble. Since μ is imposed by the surroundings and the number of molecules is not set, we have a new set of unknowns corresponding to the number of molecules in each replica. This means that, unlike in Chapter 3, where there were only two conditions to satisfy for the canonical ensemble (total number of elements in the ensemble in Eq. 3.1 and total energy of the ensemble in Eq. 3.2), we now have a third condition on the total number of molecules in the ensemble. Following the same steps as in Chapter 3, *i.e.*, identifying the distribution that maximizes the number of states W subject to the three conditions and determining the parameters through the Lagrange multipliers method, we obtain the grand partition function as

$$\Theta(\mu, V, T) = \sum_{N=0}^{\infty} Q(N, V, T) \exp(\mu N / k_B T) \tag{8.1}$$

with the product pV playing this time the role of the thermodynamic "characteristic" function directly connected to the logarithm of the ensemble partition function through

$$pV = k_B T \ln \Theta(\mu, V, T) \tag{8.2}$$

The grand partition function $\Theta(\mu, V, T)$ is related to the canonical partition functions $Q(N, V, T)$ for all N ranging from 0 to ∞. This means that we can introduce a probability distribution for the number of molecules in the system, denoted by $p(N)$ and given by

$$p(N) = \frac{Q(N, V, T) \exp(\mu N / k_B T)}{\Theta(\mu, V, T)} \tag{8.3}$$

and define average properties as a function of $p(N)$ such as, for instance, the average number of molecules $<N>$ as

$$< N >= \sum_{N=0}^{\infty} N p(N) \tag{8.4}$$

and, from there, the remaining thermodynamic properties as

$$< G >= \mu < N > \tag{8.5}$$

$$< E >= \sum_{N=0}^{\infty} < E(N,V,T) > p(N) \tag{8.6}$$

$$< S >= \frac{k_B \ln \Theta(\mu,V,T)}{< N >} + \frac{< E > - \mu}{T} \tag{8.7}$$

8.1.2 MONTE CARLO EXPLORATION

Monte Carlo simulations in the grand-canonical ensembles are a direct route to evaluate the properties of a system in equilibrium with another phase. As discussed in the previous section, the key is to allow the number of molecules in the system to change due to this chemical equilibrium. This means that, unlike in the canonical (N,V,T) ensemble, the Monte Carlo moves will not only consist of random displacements of molecules but will only include deletions of randomly selected molecules as well as insertions of additional molecules at random positions within the system. To determine the corresponding Metropolis acceptance rules, we start with the ensemble average of a phase space function B in the grand-canonical ensemble, given by

$$< B >= \frac{1}{\Theta(\mu,V,T)} \sum_{N=0}^{\infty} \frac{(Vz)^N}{N!} \int B(s) \exp(-\beta U(s)) ds^{3N} \tag{8.8}$$

in which we introduce the activity $z = \exp(\beta \mu)/\Lambda^3$ and Λ is defined as $\Lambda = (h^2/2\pi m k_B T)^{1/2}$. For simplicity, we have written the ensemble average for a system composed of atoms for which the atomic partition function only includes contributions from the translational degrees of freedom Λ^{-3}. The examples in the next chapter will discuss how the internal degrees of freedom contribute to the partition function of molecular systems and are included in the activity z. The acceptance rules derived below are thus general in z, and only the expression for z will need to be updated to reflect the nature of the molecule. We now turn to the acceptance rules for the three types of random moves, starting with the random displacement of a molecule.

The ensemble average $< B >$ shows that the probability distribution $P(\Gamma,N)$ used in the averaging of B is equal to

$$P(\Gamma,N) = \frac{1}{\Theta(\mu,V,T)} \frac{(Vz)^N}{N!} \exp(-\beta U(\Gamma)) \tag{8.9}$$

If we consider the random displacement of a molecule or, in other words, a transition from a configuration of the system Γ to a new configuration Γ' with the same number of molecules N, the acceptance rule is given by the ratio $\frac{P(\Gamma')}{P(\Gamma)}$ as

$$\mathrm{acc}(\Gamma \to \Gamma') = \min\left(1, \exp(-\beta \Delta U)\right) \tag{8.10}$$

in which $\Delta U = U(\Gamma') - U(\Gamma)$. As we can see, we recover the same acceptance rule as for a random displacement in the canonical ensemble. This is not surprising since the number of molecules does not vary during the move. In other words, despite the simulation being performed within the grand-canonical ensemble, the system remains at constant N, V, and T for this type of move. We now focus on random moves leading to a change in the number of molecules in the system. First, when a new molecule is created at a random position in the system, the corresponding change in configuration can be written as $\Gamma, N \to \Gamma', N' = N + 1$, and the acceptance rule becomes

$$\text{acc}(\Gamma, N \to \Gamma', N' = N + 1) = \min\left(1, \frac{vZ}{N+1} \exp(-\beta \Delta U)\right) \tag{8.11}$$

On the other hand, when a molecule in the system is deleted, the change in configuration is $\Gamma, N \to \Gamma', N' = N - 1$, and the acceptance rule is given by

$$\text{acc}(\Gamma, N \to \Gamma', N' = N - 1) = \min\left(1, \frac{N}{vZ} \exp(-\beta \Delta U)\right) \tag{8.12}$$

The acceptance rules from the three types of moves can be summarized into a single acceptance for a random move $\Gamma, N \to \Gamma', N'$ as

$$\text{acc}(\Gamma, N \to \Gamma', N') = \min\left(1, \left(\frac{vZ}{\max(N, N')}\right)^{\Delta N} \exp(-\beta \Delta U)\right) \tag{8.13}$$

in which $\Delta N = N' - N$ and $\max(N, N')$ denotes the greater integer between N and N'. As with Monte Carlo simulations in the canonical ensemble, properties are collected after every random move attempted. The only difference is that the number of molecules varies during the simulation, and properties on a per-mole basis need to be calculated with respect to the average number of molecules throughout the simulation. In Figure 8.2, we include a pseudo-code showing the implementation of a Monte Carlo simulation in the grand-canonical ensemble.

8.1.3 GRAND PARTITION FUNCTION DETERMINATION

Monte Carlo simulations in the grand-canonical ensemble are especially appealing since they allow one to calculate ensemble averages without ever determining the value of the grand partition function. This is because the underlying Metropolis rules rely on the value taken by probability ratios, with the grand partition functions appearing at the denominator for each probability and thus canceling out in the overall acceptance criterion. This implies that a given configuration will be sampled the correct number of times, *i.e.*, several times proportional to its probability, and that the running total for a property accumulated throughout the simulation will automatically include the correct weight for each value. In other words, the Monte Carlo simulation bypasses the need to compute the partition function for a given set (μ, V, T). What happens now if we want to determine the properties of the system for a very different μ? Can we reuse the series of configurations generated during the simulation? Since the sequence of configurations depends on the value of μ, we will need to run another grand-canonical simulation for another chemical potential. It would be unnecessary to run another simulation if we could know the value of the partition function since all thermodynamic properties can be calculated from this function. Analyzing the equation for the grand partition function in Eq. 8.1, we see that if the value for the $Q(N, V, T)$ functions is known, then the grand partition function follows for any value of μ through the simple weighted sum.

$$\Theta(\mu, V, T) = \sum_{N=0}^{\infty} Q(N, V, T) \exp(\beta \mu N) \tag{8.14}$$

To determine the value of the partition functions $Q(N, V, T)$, we carry out simulations in the grand-canonical ensemble with a Wang-Landau sampling [91,151]. To this end, we follow the reasoning developed in Section 6.3.2 to determine the corresponding biased distribution that samples evenly all values for the number of molecules N in a system with a fixed T and V. The joint distribution $p(\Gamma, N)$ is given by

$$p(\Gamma, N) = \frac{V^N \exp[\beta \mu N - \beta U(\Gamma)]}{N! \Lambda^{3N} \Theta(\mu, V, T)} \tag{8.15}$$

Algorithm 7 Monte Carlo - μVT

1: Start from a configuration Γ of potential energy $U = U(\Gamma)$ with N molecules

2: **while** i_{move} is less than the number of attempted MC moves N_{move} **do**

3:      Generate a new configuration Γ' with N' molecules through a random move. For example, choose a molecule at random and translate by a random vector, or rotate it by a random angle, or create/delete a molecule...

4:      Compute the energy for the new configuration $U' = U(\Gamma')$

5:      Compute the change in energy of the system $\Delta U = U' - U$ as a result of the random move

6:      Compute the change in number of molecules in the system $\Delta N = N' - U$

7:      Compute the transition probability $W = \left(\frac{vZ}{\max(N,N')}\right)^{\Delta N} e(-\beta \Delta U)$

8:      Generate a random number r in the interval $[0, 1]$

9:      **if** $r \leq W$ **then**

10:          Accept the new configuration

11:          Update current configuration state, *i.e.*, $\Gamma \leftarrow \Gamma'$

12:          Update number of molecules, *i.e.*, $N \leftarrow N'$

13:          Update the current value for the energy, *i.e.*, $U \leftarrow U'$

14:      **else**

15:          Reject move

16:          Keep Γ as current configuration state

17:          Keep N as current number of molecules

18:          Keep U as current value for the energy

19:      **end if**

20:

21:      Accumulate the current value for the energy in the running total for the energy

22:      Accumulate the current value for the number of molecules in the running total

23:      Accumulate running total for other quantities of interest

24: **end while**

25: Compute averages over all configurations, *i.e.*, divide running totals by N_{move}

Figure 8.2   Pseudo-code for a Monte Carlo simulation in the grand-canonical (μ, V, T) ensemble.

Wang-Landau simulations in the grand-canonical ensemble

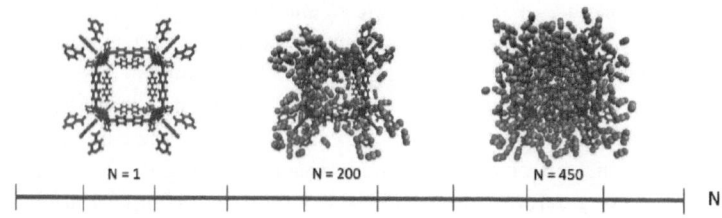

1. Set Q(N,V,T)=1 and histogram H(N)=0 for all N; choose an initial value for the modification factor f (e.g., f=e¹)

2. Perform random moves and accept/reject according to the acceptance rule

3. Update the histograms for the number of visits H(N) and for the running estimate of Q(N,V,T):
   - H(N) ← H(N)+1
   - Log Q(N,V,T) ← Log Q(N,V,T) + Log f

4. Repeat steps 2 &3 until the histogram H(N) is flat

5. Decrease f according to f ←f¹/² initialize H(N) to 0 for all N, and repeat steps 2 - 5 until f~ exp(10⁻⁸)

6. Calculate properties using final estimates for Q(N,V,T)

Figure 8.3 Summary of the steps involved in the iterative determination of the grand partition function of CO_2 adsorbed in nanoporous material IRMOF-1 via Wang-Landau simulations [92].

and the particle number distribution $p(N)$ is

$$p(N) = \frac{Q(N,V,T)\exp(\beta\mu N)}{\Theta(\mu,V,T}$$

(8.16)

Taking the biased probability distribution $p_{\text{bias}}(\Gamma,N)$ as the ratio of $p(\Gamma,N)$ to $p(N)$, we obtain

$$p_{\text{bias}}(\Gamma,N) = \frac{V^N \exp[-\beta U(\Gamma)]}{N!\Lambda^{3N}Q(N,V,T)}$$

(8.17)

In a Wang-Landau simulation, the acceptance rule from a configuration Γ with N molecules to a configuration Γ' with N' molecules is given by the ratio of the biased probabilities according to

$$\text{acc}(\Gamma,N \rightarrow \Gamma',N') = \min\left[1, \frac{p_{\text{bias}}(\Gamma',N')}{p_{\text{bias}}(\Gamma,N)}\right]$$

(8.18)

Plugging in the expression of the biased probability, we obtain the following acceptance rule.

$$\text{acc}(\Gamma,N \rightarrow \Gamma',N') = \min\left[1, \frac{Q(N,V,T)}{Q(N',V,T)} \frac{V^{N'}N!\Lambda^{3N}\exp[-\beta U(\Gamma')]}{V^N N'!\Lambda^{3N'}\exp[-\beta U(\Gamma)]}\right]$$

(8.19)

As we can see, the chemical potential does not appear in the acceptance rule. Still, it is applied later in post-simulation calculations as a weighting factor in evaluating $\Theta(\mu,V,T)$ using Eq. 8.14. The Wang-Landau simulations thus rely on determining the $Q(N,V,T)$ functions through an iterative scheme, summarized in Figure 8.3.

8.2 THERMAL AND MECHANICAL CUES

8.2.1 THE ISOTHERMAL-ISOBARIC ENSEMBLE

Another case may arise. Instead of being in chemical equilibrium with its surroundings, a system may be in mechanical equilibrium. This means that, rather than being subjected to a constraint on the chemical potential due to the exchange of matter with the surroundings, the pressure can be held constant due to the mechanical (PV) work exerted on the system by the surroundings. This leads us to introduce another statistical ensemble, the isothermal-isobaric (N,P,T) ensemble, characterized by a fixed number of molecules N, fixed pressure P, and fixed temperature T.

Before determining this ensemble's partition function and ensemble averages, let us compare the isothermal-isobaric ensemble to the canonical ensemble. The canonical ensemble consists of many replicas of a system of N molecules enclosed in a rigid vessel of volume V and held at constant temperature T. The isothermal-isobaric ensemble can derive from the canonical ensemble through a simple modification. Instead of having a rigid vessel enclosing the N molecules of a system, imagine that we now have a deformable container that allows the volume V to fluctuate in response to the pressure the surroundings apply. In the isothermal-isobaric case, the reservoir exchanges energy with the system through mechanical work through the systems' deformation. This leads to the onset of a mechanical equilibrium between a system and its surroundings. The replica that forms an isothermal-isobaric ensemble all have the same temperature and number of molecules but have different V and have, thus, different densities.

The isothermal-isobaric partition function $Q(N,P,T)$ is obtained using the same reasoning as the canonical and grand partition functions. Since P is imposed by the surroundings, and the volume for each replica is not set, we have a new set of unknowns and an additional condition to satisfy, compared to the canonical ensemble case. The third condition is now on the total volume, summed over all replicas, for the ensemble. Following the same steps as in Chapter 3, *i.e.*, identifying the distribution that maximizes the number of states W subject to the three conditions and determining the parameters through the Lagrange multipliers method, we obtain the isothermal-isobaric partition function as

$$Q(N,P,T) = \int_0^\infty Q(N,V,T)\exp(-\beta PV)dV \tag{8.20}$$

and the Gibbs free energy $G(N,P,T)$ is now the thermodynamic "characteristic" function directly connected to the logarithm of the ensemble partition function through

$$G(N,P,T) = -k_B T \ln Q(N,P,T) \tag{8.21}$$

As shown in Eq. 8.20, the isothermal-isobaric partition function $Q(N,P,T)$ is therefore related to the canonical partition functions $Q(N,V,T)$ for all V ranging from 0 to ∞. This means that we can define a probability distribution for the volume $p(V)$ as

$$p(V) = \frac{Q(N,V,T)\exp(-\beta PV)}{Q(N,P,T)} \tag{8.22}$$

and define average properties as a function of $p(V)$, starting with the average volume of the system

$$<V> = \int_0^\infty V p(V)dV \tag{8.23}$$

The remaining thermodynamic properties can be calculated as, for instance,

$$<E> = \int_0^\infty <E(N,V,T)> p(V)dV \tag{8.24}$$

$$<S> = \frac{k_B \ln Q(N,P,T)}{N} + \frac{<E>+P<V>/N}{T} \tag{8.25}$$

8.2.2 PROPERTIES CALCULATIONS

Monte Carlo simulations in the isothermal-isobaric allow for the direct calculation of the properties of a system in mechanical and thermal equilibrium with its surroundings. Since the volume of the system is allowed to change unlike in the canonical (N, V, T) ensemble, the Monte Carlo moves will consist of random volume changes in addition to random displacements of molecules. To determine the Metropolis acceptance rules, we begin with the ensemble average $< B >$ of a phase space function B in the isothermal-isobaric ensemble, defined as

$$< B >= \frac{1}{Q(N, P, T)} \int_{V=0}^{\infty} V^N \exp(-\beta PV) \int_s B(s) \exp(-\beta U(s)) ds^{3N} \qquad (8.26)$$

The probability distribution $P(\Gamma, V)$ can used in the averaging of B is equal to

$$P(\Gamma, V) = \frac{V^N}{Q(N, P, T)} \exp(-\beta(U(\Gamma) + PV)) \qquad (8.27)$$

If we consider the random displacement of a molecule or, in other words, a transition from a configuration of the system Γ to a new configuration Γ' with the same number of molecules N, the acceptance rule is given by the ratio $\frac{P(\Gamma')}{P(\Gamma)}$ as

$$\mathrm{acc}(\Gamma \to \Gamma') = \min[1, \exp(-\beta \Delta U)] \qquad (8.28)$$

with $\Delta U = U(\Gamma') - U(\Gamma)$. As previously discussed in the case of the grand-canonical ensemble, we recover the same acceptance rule for a random displacement as in the canonical ensemble. This is expected since, during this type of random move, the system remains at constant N, V, and T. We now turn to random changes in volume. Considering a change from configuration $\Gamma, V \to \Gamma', V'$, the acceptance rule is given by

$$\mathrm{acc}(\Gamma, V \to \Gamma', V') = \min\left[1, \left(\frac{V'}{V}\right)^N \exp(-\beta(\Delta U + P\Delta V))\right] \qquad (8.29)$$

in which $\Delta V = V' - V$ and $(\Delta U + P\Delta V)$ stands for the enthalpy change during the rencom volume change.

The acceptance rules from the two types of moves can be summarized into a single acceptance for a random move $\Gamma, V \to \Gamma', V'$ as

$$\mathrm{acc}(\Gamma, V \to \Gamma', V') = \min\left[1, \left(\frac{V'}{V}\right)^N \exp(-\beta(\Delta U + P\Delta V))\right[\qquad (8.30)$$

Similarly to Monte Carlo simulations in the canonical and grand-canonical ensemble, properties are collected after every random move attempted. In Figure 8.4, we include a pseudo-code outlining how a Monte Carlo simulation in the isothermal-isobaric ensemble can be implemented.

8.2.3 PARTITION FUNCTION COMPUTATION

Monte Carlo simulations in the isothermal-isobaric ensemble share the same advantages and drawbacks as Monte Carlo simulations in the other ensembles. They generate a chain of configurations that correctly sample the volume occupied by the system for a given set (N, P, T) and, as such, provide reliable ensemble averages of the thermodynamic properties of the system. However, suppose we need to evaluate the properties of the system at, for instance, a much higher pressure. In that case, the chain of configurations generated during the simulations will not provide an accurate picture of the system for this new pressure value. In other words, the volumes sampled at high pressure will be significantly smaller than those sampled at low pressure. We must start from scratch and run another simulation at high pressure. As with the grand-canonical ensemble, the alternative is to use Wang-Landau sampling to sample the entire volume range evenly, evaluate the isothermal-isobaric partition function, and calculate all thermodynamic properties from the partition function.

Algorithm 8 Monte Carlo - NPT

1: Start from a configuration Γ of potential energy $U = U(\Gamma)$ and volume V

2: **while** i_{move} is less than the number of attempted MC moves N_{move} **do**

3:     Generate a new configuration Γ' and volume V' through a random move. For example, choose a molecule at random and translate by a random vector, rotate it by a random angle, or perform a random volume change...

4:     Compute the energy for the new configuration $U' = U(\Gamma')$

5:     Compute the change in energy of the system $\Delta U = U' - U$ as a result of the random move

6:     Compute the change in volume $\Delta V = V' - V$ as a result of the random move

7:     Compute the transition probability $W = \left(\frac{V'}{V}\right)^N e^{-\beta(\Delta U + P \Delta V)}$

8:     Generate a random number r in the interval $[0, 1]$

9:     **if** $r \leq W$ **then**

10:         Accept the new configuration

11:         Update current configuration state, *i.e.*, $\Gamma \leftarrow \Gamma'$

12:         Update current volume, *i.e.*, $V \leftarrow V'$

13:         Update the current value for the energy, *i.e.*, $U \leftarrow U'$

14:     **else**

15:         Reject move and keep Γ as current configuration state

16:         Keep V as current value for the volume

17:         Keep U as current value for the energy

18:     **end if**

19:     Accumulate the current value for the energy in the running total for the energy

20:     Accumulate the current value for the volume in the running total for the volume

21:     Accumulate running total for other quantities of interest

22: **end while**

23: Compute averages over all configurations, *i.e.*, divide running totals by N_{move}

Figure 8.4 Pseudo-code for a Monte Carlo simulation in the isothermal-isobaric (N, P, T) ensemble.

As we will see shortly, this approach hinges on the iterative determination of the partition functions $Q(N, V, T)$ over the volume range and the evaluation during post-simulation processing of $Q(N, P, T)$ according to

$$Q(N, P, T) = \int_0^\infty Q(N, V, T) \exp(-\beta P V) dV \tag{8.31}$$

To ensure a uniform volume sampling, we must determine the biased distribution for the Wang-Landau sampling scheme [90,151]. We start with the joint distribution $p(\Gamma, V)$ as

$$p(\Gamma, V) = \frac{V^N \exp[-\beta P V - \beta U(\Gamma)]}{N! \Lambda^{3N} Q(N, P, T)} \tag{8.32}$$

The volume distribution $p(V)$ is obtained by integrating $p(\Gamma, V)$ over the configuration space, which gives

$$p(V) = \frac{Q(N, V, T) \exp(-\beta P V)}{Q(N, P, T)} \tag{8.33}$$

The biased probability distribution $p_{bias}(\Gamma, V)$ is, in this case, equal to the ratio of $p(\Gamma, V)$ to $p(V)$ and is equal to

$$p_{bias}(\Gamma, V) = \frac{V^N \exp[-\beta U(\Gamma)]}{N! \Lambda^{3N} Q(N, V, T)} \tag{8.34}$$

Wang-Landau sampling in the isothermal-isobaric ensemble

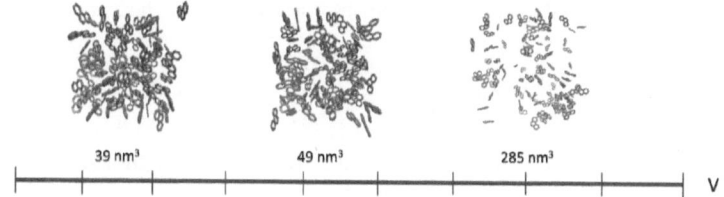

39 nm³   49 nm³   285 nm³

V

1. Set Q(N,V,T)=1 and histogram H(V)=0 for all V intervals; choose an initial value for the modification factor f (e.g., f=e¹)

2. Perform random moves and accept/reject according to the acceptance rule

3. Update the histograms for the number of visits H(V) and for the running estimate of Q(N,V,T):
   • H(V) ← H(V)+1
   • Log Q(N,V,T) ← Log Q(N,V,T) + Log f

4. Repeat steps 2 &3 until the histogram H(V) is flat

5. Decrease f according to f ← f¹ᐟ² initialize H(V) to 0 for all V intervals, and repeat steps 2 - 5 until f~exp(10⁻⁵)

6. Calculate properties using final estimates for Q(N,V,T)

Figure 8.5 Summary of the steps involved in the iterative determination of the isothermal-isobaric partition function of a polycyclic aromatic hydrocarbon molecules system via Wang-Landau simulations.

The acceptance rule from a configuration Γ with volume V to a configuration Γ' with volume V' is given by

$$\mathrm{acc}(\Gamma,V \to \Gamma',V') = \min\left[1, \frac{p_{\mathrm{bias}}(\Gamma',V')}{p_{\mathrm{bias}}(\Gamma,V)}\right] \tag{8.35}$$

which, when plugging in the expression of the biased probability, gives

$$\mathrm{acc}(\Gamma,V \to \Gamma',V') = \min\left[1, \frac{Q(N,V,T)}{Q(N,V',T)} \frac{V'^N \exp[-\beta U(\Gamma')]}{V^N \exp[-\beta U(\Gamma)]}\right] \tag{8.36}$$

Pressure does not appear in the acceptance rule. Still, it is applied later in post-simulation calculations as a weighting factor in evaluating $Q(N,P,T)$ using Eq. 8.31. As in the case of the grand-canonical ensemble, Wang-Landau simulations in the isothermal-isobaric ensemble rely on the iterative determination of the $Q(N,V,T)$ functions summarized in Figure 8.5.

8.3 VARIATIONS AND APPLICATIONS

8.3.1 MULTI-COMPONENT SYSTEMS AND SEMI-GRAND APPROACH

Systems in nature rarely contain a single component. How do we account for the presence of different identities within the group? Thus, the next step is generalizing this approach to multiple identities and components. We see below how we can extend the equations to more than one type of molecule in the case of a binary system. If we consider an ideal (non-interacting) binary system composed of distinguishable molecules, we can apply the same reasoning that led us to obtain Eqs. 3.25 and 3.26. Since the total energy for the entire system is simply the sum of the energy for the two subsystems of N_1 molecules and N_2 molecules, we have, for the ideal mixture,

$$Q(N,V,T) = Q(N_1,V,T) \times Q(N_2,V,T)$$
$$= \frac{q_1^{N_1} q_2^{N_2}}{N_1! N_2!} \tag{8.37}$$

As discussed in Chapter 4, q_1 and q_2 can take different forms depending on the nature of the components in the system with for atoms, $e.g.$, for q_1, $q_1 = \frac{V}{\Lambda_1^3}$, in which $\Lambda = \sqrt{\frac{h^2}{2\pi m_1 k_B T}}$, and for a polyatomic molecule, $q_1 = \frac{V}{\Lambda^3} q_{int,1}$, in which $q_{int,1}$ encompasses the internal degrees of freedom of rotation and vibration (see Eqs. 4.16–4.21).

Generalizing this to a system of interacting atoms, we obtain the canonical partition function $Q(N_1, N_2, V, T)$

$$Q(N_1, N_2, V, T) = \frac{V^{(N_1+N_2)}}{N_1! \Lambda_1^{3N_1} N_2! \Lambda_2^{3N_2}} \int \exp\left(-\beta U(\Gamma)\right) d\Gamma \tag{8.38}$$

As discussed above, this equation can be extended to the case of molecules by including in the numerator the factor $q_{int,1}^{N_1} q_{int,2}^{N_2}$. For conciseness, we provide the partition functions in the other ensembles on the example of atoms below.

Building on the canonical partition function for the mixture, we have the following equation for the grand-canonical partition function $\Theta(\mu_1, \mu_2, V, T)$

$$\Theta(\mu_1, \mu_2, V, T) = \sum_{N_1} \sum_{N_2} Q(N_1, N_2, V, T) \exp(\beta \mu_1 N_1) \exp(\beta \mu_2 N_2) \tag{8.39}$$

in which μ_i ($i = 1, 2$) the chemical potential of atoms/molecules of type i. Introducing the activity $\lambda_i = \exp(\mu_i/k_B T)$, we define the number distribution $p(N_1, N_2)$ for the binary mixture as

$$p(N_1, N_2) = \frac{Q(N_1, N_2, V, T) \lambda_1^{N_1} \lambda_2^{N_2}}{\sum_{N_1} \sum_{N_2} Q(N_1, N_2, V, T) \lambda_1^{N_1} \lambda_2^{N_2}} \tag{8.40}$$

and evaluate all properties by following the same steps as for the single-component case. For instance, the average numbers of molecules are given by $< N_i > = \sum_{N_1} \sum_{N_2} N_i \times p(N_1, N_2)$. We give below the equations for several thermodynamic properties:

$$< P > = \frac{k_B T \ln \Theta(\lambda_1, \lambda_2, V, T)}{V} \tag{8.41}$$

$$< \rho > = \sum_{N_1} \sum_{N_2} \frac{(m_1 N_1 + m_2 N_2)}{V} p(N_1, N_2) \tag{8.42}$$

$$< E > = \sum_{N_1} \sum_{N_2} \left(E_{pot}(N_1, N_2) + \frac{n_{f,1}}{2} N_1 k_B T + \frac{n_{f,2}}{2} N_2 k_B T \right) p(N_1, N_2) \tag{8.43}$$

where $n_{f,i}$ ($i = 1, 2$) denotes the number of degrees of freedom for molecule i. We recall $n_{f,i} = 3$ if i is an atom, 5 if i is a linear molecule and 6 in the general case. Finally, the enthalpy and entropy are given by

$$< H > = < E > + k_B T \ln \Theta(\lambda_1, \lambda_2, V, T) \tag{8.44}$$

$$< S > = k_B \ln \Theta(\lambda_1, \lambda_2, V, T) + \frac{(< E > - \mu_1 < N_1 > - \mu_2 < N_2 >)}{T} \tag{8.45}$$

Determining the properties of multi-component systems has also inspired scientists to derive ensembles from the three isothermal ensembles we have discussed to calculate specific properties. We will discuss in the next section one such example, the Gibbs ensemble Monte Carlo (GEMC) approach [281,282], that was tailored for the determination of vapor-liquid coexistence properties. We mention here another example, the semi-grand ensemble [210]. The semi-grand ensemble can be considered a hybrid between the canonical and grand canonical. In its initial derivation, the idea was to use a constraint on either the total number of molecules in the system, i.e., maintaining $N = N_1 + N_2$ constant rather than letting N_1 and N_2 fluctuate freely. In this case, the insertion of new molecules or the deletion of existing molecules was replaced by identity changes, a molecule of type 1 being, for instance, switched into a molecule of type 2 during a Monte Carlo move. Other implementations have focused on determining the solubility [314] and kept the number of molecules of type 1 constant (for instance, a bulky solute). In contrast, the number of molecules of type 2 (the solvent in this example) can change. More recently, the semi-grand approach has been applied to study the phase behavior of multicomponent systems [120], the impact of polydispersity on thermodynamic properties [26], and, in recent years, the properties of mixed biological membranes [63].

8.3.2 A FIRST STEP TOWARD COEXISTENCE: GIBBS ENSEMBLE MONTE CARLO METHOD

We have seen how we could determine the properties of coexisting phases with simulations in, for instance, the grand-canonical ensemble. To achieve this, we need to carry out a series of simulations for each of the two phases and find the state points for which the two phases share the same T, P, and μ, *i.e.*, are in thermal, mechanical, and chemical equilibrium. This requires a significant number of independent simulations to obtain the full coexistence boundary between the two phases. Can we reduce the number of simulations, perhaps requiring only a single simulation run per coexistence point? A first solution could be to simulate the interface between the two phases directly. There are a few difficulties involved, however, as such a simulation actually contains three subsystems, that are the interface, phase I and phase II. This approach will thus require simulating a system with a large N and will not be the most efficient if we are not interested in the properties of the interfacial region and need to determine the properties of phases I and II only.

Panagiotopoulos [281,282] proposed an efficient solution to address this challenge by introducing the Gibbs Ensemble Monte Carlo (GEMC) method. The GEMC approach combines the canonical, grand-canonical, and isothermal-isobaric ensembles to achieve this efficiency. Indeed, in its implementation for single-component systems, the system consists of two physically separated boxes at the same temperature coupled through Monte Carlo moves that ensure the two boxes have the same pressure and chemical potential. The total number of molecules N is split between the two boxes, $N = N^I + N^{II}$, and the total volume V is the sum of the volume for the two boxes, $V = V^I + V^{II}$. A (N,V,T) GEMC simulation thus consists of three types of random moves, starting with the random displacement of a molecule within the box it originates from according to the acceptance rule provided below.

$$\mathrm{acc}(\Gamma_i \to \Gamma_i') = \min\left[1, \exp\left(-\beta \Delta U_i\right)\right] \tag{8.46}$$

in which $i = I$ or II. The second type of random move consists of a random volume change, with the volume of, for instance, box I increasing by ΔV and, correspondingly, box II decreasing by ΔV. This leads to the following acceptance rule, calculated as the product of the probabilities for two volume change moves for boxes I and II at the same pressure P.

$$\mathrm{acc}(\Gamma_i \to \Gamma_i') = \min\left[1, \exp\left(-\beta\left[\Delta U_I + \Delta U_{II} - N_I k_B T \ln \frac{V^I + \Delta V}{V^I} - N_{II} k_B T \ln \frac{V^{II} + \Delta V}{V^{II}}\right]\right)\right] \tag{8.47}$$

The third type of move involves the transfer of a molecule from one box to the other. If the molecule is transferred from box *II* to box *I*, the transfer can be split into two familiar Monte Carlo moves corresponding to creating a molecule in box *I* and deleting a molecule in box *II*. This yields the following acceptance rule.

$$\mathrm{acc}(\Gamma_i \to \Gamma_i') = \min\left[1, \exp\left(-\beta\left[\Delta U_I + \Delta U_{II} + k_B T \ln \frac{V^{II}(N^I + 1) + \Delta V}{V^I N^{II}}\right]\right)\right] \tag{8.48}$$

The various types of moves involved in a GEMC simulation are summarized in Figure 8.6. Phase equilibria in binary mixtures can be simulated by carrying out (N,P,T) GEMC simulations, *i.e.* by fixing an additional intensive thermodynamic property (P) to fully determine the state of the system

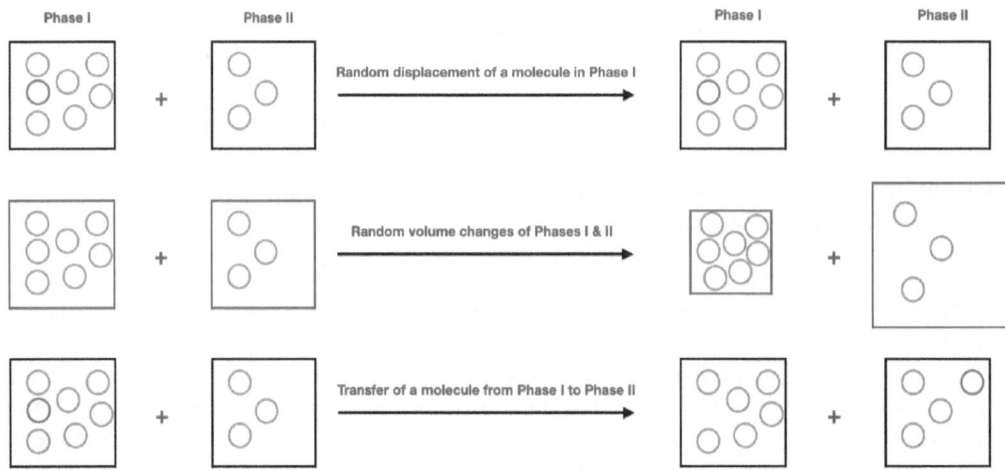

Figure 8.6 Monte Carlo moves involved in a (N, V, T) GEMC simulation.

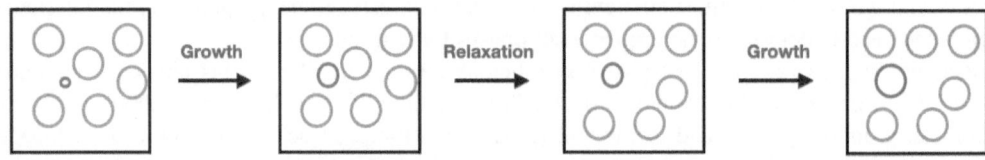

Figure 8.7 Insertion of a molecule in the expanded ensemble approach. The newly inserted molecule on the left grows gradually from a small sized molecule to a full-size molecule. As growth occurs, the rest of the system relaxes to accommodate this growth.

in accord with Gibbs' phase rule. Implementations of the GEMC method apply to various types of phase equilibria, including liquid-vapor and solid-vapor equilibria [53].

As with all approaches based on the grand-canonical ensemble, the insertion of molecules at a randomly selected position in a dense phase is fraught with a low acceptance probability and low success rates. This can be alleviated using alternate strategies, such as the Configurational Bias Monte Carlo (CBMC) or Expanded Ensemble method. In the CBMC method, a chain molecule is created bead by bead, with the location of the next bead being drawn from a set of energetically reasonable positions [332]. In the Expanded Ensemble approach, a molecule is grown gradually from a minute version of itself to a full-size molecule while allowing the rest of the system to relax around the growing molecule [121,232,264]. A schematic of the expanded ensemble approach is shown in Figure 8.7. Such approaches have enabled the simulations of large molecules in GEMC simulations [9,279,296,331,372,393].

8.3.3 RECYCLING AND REWEIGHTING

A data analysis approach can also be applied to recycle and retreat from Monte Carlo simulations. The positive outcome from such an analysis is that the results obtained for a given set of conditions can be extrapolated to a slightly different, neighboring set of conditions. This is achieved through a technique known as histogram reweighting [132–134,200]. The concept of histogram relies on the simple scaling relation that can be introduced between probability distributions obtained for two different temperatures. For instance, considering the partition function for several molecules N, a temperature T_1, and a volume V with, for simplicity, a discrete distribution of energy levels.

$$Q(N,V,T_1) = \int \Omega(N,V,E) \exp\left(-\frac{E}{k_B T_1}\right) dE \propto \int P_{N,V,T_1}(E) dE \qquad (8.49)$$

by introducing $P_{N,V,T_1}(E) \propto \Omega(N,V,E) \exp\left(-\frac{E}{k_B T_1}\right)$. Now, if we would like to obtain the partition function for a temperature T_2, we may write

$$P_{N,V,T_2}(E) \propto \Omega(N,V,E) \exp\left(-\frac{E}{k_B T_2}\right) = \Omega(N,V,E) \exp\left(-\frac{E}{k_B T_1}\right) \exp\left[-E\left(\frac{1}{k_B T_2} - \frac{1}{k_B T_1}\right)\right]$$
$$(8.50)$$

This means that $P_{N,V,T_2}(E)$ can be, in principle, obtained from $P_{N,V,T_1}(E)$ through a simple reweighting process according to

$$P_{N,V,T_2}(E) = P_{N,V,T_1}(E) \exp\left[-E\left(\frac{1}{k_B T_2} - \frac{1}{k_B T_1}\right)\right] \qquad (8.51)$$

This approach can be leveraged to extrapolate Monte Carlo simulation results obtained at a given temperature to temperatures that are close by. However, the temperature range over which simulation results can be extrapolated reliably is limited. This is due to the statistical errors in determining the histogram for $P_{N,V,T_1}(E)$. Indeed, the statistical errors become very significant in the tails of the histogram, and reweighting to a temperature T_2 that significantly departs from T_1 would require an accurate determination of the tails of $P_{N,V,T_1}(E)$. This range is, however, large enough to determine a sizeable fraction of the phase coexistence curve by histogram reweighting in many cases [119, 295,383].

9 Collective Properties from Partition Functions

"The relation due to Boltzmann between entropy and "thermodynamic probability" is enunciated in a precise form. (. . .) each of the other thermodynamic potentials is related (. . .) to a "thermodynamic probability", for which a more suitable name is a "partition function"."

Edward A. Guggenheim,
Grand partition functions and so-called "thermodynamic probability" [171]

The observation made by Guggenheim captures the interplay between thermodynamics and statistics, that is between entropy and probability through $S(N,V,E) = k_B \ln \Omega(N,V,E)$ in the microcanonical ensemble, or, more generally, between thermodynamic potentials and partition functions. As discussed in the previous chapters, the thermodynamic potentials for a given set of fixed thermodynamic properties are related to the logarithm of the partition function in the corresponding ensemble. Such relations apply to adiabatic and isothermal ensembles alike. The other thermodynamic state functions can then be deduced from the partition function, via equations that ensure consistency with macroscopic thermodynamics. The partition functions thus contain all the necessary information to fully characterize in a single function all the statistical information gleaned from a system that evolves on a $3N$-dimensional configurational space, where N stands for the number of atoms. While this information has been summarized in mathematical equations until now, it can be numerically determined through the calculations of the integrals involved in the formula. The determination of partition functions and their practical use in statistical mechanics equations are the topics covered in this chapter. We mentioned in Chapters 6 and 8 how Monte Carlo simulations, performed according to the Wang-Landau sampling scheme, can lead formally to the iterative determination of partition functions. We discuss below the practical applications of these methods, as well as additional integration methods providing numerical estimates for partition functions. We also examine several test cases, including the use of the so obtained partition functions to locate phase transitions or determine the thermodynamic properties of adsorption.

9.1 GENERATING DATA ON PARTITION FUNCTIONS

9.1.1 STARTING FROM A

The most straightforward approach to determining the numerical value of a partition function is start from the beginning, *i.e.*, the relation between the Helmholtz free energy A and the canonical partition function $Q(N,V,T)$. Since $A(N,V,T)$ is the thermodynamic potential for the canonical ensemble (N,V,T), we have the following equation for $Q(N,V,T)$

$$Q(N,V,T) = \exp\left(-\frac{A(N,V,T)}{k_B T}\right) \tag{9.1}$$

This means we can determine $Q(N,V,T)$ by taking an indirect route involving the prior evaluation of $A(N,V,T)$. Fortunately, thermodynamic relations provide a simple way to estimate $A(N,V,T)$ using conventional Monte Carlo simulations. This is, for instance, the case if we consider the partial derivative of A with respect to the volume, given by

DOI: 10.1201/9781003006411-11

$$\left(\frac{\partial A}{\partial V}\right)_{N,T} = -P \tag{9.2}$$

This relation provides the basis for a very efficient method for calculating A, known as thermo-dynamic integration [143,257]. Indeed, the Helmholtz free energy resulting from a change in the volume of the system can be evaluated by integrating $-P$ over the change as

$$A(N,V,T) = A(N,V_0,T) - \int_{V_0}^{V} PdV' \tag{9.3}$$

If the volume V_0 is taken to be very large, then the intermolecular interactions vanish, giving rise to an ideal gas. The Helmholtz free energy $A(N,V_0,T)$ can then be taken to be equal to that of the ideal gas $A_{\text{id}}(N,V_0,T)$ which, for instance, in the case of a system of N atoms is equal to

$$A_{\text{id}}(N,V_0,T) = -Nk_BT\ln\left[\left(\frac{2\pi mk_BT}{h^2}\right)^{3/2}\frac{V_0e}{N}\right] \tag{9.4}$$

We recall that the general equation for A_{id} in the case of a polyatomic molecular fluid is provided in Eq. 4.35. However, this equation contains two logarithmically divergent terms as $V \to \infty$. $A_{\text{id}}(N,V_0,T)$ is a logarithmically divergent function of V_0. Rearranging this equation to separate terms that depend on V_0 from terms that are independent of V_0 and introducing the density as $\rho_0 = N/V_0$, we find

$$\frac{A_{\text{id}}(\rho_0,T)}{Nk_BT} - \ln\rho_0 = -\ln\left[\left(\frac{2\pi mk_BT}{h^2}\right)^{3/2}e\right] \tag{9.5}$$

The Helmholtz free energy of the ideal gas at a density $\rho = N/V$ can thus be written as a function of $A_{\text{id}}(\rho_0,T)$ via

$$\frac{A_{\text{id}}(\rho,T)}{Nk_BT} = \frac{A_{\text{id}}(\rho_0,T)}{Nk_BT} - \ln\left(\frac{\rho_0}{\rho}\right) \tag{9.6}$$

Plugging in $d\rho = -\frac{\rho^2}{N}dV$ and recognizing the ideal gas pressure as $P_{\text{id}} = \rho k_BT$, the equation for $A(\rho,T)$ becomes a more easily manageable expression from a computational standpoint as

$$\frac{A(\rho,T)}{Nk_BT} = \frac{A_{\text{id}}(\rho,T)}{Nk_BT} - \int_{\rho_0}^{\rho}\frac{1}{\rho'}\left(\frac{P}{\rho'k_BT} - 1\right)d\rho' \tag{9.7}$$

Here we use the fact that the integral of $1/\rho'$ over the $[\rho_0,\rho]$ interval equals $\ln(\rho/\rho_0)$. Furthermore, the integrand in the above equation converges at low density and takes the value of the second virial coefficient B_2, as shown in the well-known virial equation

$$\frac{1}{\rho'}\left(\frac{P}{\rho'k_BT} - 1\right) = B_2(T) + B_3(T)\rho' + \mathcal{O}(\rho'^{\in}) \tag{9.8}$$

The second virial coefficient can be obtained from the intermolecular potential as

$$B_2(T) = -2\pi\int_{0}^{\infty}[\exp(-\beta V(r)) - 1]r^2dr \tag{9.9}$$

This provides a numerical estimate for the value of the integrand at low density. Now that we have rearranged the equation so that all terms are now convergent as $\rho \to 0$ (or, equivalently, $V \to \infty$), we obtain

$$\frac{A(\rho,T)}{Nk_BT} = \frac{A_{\text{id}}(\rho,T)}{Nk_BT} - \int_{0}^{\rho}\frac{1}{\rho'}\left(\frac{P}{\rho'k_BT} - 1\right)d\rho' \tag{9.10}$$

The integral can now be evaluated numerically over a series of Monte Carlo simulations. To this end, we vary the system's density and use Monte Carlo simulations to obtain an equation of state for P as a function of ρ. This can be done either during simulations in the (N,P,T) ensemble by averaging the average density $<\rho> = \frac{N}{<V>}$ as a function of the input P during simulations in the (N,V,T) ensemble using the virial theorem to evaluate $<P>$ as a function of the input $\rho = N/V$ density according to

$$<P> = \rho k_B T + \frac{1}{V} < \sum_{i=1}^{N} \mathbf{r}_i . \mathbf{f}_i >$$

(9.11)

in which $\mathbf{r}_i$ and $\mathbf{f}_i$ denote the position of and force acting on molecule i.

9.1.2 FROM DILUTE TO CONDENSED PHASES

The path we followed in the previous section shows that free energy differences can be obtained fairly easily from Monte Carlo simulations. Indeed, through the integration process, we calculate the above free energy differences between systems of decreasing density until we get to a low enough density for which we know that the integrand is equal to $B_2(T)$. This allows us to obtain the glute value for the Helmholtz free energy, which is required to obtain the value for the canonical partition function. This will be a common theme to all determinations of partition functions. One *always* needs to connect to a state (density, temperature,...) for which we know exactly what the partition function and absolute Helmholtz free energy are equal to [191]. Such a state can be the ideal gas (low density state) as discussed above or, in the case of crystals, the Einstein crystal [142]. Early in the 20th century, Einstein proposed a simple model for an atomic crystal according to which each atom vibrates about its equilibrium position. In the Einstein picture, the crystal is essentially a set of $3N$ harmonic oscillators which all have the same frequency ν, with the $3N$ oscillators corresponding to the N atoms vibrating independently from each other along the three spatial directions x, y, and z. This simple model proved instrumental in the early successes of statistical mechanics. This was indeed the first qualitative explanation for the observation by two physicists, Dulong and Petit, that the heat capacity of crystals tends to $3Nk_B$ when $T \to \infty$.

To calculate the free energy of a crystal, we use thermodynamic integration, but this time determine the free energy change along a path that connects the "real" crystal for which we want to determine the absolute Helmholtz free energy to the Einstein crystal with the same structure. In the Frenkel-Ladd approach [142], this is achieved by gradually "switching on" the harmonic springs that tether the atoms to their crystallographic sites and, at the same time, "switching off" gradually the intermolecular interactions. We thus need to change the Hamiltonian accordingly by introducing a switching parameter λ that spans the path connecting the real crystal ($\lambda = 0$) to the Einstein crystal ($\lambda = 1$). The λ-dependent Hamiltonian for the system $\mathcal{H}(\lambda)$ is given by

$$\mathcal{H}(\lambda) = \mathcal{H}_0 + \lambda \sum_{i=1}^{N} \left(\mathbf{r}_i - \mathbf{r}_i^0\right)^2 + (1-\lambda)U(\mathbf{r}_1,\ldots,\mathbf{r}_N)$$

(9.12)

in which $\mathbf{r}_i$ denotes the position of atom i, $\mathbf{r}_i^0$ its lattice position, and U is the potential energy for the "real" crystal. This allows us to obtain $\frac{\partial A}{\partial \lambda}$ as

$$\frac{\partial A}{\partial \lambda} = \left\langle \frac{\partial \mathcal{H}(\lambda)}{\partial \lambda} \right\rangle$$

(9.13)

The change in Helmholtz free energy is thus given by

$$A(N,V,T,\lambda = 0) = A(N,V,T,\lambda = 1) - \int_0^1 \left\langle \frac{\partial \mathcal{H}(\lambda)}{\partial \lambda} \right\rangle d\lambda$$

(9.14)

In practice, the Frenkel-Ladd integration needs to be carried out with the center-of-mass (CM) of the systems ("real" crystal and Einstein crystal) fixed [293] and the path involves the following steps [208]: ("real" crystal) → ("real" crystal with CM constraint) → (Einstein crystal with CM constraint) → (Einstein crystal). The Helmholtz free energy of a three-dimensional atomic crystal

$$\frac{\beta A}{N} = -\frac{3}{2}\ln\left[\frac{4\pi^2 m}{\alpha\beta^2}\right] - \frac{\beta}{N}\int_0^1 \left\langle\frac{\partial \mathscr{U}^{CM}(\lambda)}{\partial\lambda}\right\rangle d\lambda - \frac{3}{2N}\ln\left(\frac{\alpha\beta}{2\pi}\right) - \frac{3\ln N}{2N} + \frac{\ln\rho}{N} \tag{9.15}$$

The approach discussed here in the case of an atomic solid has also been extended to molecular systems [379].

9.1.3 DIRECT DETERMINATION OF PARTITION FUNCTIONS

After discussing the indirect A-route to determining partition functions, we now focus on leveraging the Wang-Landau sampling scheme to obtain the partition function directly. We recall that the two examples of Wang-Landau simulations covered in Chapter 8 lead to the iterative determination of $Q(N,V,T)$ by either sampling evenly the number of molecules over a range $[0,N_{max}]$, or sampling uniformly the volume over the range $[V_{min},V_{Max}]$. The grand partition function $\Theta(\mu,V,T)$ is then obtained by summing the weighted contributions of the canonical partition functions $Q(N,V,T)$ according to

$$\Theta(\mu,V,T) = \sum_{N=0}^{N_{max}} Q(N,V,T)\exp(\beta\mu N) \tag{9.16}$$

In practice, N_{max} must be chosen large enough so that the contributions of $Q(N,V,T)$ for larger N are negligible. Examining the behavior of the number distribution $p(N)$ allows us to assess how large N_{max} needs to be, with $p(N) \to 0$ as N approaches N_{max}.

The isothermal-isobaric partition function $Q(N,P,T)$ is evaluated by integrating the weighted contributions of the canonical partition functions $Q(N,V,T)$ via

$$Q(N,P,T) = \int_{V_{min}}^{V_{max}} Q(N,V,T)\exp(-\beta PV)dV \tag{9.17}$$

As in the previous case, V_{min} and V_{max} need to be chosen so that the contributions from $Q(N,V,T)$ vanish beyond the two bounds. This can be assessed by checking the volume distribution $p(V) \to 0$ at both ends of the volume range.

Thermodynamic integration calculates free energy differences and needs a connection to provide absolute free energies. Similarly, the iterative updating process carried out in the Wang-Landau sampling gives a histogram that adequately quantifies ratios of partition functions for two different N values, in the case of the grand-canonical ensemble, or for other volumes V, in the case of the isothermal-isobaric ensemble. To provide absolute values for the partition functions, Wang-Landau simulations also need a connection with a reference state of the known partition functions. For instance, in the (N,P,T) ensemble, we can use the same reference state for the first example of thermodynamic integration we discussed, *i.e.*, the ideal gas. This can be done by simply setting the volume range sampled evenly during the (N,P,T) Wang-Landau simulations so that it covers tremendous V values and provides a value of the partition function equal to that of the ideal gas. For such volumes, we have $Q(N,V,T) = \frac{q^N}{N!}$ in which q is the molecular partition function. The situation is, however, much simpler for the grand-canonical ensemble. Indeed, a state with a known partition function is always sampled over the range $[0,N_{max}]$ whatever the system or conditions are since for $N = 0$; we have $Q(0,V,T) = 1$ or, in other words, $\ln Q(N,V,T) = 0$ as shown in Figure 9.1 in the case of a system of Argon atoms [91]. Resetting the value of $Q(0,V,T) = 1$ in the histogram gives the absolute value for $Q(N,V,T)$ over the entire range shown in Figure 9.1b. From there, we can

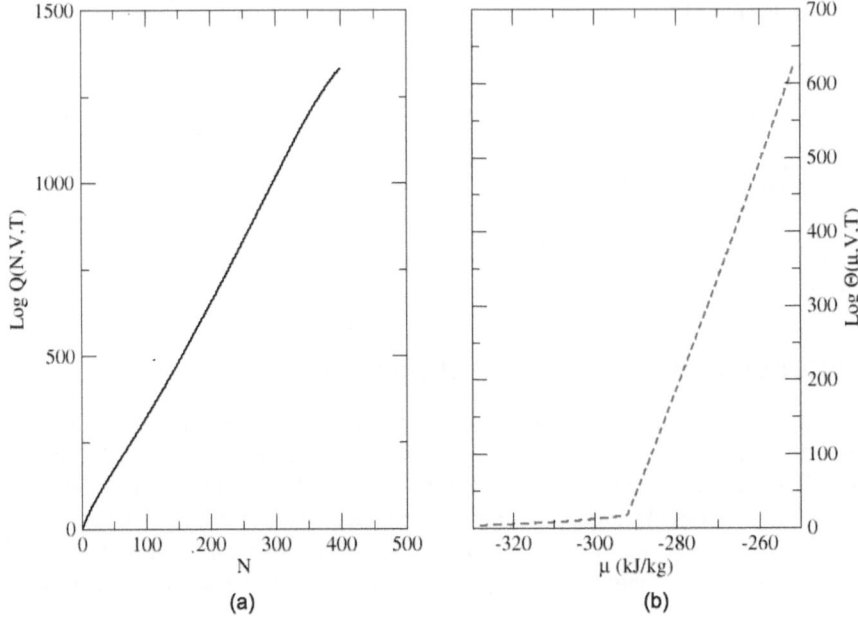

Figure 9.1 The canonical partition function $Q(N,V,T)$ is shown on the panel (a) for a system of Argon atoms at $T = 122.9$ K. Using this data, the grand partition function $\Theta(\mu,V,T)$ can be directly evaluated for any value of the chemical potential μ (b).

evaluate $\Theta(\mu,V,T)$ for any value of the chemical potential μ and plot the results in Figure 9.1. The plot shows a sharp change in the behavior of the partition function around -292.7 kJ/kg.

To better understand the implications of this change in behavior of $\Theta(\mu,V,T)$, we calculate the number distribution for two different chemical potentials that closely bracket the value of μ for which the sharp change in $\Theta(\mu,V,T)$ is observed. We recall that the number distribution $p(N)$ is given by

$$p(N) = \frac{Q(N,V,T)\exp(\beta,\mu,N)}{\Theta(\mu,V,T)} \tag{9.18}$$

Results are shown in Figure 9.2. Despite the very small change in μ, the results reveal that the value of N for which the maximum in $p(N)$ is reached has changed dramatically from a very low N value on the left panel to a much larger N value on the right panel. This points to a sudden change in the system's density and, thus, to the onset of an emergent phenomenon known as phase transition. We will characterize this phenomenon in the next section.

9.2 THE CASE OF PHASE TRANSITIONS

9.2.1 MATCHING PROBABILITIES

The window of chemical potentials during which the sudden change in density occurs is relatively narrow (see Figure 9.2). To explain how to locate the thermodynamic conditions leading to a phase transition, we examine more closely the behavior of the number distribution $p(N)$ as a function of μ. Choosing a chemical potential between the two values listed in the caption of Figure 9.2 results in a $p(N)$ which now exhibits two peaks, a first peak for a low N value or, equivalently, at low density, and a second peak for a significant N value or, equivalently, at high density. We show the number distribution so obtained in Figure 9.3. This means that, for such a chemical potential, two highly probable phases are associated with the two peaks.

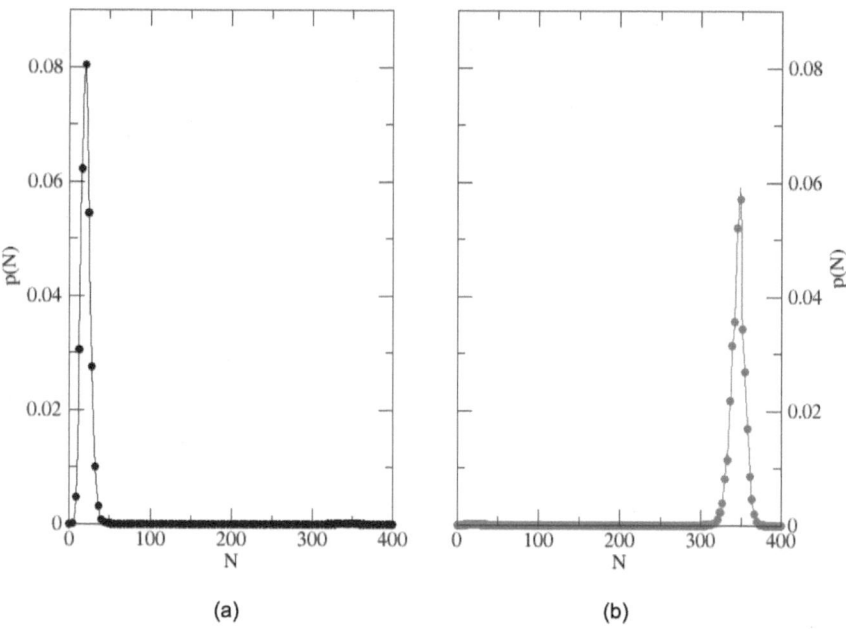

Figure 9.2 The number distribution is shown for Argon at $T = 122.9$ K for two different values of the chemical potential: $\mu = -292.76$ kJ/kg (a) and $\mu = -291.61$ kJ/kg (b).

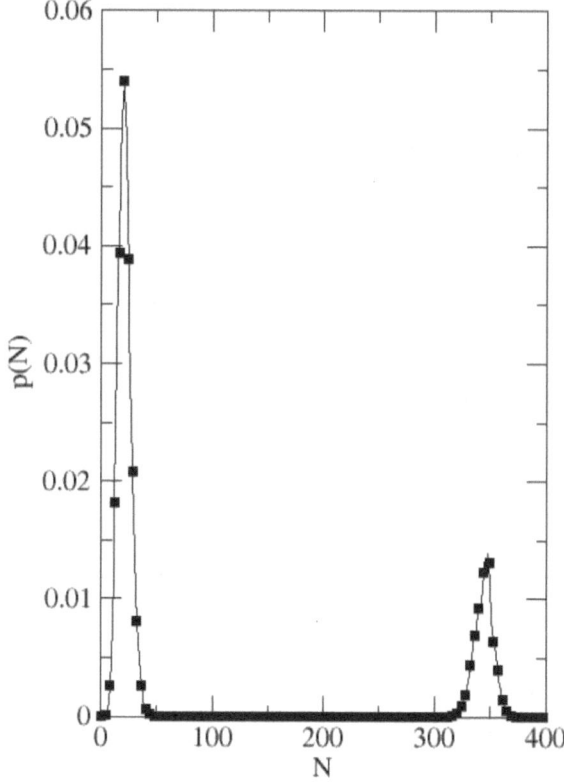

Figure 9.3 The number distribution is shown for Argon at $T = 122.9$ K for $\mu = -291.25$ kJ/kg. The two peaks correspond to the two high-probability phases, one at low density (peak on the left) and the other at high density (peak on the right).

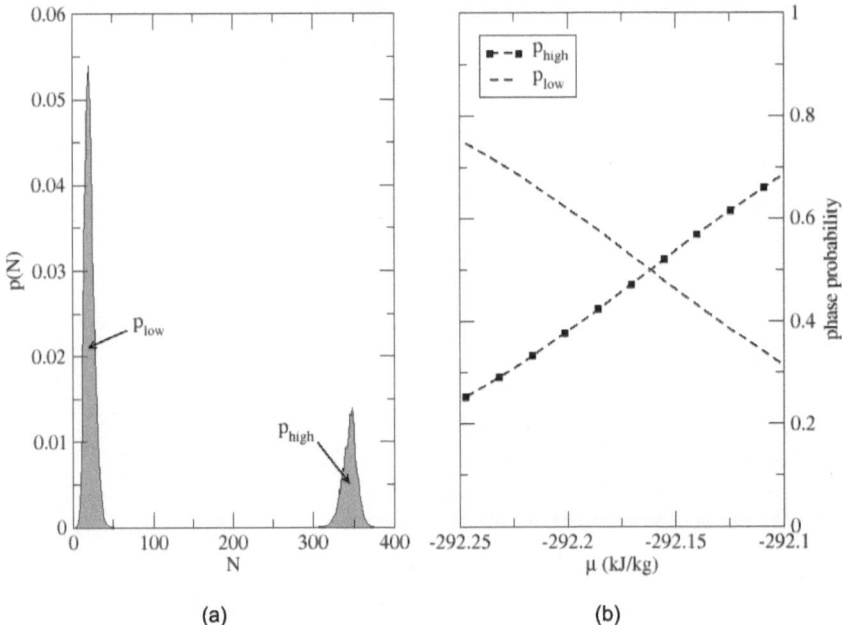

(a)   (b)

Figure 9.4 Panel (a) the number distribution exhibits two peaks, and the probability associated with each of the two phases is shown in blue below each of the two peaks. Panel (b) Variation of the probabilities for the low-density phase p_{low} and for the high-density phase p_{high} as a function of μ over the $[-292.76, -291.61]]$ range (in kJ/kg).

The chemical potential at coexistence μ_{coex} will thus be obtained at the exact point where the two phases are equally probable. Mathematically, this point is reached when the areas (integrals) under the two peaks are equal. In the case of the grand-canonical ensemble, since the number of molecules N is a discrete quantity, we have the following condition.

$$\sum_{N=0}^{N_b} p(N) = \sum_{N=N_b}^{N_{max}} p(N) \tag{9.19}$$

In this equation, we introduce the boundary value, N_b, which corresponds to the value of N for which $p(N)$ reaches a minimum between the two maxima. The left-hand side of this equation corresponds to the probability of the low-density phase, p_{low}. In contrast, the right-hand side denotes the probability associated with the high-density phase, p_{high}. We sketch in Figure 9.4 the areas associated with p_{low} and p_{high}.

As μ increases, the probability for the low-density phase p_{low} decreases, while the probability for the high-density phase increases. As shown in Figure 9.4, the two functions intersect for a chemical potential $\mu_{coex} = -292.16$ kJ/kg when both probabilities equal 0.5. At μ_{coex}, the two phases are thus equally probable, which means that we have identified a state point at the phase boundary, and the (μ_{coex}, T) conditions for which phase coexistence takes place.

9.2.2 FEATURES OF COEXISTENCE

Once we have determined the (μ_{coex}, T) conditions for phase coexistence, we can use the knowledge we gathered on μ_{coex}, $Q(N, V, T)$, and $\Theta(\mu_{coex}, V, T)$ to determine the properties of the coexisting phases. We recall that, in the grand-canonical ensemble, PV is the thermodynamic potential given by

$$PV = k_B T \log \Theta(\mu, V, T) \tag{9.20}$$

It follows that the pressure at coexistence P_{coex} is simply

$$P_{coex} = \frac{k_B T \log \Theta(\mu_{coex}, V, T)}{V} \tag{9.21}$$

Now let us turn to evaluating the properties of the two coexisting phases. We have seen previously that the average number of molecules in the system can be evaluated through

$$<N> = \sum_0^{N_{max}} N \times p(N) \tag{9.22}$$

Applying this equation to a system at coexistence will consider the two peaks of $p(N)$ and average the number of molecules over the two phases. In other words, $<N>$ would be equal to the average between $<N>_I$, the average number of molecules in phase I, and $<N>_{II}$, the average number of molecules in phase II. Since we are interested in the properties of the two phases I and II at coexistence, we must carry two separate sums for each. Specifically, we will have for $<N>_I$

$$<N>_I = \sum_0^{N_b} N \times p(N) \tag{9.23}$$

in which N_b denotes the minimum observed for $p(N)$ between the two peaks, and for $<N>_{II}>$, is given by

$$<N>_{II} = \sum_{N_b}^{N_{max}} N \times p(N) \tag{9.24}$$

The density of the two coexisting phases is thus calculated as $<\rho>_I = \frac{<N>_I}{V}$ and $<\rho>_{II} = \frac{<N>_{II}}{V}$. We provide in Figure 9.5 a plot of the coexisting densities obtained for Argon at the vapor-liquid phase transition using Wang-Landau simulations [91].

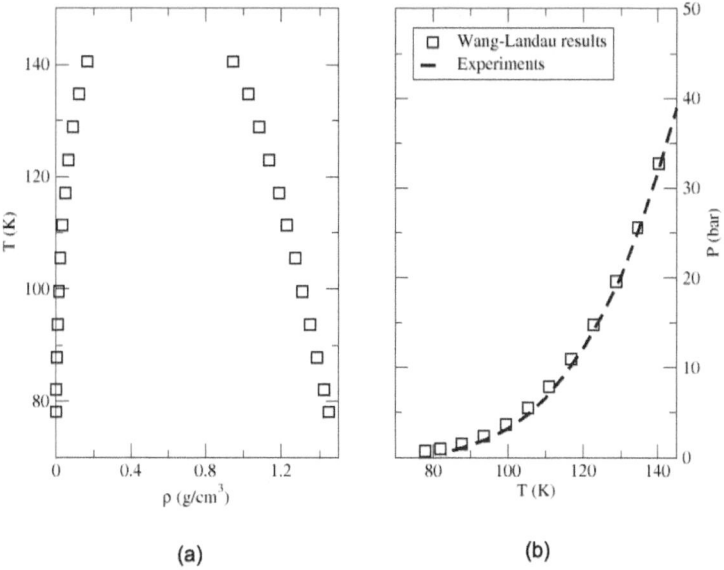

(a)   (b)

Figure 9.5 Vapor-liquid phase coexistence for Argon from Wang-Landau simulations. We show the densities at coexistence for the two phases on the panel (a) and compare the Wang-Landau results and the experimental data [378] for pressure (b).

The thermodynamic properties of the two phases can be evaluated by following this approach, *i.e.*, by carrying out two separate sums over each of the peaks for $p(N)$. The molar internal energies for phases I and II are given by

$$E_{\mathrm{I}} = \frac{\sum_{N<N_b} \left(U(N) + \frac{3}{2}k_B T\right) p(N)}{\sum_{N<N_b} p(N)} \tag{9.25}$$

$$E_{\mathrm{II}} = \frac{\sum_{N>N_b} \left(U(N) + \frac{3}{2}k_B T\right) p(N)}{\sum_{N>N_b} p(N)} \tag{9.26}$$

in which $U(N)$ is the average potential energy per Argon atom collected during the Wang-Landau simulations.

The molar entropies for phases I and II can then be evaluated as

$$S_{\mathrm{I}} = \frac{k_B \ln \Theta(\mu_{\mathrm{coex}}, V, T)}{<N>_{\mathrm{I}}} + \frac{(E_{\mathrm{I}} - \mu)}{T} \tag{9.27}$$

$$S_{\mathrm{II}} = \frac{k_B \ln \Theta(\mu_{\mathrm{coex}}, V, T)}{<N>_{\mathrm{II}}} + \frac{(E_{\mathrm{II}} - \mu)}{T} \tag{9.28}$$

9.2.3   EXTENSION TO MULTI-COMPONENT SYSTEMS

This approach to the determination of the partition function of a system can be extended to the case of a binary, ternary, or multi-component mixture. For instance, the grand-canonical ensemble for a binary mixture is denoted by (μ_1, μ_2, V, T), in which N_1 and N_2 indicate the number of molecules for each of the two components, labeled with the indices 1 and 2. This means that the number of molecules of each component will fluctuate to ensure that chemical equilibrium is provided and that the chemical potentials for the two species are the same in the system and the reservoir. The grand partition function is given in this case by

$$\Theta(\mu_1, \mu_2, V, T) = \sum_{N_1=0}^{\infty} \sum_{N_2=0}^{\infty} Q(N_1, N_2, V, T) \exp(\beta \mu_1 N_1) \exp(\beta \mu_2 N_2) \tag{9.29}$$

in which $Q(N_1, N_2, V, T)$ are the canonical partition functions for different sets of (N_1, N_2).

If, for instance, we are interested in a mixture containing two different types of atoms, $Q(N_1, N_2, V, T)$ is the canonical partition function given by

$$Q(N_1, N_2, V, T) = \frac{V(N_1+N_2)}{N_1! \Lambda_1^{3N_1} N_2! \Lambda_2^{3N_2}} \int \exp\left(-\beta U(\Gamma)\right) d\Gamma \tag{9.30}$$

where Γ denotes a specific configuration of the system and Λ_i $(i=1,2)$ is the De Broglie wavelength for an atom of type i. Introducing the activity $\lambda_i = \exp(\mu_i/k_B T)$ $(i=1,2)$, we can recast grand partition function as

$$\Theta(\lambda_1, \lambda_2, V, T) = \sum_{N_1=0}^{\infty} \sum_{N_2=0}^{\infty} Q(N_1, N_2, V, T) \lambda_1^{N_1} \lambda_2^{N_2} \tag{9.31}$$

Wang-Landau simulations can provide a numerical estimate for $\Theta(\lambda_1, \lambda_2, V, T)$. To this end [93], in line with the single-component case, we start by defining the joint Boltzmann distribution

$$p(\Gamma, N_1, N_2) = \frac{V^{N_1} V^{N_2} \lambda_1^{N_1} \lambda_2^{N_2} \exp\left(-\beta U(\Gamma)\right)}{N_1! \Lambda_1^{3N_1} N_2! \Lambda_2^{3N_2} \Theta(\lambda_1, \lambda_2, V, T)} \tag{9.32}$$

The number distribution $p(N_1, N_2)$ can then be estimated from Eq. 9.32 as

$$p(N_1, N_2) = \int p(\Gamma, N_1, N_2) d\Gamma = \frac{Q(N_1, N_2, V, T) \lambda_1^{N_1} \lambda_2^{N_2}}{\Theta(\lambda_1, \lambda_2, V, T)} \tag{9.33}$$

The bias distribution $p_{\text{bias}}(\Gamma, N_1, N_2)$ is given by

$$p_{\text{bias}}(\Gamma, N_1, N_2) = \frac{p(\Gamma, N_1, N_2)}{p(N_1, N_2)} = \frac{V^{N_1} V^{N_2} \exp(-\beta U(\Gamma))}{N_1! \Lambda_1^{3N_1} N_2! \Lambda_2^{3N_2} Q(N_1, N_2, V, T)} \tag{9.34}$$

The acceptance rule for the random move $\Gamma, N_1, N_2 \rightarrow \Gamma', N_1', N_2'$ can then be obtained from the ratio of the biased probabilities as

$$\text{acc}(\Gamma, N_1, N_2 \rightarrow \Gamma', N_1', N_2') = \min \left[1, \frac{p_{\text{bias}}(\Gamma', N_1', N_2')}{p_{\text{bias}}(\Gamma, N_1, N_2)} \right] \tag{9.35}$$

As with single-component systems, thermodynamic properties can be directly calculated using the number distribution $p(N_1, N_2)$. Similarly, the location of the phase boundaries can be obtained by the appropriate sums of $p(N_1, N_2)$ over regions corresponding to each phase. More specifically, to determine the conditions of phase coexistence in the case of binary mixtures, we find numerically the (λ_1, λ_2) sets for which the probability for the high-density phase is equal to the probability for the low-density phase. This is captured by the equation below.

$$\sum_{N_1 > N_{1,b}} \sum_{N_2 > N_{2,b}} p(N_1, N_2) = \sum_{N_1 < N_{1,b}} \sum_{N_2 < N_{2,b}} p(N_1, N_2) \tag{9.36}$$

In this equation, the term on the left-hand-side is equal to p_{high} and the term on the right-hand-side is equal to p_{low}, with $N_{b,1}$ and $N_{b,2}$ corresponding to the (N_1, N_2) values for which $p(N_1, N_2)$ reaches a minimum between the two peaks. Once $(\lambda_{1,\text{coex}}, \lambda_{2,\text{coex}})$ sets have been identified, the thermodynamic properties of the two coexisting phases can be calculated using the same reasoning as for single-component systems.

9.3 GAS STORAGE AND SEPARATION APPLICATIONS

9.3.1 PARTITION FUNCTIONS FOR ADSORBED FLUIDS

As we have seen in the previous chapter, the grand-canonical ensemble is ideally suited to determine the properties of a system for which the only information we have is the knowledge that this system is in chemical equilibrium with its surroundings. A typical example is a molecular fluid adsorbed in a porous material. In this case, the adsorbed molecules constitute a confined, or adsorbed, phase that is in equilibrium with a vapor or a liquid composed of molecules of the exact nature. Adsorption in a porous material is central to a wide range of phenomena in chemistry, physics, biology, and engineering, and, as a result, the thermodynamic characterization of adsorption is crucial for many industrial applications. Examples of porous materials include, among others, aluminosilicate materials known as zeolites. Zeolites (which corresponds to "boiling stones" in Greek) were discovered during the 18th century by Alex Cronstedt, who proposed to classify minerals according to their chemical composition rather than based on their physical properties [246]. In 1756, Cronstedt heated a natural zeolite mineral and found that bubbles formed in the solid material and that the zeolite released significant amounts of steam from its pores. These materials, whether natural or synthesized in laboratories, quickly proved to be very potent materials for applications involving the storage of molecules, separation processes, and catalysis. In recent decades, another class of nanoporous materials, known as Metal-Organic Frameworks (MOFs) or Covalent Organic Frameworks (COFs), has emerged as remarkably efficient nanoporous materials [12,117,225,387,395].

Such structures consists of metal oxides subunits located on the corners of the cubic structure, with organic aromatic linkers connecting the corners of the cube. As a result, a wide cavity in the cube's center can accommodate "guest" molecules and thus adsorb, *e.g.*, large quantities of gas. In Figure 9.6a, we show the unit cell of the isoreticular MOF IRMOF-1 with a few CO_2 molecules adsorbed inside the pore. The snapshot shows that the CO_2 molecules decorate the "walls" of the cavity due to the strong intermolecular interactions between the guest molecules (here, CO_2) and the atoms that make up the MOF structure.

This means that, when calculating the system's potential energy, we will need to account not only for the interactions between guest molecules but also for the interactions between guest molecules and atoms of the structures. The latter has been most often modeled using Lennard-Jones potentials and/or point charges [152,228,334,389], with the parameters for these interactions being taken from multi-purpose force fields such as the DREIDING force field [249]. In most cases, a good approximation consists of considering the porous material to be rigid. This implies that the coordinates of the atoms constituting are fixed and that, from the standpoint of the intermolecular interactions, the atoms form an interaction field in which the guest molecules can move inside the pores of the material, enter the pore from the surrounding gas phase, or leave the pore to return to the surrounding gas phase. These correspond to the typical three types of random moves we have discussed for Monte Carlo simulations in the grand-canonical ensemble in the previous chapter. Therefore, to obtain the grand partition function of the adsorbed phase, *i.e.*, the "guest" molecules contained within the nanoporous material, we will need to carry out Wang-Landau simulations in the grand-canonical ensemble with the same acceptance rules as we employed so far in the absence of the adsorbent. We will thus have the following acceptance rule for a random move $\Gamma, N \to \Gamma', N'$.

$$\mathrm{acc}(\Gamma, N \to \Gamma', N') = \min\left(1, \left(\frac{Vz}{\max(N,N')}\right)^{\Delta N} \exp(-\beta \Delta U)\right) \qquad (9.37)$$

in which ΔU denotes the change in potential energy of the system, z the activity, $\Delta N = N' - N$ the difference in the number of molecules adsorbed in the nanoporous material, and $\max(N,N')$ the greater integer between N and N'. ΔU includes the contribution from the interactions of guest molecules and atoms from the nanoporous material. We show in the Figure 9.6b the canonical

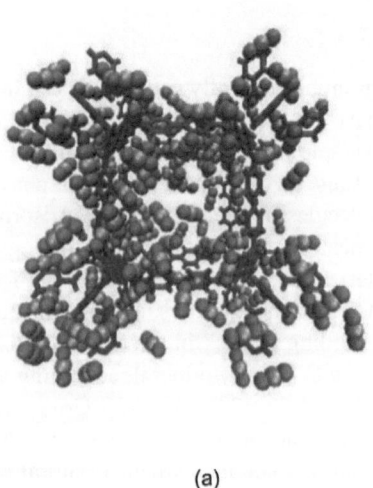

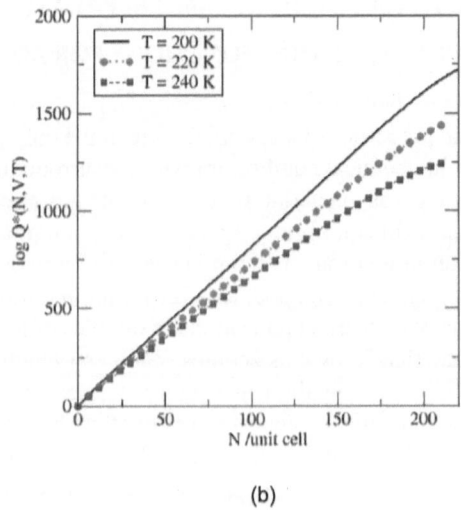

(a)   (b)

Figure 9.6 (a) Snapshot of CO_2 adsorbed in a nanoporous material known as IRMOF-1. The snapshot shows a single unit cell of the nanoporous material. (b) Variation of the canonical partition function for the confined fluid as a function of N for different temperatures.

partition functions obtained [92] for the adsorption of CO_2 in IRMOF-1 for temperatures ranging from $T = 200$ K to $T = 240$ K.

9.3.2 THERMODYNAMIC PROPERTIES OF ADSORPTION

Once we have the adsorbed fluid's canonical partition functions, adsorption's thermodynamic properties can be evaluated according to the same procedure for a "regular", unconfined fluid. We start with the number distribution $p(N)$ defined as

$$p(N) = \frac{Q(N,V,T)\exp\beta\mu N}{\Theta(\mu,V,T)} \tag{9.38}$$

Here N denotes the number of guest molecules adsorbed in the porous nanomaterial. A subtle difference here is that the value of μ is imposed by the surroundings, *i.e.*, the external phase in equilibrium with the adsorbed phase. Since μ is not a quantity that can be measured directly in experiments, the adsorption properties are generally calculated as a function of the pressure of the external phase. The relation between μ and P can be obtained from, *e.g.*, an ideal gas (for pressures below 1 bar), an equation of state, or an independent simulation in the grand-canonical ensemble for the "regular" fluid. Once $p(N)$ is known, the adsorption isotherm, *i.e.*, the amount of the adsorbed fluid per "unit" of the porous material can be evaluated as

$$N_{\text{ads}} = \sum_{0}^{N_{\text{max}}} N p(N) \tag{9.39}$$

The adsorption isotherm or loading can be calculated as several molecules of guest molecules per unit cell of the material, as shown in Figure 9.7a, or as a ratio of the weight of adsorbed fluid per unit weight of porous material. For such temperatures, the adsorption isotherms show a step, a behavior akin to a gas-liquid transition in the confined phase [69].

The partition function can also determine the internal energy E_{ads} and entropy S_{ads} of the adsorbed fluid. To this end, we need to provide more information on the model chosen to simulate the CO_2 molecules. Specifically, in the case of CO_2, a reasonable approximation is to consider

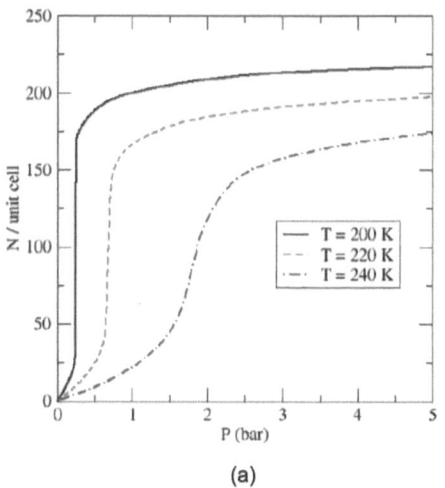

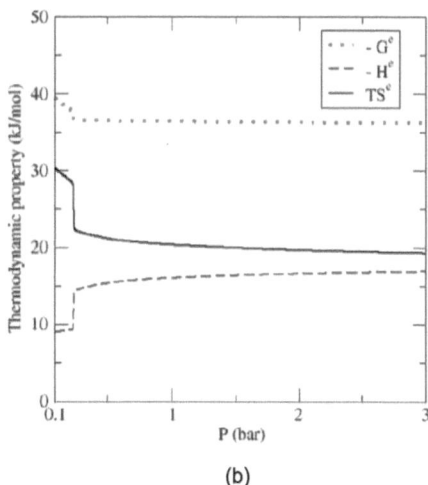

(a) (b)

Figure 9.7 (a) Adsorption isotherms obtained from Wang-Landau simulations for CO_2 in IRMOF-1 at $T = 200$ K, 220 K, and 240 K. (b) $T = 200$ K. Variation of the excess Gibbs free energy, enthalpy, and entropy of CO_2 adsorbed in IRMOF-1 as a function of pressure [92].

the molecule linear, rigid, and polyatomic. This means that the number of degrees of freedom equals 5 per molecule. Out of these 5 degrees of freedom, 3 are translational degrees, and the remaining two are rotational degrees of freedom. This means that the internal energy for the system is

$$E_{\text{ads}} = \sum \left(U(N) + \frac{5}{2}k_B T \right) p(N) \tag{9.40}$$

where we have added to the potential energy U the contribution of $5/2k_B T$ arising from the 5 degrees of freedom per CO_2 molecule.

The entropy for the adsorbed phase can then be evaluated according to

$$S_{\text{ads}} = \frac{k_B \ln \Theta(\mu, V, T) + E_{\text{ads}} - N_{\text{ads}}\mu}{T} \tag{9.41}$$

From there, the enthalpy of the adsorbed phase H_{ads} can be calculated from E_{ads} by adding the product $P_g V$, in which P_g denotes the pressure of the external gas phase in equilibrium with the adsorbed phase. The Gibbs free energy follows from, for instance, $G_{\text{ads}} = H_{\text{ads}} - T S_{\text{ads}}$. For comparison with experiments, it is also convenient to give results regarding excess properties, $i.e.$, relative to a system of the external gas phase occupying the same volume as the adsorbed fluid [266]. We show in Figure 9.7b the variations of the excess Gibbs free energy G^e, entropy S^e, and enthalpy H^e. The plot shows that the steep increase in the number of molecules adsorbed per unit cell correlates with a sharp decrease in entropy that results from the increased density of the adsorbed phase and with an increase in $-H^e$, that corresponds to the increased number of attractive interactions taking place between guest molecules.

9.3.3 ENVIRONMENTAL AND ENERGY APPLICATIONS

Statistical mechanics and simulations in the grand-canonical ensemble have shed light on various adsorption phenomena over the past decades [155]. In recent years, the emergence of new materials such as MOFs and COFs has resulted in a renewed interest in the application of molecular simulation methods to determine the thermodynamic properties of adsorption and screen potential candidates for specific applications. MOFs are highly tailorable since the metal subunit and the organic linker can be modified to optimize the intermolecular interactions between the guest molecules and the host material. Such materials can be designed to be highly specific toward a given adsorbent. In other words, they can serve as sensors or selectively capture environmental contaminants [170]. The capture and sequestration of carbon dioxide have drawn considerable interest, and several MOF structures have proved to be incredibly efficient porous materials for the adsorption of CO_2 [69,92, 113,389,405].

MOFs and COFs have also been increasingly evaluated for several energy applications. This is the case, for instance, of the storage of methane in MOFs, to design a high-capacity fuel tank for vehicles [149]. COFs are also promising materials for the storage of hydrogen [44,157,175]. This is because, instead of relying on high-atomic-weight elements such as Zn like MOFs, COFs are built with lightweight elements, such as Boron. This is especially important when designing a porous material to store a low molecular weight fuel like Hydrogen. Indeed, COFs exhibit remarkable loading in grams of hydrogen stored per gram of COFs, an essential parameter for practical applications. We add that reliable classical models have been parameterized based on a Morse potential for the interactions between guest molecules (Hydrogen) and the atoms from the COF structure. Furthermore, their use in grand-canonical Monte Carlo and Wang-Landau simulations leads to accurate predictions for the adsorption isotherms [209,259].

Finally, grand-canonical Monte Carlo and Wang-Landau simulations have also proved to be instrumental in the case of mixture adsorption. Indeed, they can shed light on the interplay between the features of the MOFs/COFs and their selectivity toward a specific adsorbent. This is significant in separation applications when MOFs and COFs are used as membranes. The mixtures studied using these methods include the industrially relevant CO_2-CH_4 mixture [207], as well as mixtures of compounds with very similar chemical structures such as Xenon and Krypton, or ethylene and acethylene [165]. Given the high tailorability of MOFs and COFs, data-driven approaches and machine learning are poised to play an increasing role in the design of nanoporous materials and the efficient characterization of their thermodynamic properties.

10 Machine Learning Molecular Trends

"To answer the questions of how information about the physical world is sensed, (...), and how does information retained in memory influence recognition and behavior, a theory is developed for a hypothetical nervous system called a perceptron."

Frank Rosenblatt,
The perceptron [313]

We are now equipped with a series of simulation methods that can efficiently predict, in a wide range of external settings, the behavior of molecular systems, through Metropolis Monte Carlo simulations or the policies governing their behavior, through Wang-Landau simulations. However, to provide accurate predictions for behaviors and policies, simulations must handle vast amounts of data, both as input and output. The part of the statement focusing on the "output" is quite straightforward. For instance, the number of configurations sampled using flat-histogram methods, like the Wang-Landau method, is of the order of 10^9. This means that if we consider a system with 100 atoms, the simulation will have generated 3×10^{11} numbers for (just) the coordinates of this relatively small system. On the other hand, the part of the statement regarding the "input" has been hidden so far because we had accurate classical force fields available for the systems discussed in the previous chapters. For instance, for the results obtained on CO_2 in Chapter 9, we used the TraPPE model for CO_2 [296]. In this case, the parameters for the Lennard-Jones potentials and the point charges had been extensively tested and could be used reliably in the simulations. When starting a study with a new compound, there might not be a thoroughly tested force field available. Similarly, suppose we are now working with a compound subjected to unusual conditions. In that case, the conditions might be outside the range for which the model was parametrized, and the parameters might need to be reevaluated. In those two cases, we must (re-)parametrize the force field to obtain accurate results. The parameterization process will require much data, either from experiments or from *ab initio* calculations, as input for the simulation model. We cover in this chapter how Machine Learning can address many outstanding challenges in simulations.

10.1 LEARNING INTERMOLECULAR INTERACTIONS

10.1.1 STARTING FROM EMPIRICAL DATASETS

Generating billions of system configurations during a simulation requires evaluating an even more significant number of distances between pairs of atoms, angles between triplets or quadruplets of atoms, and, ultimately, interaction energies. To simulate larger systems and/or gather statistics over longer trajectories. This challenge will be around for many decades; scientists proposed calculating the interaction energy at specific locations and interpolating the energy value between grid points with a spline fit. In practice, this means evaluating exactly the interaction energy at grid points on 2D and 3D grid models of the space over, for instance, the surface of a catalyst or inside a porous material. It was during the 1990s that the idea of replacing the cubic spline fit with the predictions from a neural network. One of the first applications [24] allowed for the prediction of the potential energy of a CO molecule over a Ni (111) surface, with the surface atoms being fixed in their equilibrium positions. This initial study considered up to three degrees of freedom for the CO molecule: (i) the lateral position x of the center-of-mass of the CO molecule, (ii) the angle θ

DOI: 10.1201/9781003006411-12

between the molecular axis and the normal to the surface, and (iii) the position of the center-of-mass of the CO molecule over the surface in the z direction that is normal to the surface. The neural network models included either two degrees of freedom (x and θ) or all three degrees of freedom (x, θ, and z) as input nodes, a hidden layer with between 6 and 15 neurons, and a single output neuron corresponding to the potential energy of the molecule of CO over the Ni (111) surface. The network was trained using several hundred data points from an empirical potential. We show in Figure 10.1 the architecture of one of these neural networks.

It soon became apparent that neural networks could outperform empirical potentials for modeling contributions arising from many-body interactions. For instance, in 1999 [185], neural networks were employed to predict the many-body term $\bar{b}_{ij}$ (also called a bond-order term [357]) in the Brenner potential [35] for hydrocarbons. Indeed, according to this model, the interaction energy between two atoms i and j is calculated as

$$V(r_{ij}) = f_c(r_{ij})\left[V_R(r_{ij}) + \bar{b}_{ij}V_A(r_{ij})\right] \tag{10.1}$$

in which $V_R(r_{ij}) = A_{ij}\exp(-\lambda_{ij}r_{ij})$ and $f_A(r) = -B_{ij}\exp(-\mu_{ij}r_{ij})$ are 2-body terms (the first being repulsive and the second attractive), and $f_c(r_{ij}) = 0.5\left[1 + \cos\left(\frac{r_{ij}-R}{S-R}\right)\right]$ is a cutoff function. A_{ij}, B_{ij}, λ_{ij}, μ_{ij}, S and R are Brenner potential parameters that are kept exactly as in the Brenner model [35]. The cutoff function f_c is set to 1 if $r_{ij} < R$ and to 0 if $r_{ij} > S$.

However, determining the bond order term $\bar{b}_{ij}$ is more involved. It is a geometry-dependent, many-body term obtained by averaging the terms b_{ij} and b_{ji} associated with the two atoms i and j plus a corrective term. b_{ij} is given by

$$b_{ij} = \left(1 + \sum_{k\neq i,j} f_c(r_{ik})g(\theta_{ijk})\exp^{\alpha_{ijk}[(r_{ij}-R_{ij}^{(e)})-(r_{ik}-R_{ik}^{(e)})]} + H_{ij}(N_i^H, N_i^C)\right)^{-\delta_i} \tag{10.2}$$

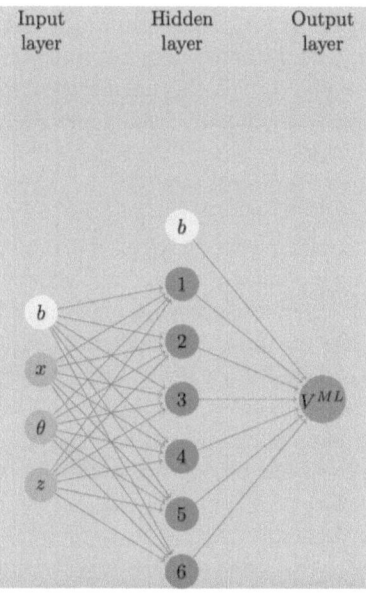

Figure 10.1 Example of a neural network used to predict the potential energy V of a CO molecule over a Ni (111) surface. The center-of-mass of the CO molecule has a lateral location x between two surface sites and is at a distance z in the direction normal to the surface. The molecular axis makes an angle θ with the normal to the surface.

The last term in Eq. 10.2 depends on the numbers of H and C atoms "connected" to i, obtaining by summing $f_c(r_{ij})$ over the j neighbors of i for each element, and with the function $g(\theta_{ijk})$ evaluated as

$$g(\theta_{ijk}) = 1 + \frac{c^2}{d^2} - \frac{c^2}{\left[d^2 + (1 + \cos\theta_{ijk})^2\right]} \tag{10.3}$$

where β, c, d and h are potential parameters and θ_{jik} denotes the angle between vectors $\mathbf{r}_{ij}$ and $\mathbf{r}_{ik}$.

The idea was thus to replace this series of mathematical operations with a series of inputs fed to a neural network to obtain directly the parameter $\bar{b}_{ij}$. Networks used for this application included five inputs, six hidden nodes in a single hidden layer, and a single output for $\bar{b}_{ij}$. Input neurons $y_1, \dots, y_5$ fed the information relative to each triplet (i, j, k) of atoms. y_1 and y_2 focus on the atom types involved in the bond. Since neural networks handle numbers, we must convert this chemical and geometric information into numbers. More specifically, y_1 provides information on the $i - j$ bond ($y_1 = 0.9$ for a C-C bond or $y_1 = 0.1$ for a C-H bond), while y_2 provides information on the $i - k$ bond. y_3 encodes the information about the angle θ_{ijk} with $y_3 = (1 + \cos\theta_{ijk})^2$. y_4 encodes the connectivity as the sum of the numbers of connected/neighboring H and C atoms as $y_4 = N_i^H + N_i^C$. Finally, the fifth input y_5 encodes information on the second order connectivity, using sums over only C atoms to determine if atoms i and j are part of a conjugated system (we refer the reader to the original paper for complete details on this term [185]). This work proved the feasibility of neural network potentials for the simulation of systems with many degrees of freedom, such as those encountered in Monte Carlo and Wang-Landau simulations.

10.1.2 TRAINING ON TIGHT-BINDING DATA

The development of neural network potentials started eight years later, with the first development of a complete potential predicted by an artificial neural network. In a study devoted to Silicon [23], the idea was taken two steps further with (i) a complete neural network formulation of the potential and (ii) training on thousands of data points taken from a single quantum (tight-binding) source. The training was much more extensive than in previous attempts and accounted for configurations with many different coordination numbers and nonequilibrium structures. This allowed us to capture the dependence of the potential energy on the environment surrounding each atom. Furthermore, since the training dataset included quantum energies, this opened the door to developing machine-learned force fields that matched the accuracy of computationally intensive, highly accurate quantum calculations at the *ab initio* DFT level later.

This first neural network potential is used as input for the information gathered on chains of four atoms. This input information was fed to the neural network as a series of nine input neurons, with the geometric information put under the form of a series of real numbers to be conveniently handled by the neural network. The first six input neurons are direct geometric data on a chain of four atoms labeled as $i - j - k - l$ with each letter corresponding to a different atom. More specifically, they include two-body information under the form of distances (r_{ij}, r_{jk}, and r_{ik}), three-body information under the form of "bond" angles (θ_{ijk} and θ_{jkl}), and four-body information through a dihedral angle (τ_{ijkl}). The remaining three input neurons cover information on the amount of "screening" occurring over parts of the chain, either around atom i (N_i) or atom j (N_j), or over the entire chain (N_{ijkl}). "Screening" is a concept drawn from many-body potentials. It takes the form of a number S_{ij} for a pair of atoms i and j and quantifies whether two atoms are shielded from interacting by another atom placed in-between i and j. Mathematically, S_{ij} is given by the following product over triplets (i, j, k)

$$S_{ij} = \Pi_{k \neq i, j} S_{ikj} \tag{10.4}$$

S_{ikj} is found by finding the prameter C in the equation of an ellipse that goes through i, j, and k

$$x^2 + \frac{y^2}{C} = \frac{r_{ij}^2}{4} \tag{10.5}$$

The parameter C is thus given by Ref. [221]

$$C = \frac{2(X_{ik} + X_{kj}) - (X_{ik} + X_{kj})^2 - 1}{1 - (X_{ik} + X_{kj})^2} \tag{10.6}$$

in which $X_{ik} = (R_{ik}/R_{ij})^2$ and $X_{kj} = (R_{kj}/R_{ij})^2$, leading to the screening factor S_{ikj} according to

$$S_{ikj} = f_c \left[\frac{C - C_{\min}}{C_{\max} - C_{\min}} \right] \tag{10.7}$$

The geometric information is passed through the input neurons, onto two hidden layers with 11 neurons, and then to a single output neuron that gives the value for the potential energy [23]. The dataset consists of energies calculated using a tight-binding model over various geometries, with $23,050$ data points used for training the neural network and 2050 for validations. Tests show that designing a neural network potential is feasible since the approach provides an accuracy similar to a complete tight-binding calculation at a fraction of the computational cost.

10.1.3   NEURAL NETWORK POTENTIALS

These early successes imply that neural network potentials (NNPs) are up-and-coming candidates for developing high-performance force fields that deliver an accuracy comparable to quantum density functional theory (DFT) results at a fraction of the cost of DFT calculations. There needs, however, to be a set of guiding principles when building and parametrizing the model [16,17]. The critical starting point for a successful NNP consists of splitting the potential energy of the system into a sum of atomic contributions as

$$U_{\text{tot}} = \sum_{i=1}^{N} U_i \tag{10.8}$$

in which U_i is equal to the NNP value evaluated for each atom i based on their local atomic environment. Symmetry functions factor in the atoms surrounding i by taking into account the distance for each neighbor and its location relative to atom i and its other neighbors. Such symmetry functions are calculated over distances up to a cutoff radius R_c, whose value is set between 6 and 10 depending on the application considered. The symmetry functions thus capture the many-body contributions to the potential energy and provide a structural "fingerprint" of the modeled system [17]. The atomic energy is then predicted by an atomic neural network whose input neurons contain the information (numbers) provided by tens of symmetry functions (typically between 20 and 100) and whose output is a single number, the atomic energy. The atomic neural network is specific to each element, and the same atomic neural network is used for all atoms of the same element. The latter naturally returns an energy value that varies according to the local environment of the atom i.

We now discuss in greater detail what symmetry functions. In this case, symmetry does not refer to a specific repeating pattern in the system or space group. It relates to the concept of structurally equivalent representations. Let us consider a system that contains, for instance, two Oxygen atoms. If we exchange the positions between the two Oxygen atoms, we obtain a new system configuration that looks the same as the old one. In other words, we have here two structurally equivalent representations of the system. The two equivalent configurations have the same energy, and, as a result, all functions that serve as input to the atomic neural network should be "symmetric" with respect to the exchange of atoms of the same element and take the same value for both configurations. Symmetry functions can be broadly defined as belonging to two classes: radial symmetry and angular symmetry. Radial symmetry functions depend on interatomic distances and can take the following generic expression.

$$G_i^R = \sum_{j=1}^{N_n} e^{-\eta \left(r_{ij} - r_0 \right)^2} \times f_c(r_{ij}) \tag{10.9}$$

in which η and r_0 are parameters defining the width and center for the Gaussian, respectively. η provides a means to control the radial extension of the symmetry function, while a value of r_0 differing from 0 allows us to emphasize the importance of specific spatial regions in the coordination shell of atom i. As we have seen in many-body force fields in this chapter, it is advantageous to use a cutoff function f_c that smoothly goes to 0 at the cutoff radius R_c. This avoids jumps in the potential energy as an atom enters/exits the spherical shell of radius R_c around atom i. As we will see in the next chapter, it is also preferable that the gradient of f_c vanish at the cutoff radius to ensure any discontinuity in the forces for molecular dynamics applications. There are many possible choices for f_c such as, among others, the following function for $r_{ij} \leq R_c$

$$f_c(r_{ij}) = \frac{1}{2}\left[\cos\left(\frac{\pi r_{ij}}{R_c}+1\right)\right] \tag{10.10}$$

and $f_c(r_{ij}) = 0$ for $r_{ij} > R_c$. We show in Figure 10.2 the behavior of f_c and its derivative with respect to r.

On the other hand, angular symmetry functions include information on triplets of atoms (i,j,k) as shown in the equation below

$$G_i^A = 2^{1-\zeta}\sum_{j\neq i}\sum_{k\neq i,j}\left[(1+\lambda\cos(\theta_{ijk}))^{\zeta}e^{-\eta\left(r_{ij}^2+r_{ik}^2+r_{jk}^2\right)} \times f_c(r_{ij})f_c(r_{ik})f_c(r_{jk})\right] \tag{10.11}$$

in which ζ and η are parameters of the symmetry function, θ_{ijk} denotes the angle for the (i,j,k) triplet, and r_{ij}, r_{ik}, and r_{jk} distances between pairs of atoms.

We show in Figure 10.3 a summary of the behaviors observed for each type of symmetry function. We can then define a basis set of radial functions by employing different values for the parameters of the radial and angular symmetry functions. Each symmetry function can then be considered an input neuron in the input layer of the atomic neural network. We recall that the output layer will contain a single neuron, giving the value of the atomic energy. The neural network can then be trained on, for instance, configurations for which the energy is obtained using a high-accuracy method,

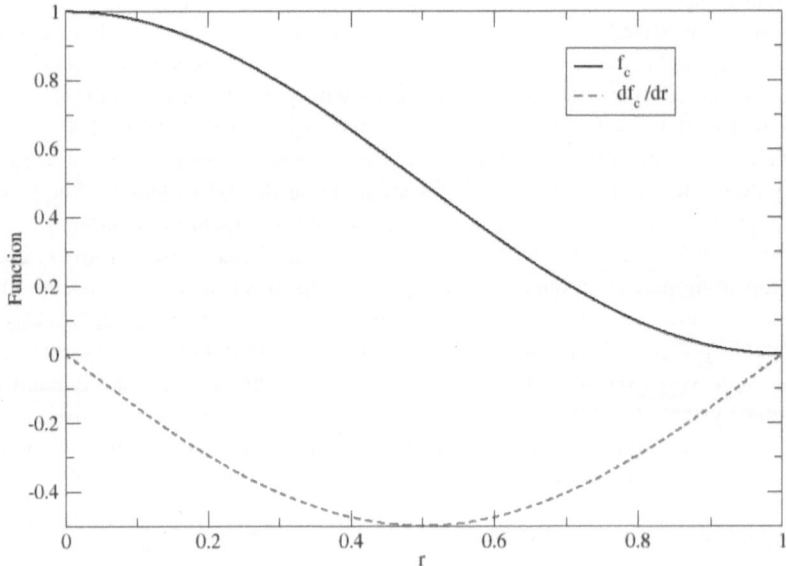

Figure 10.2 Plot of the cutoff function $f_c(r)$ controlling how much atoms in the spherical shell of radius R_c around atom i contribute to the symmetry functions centered in i.

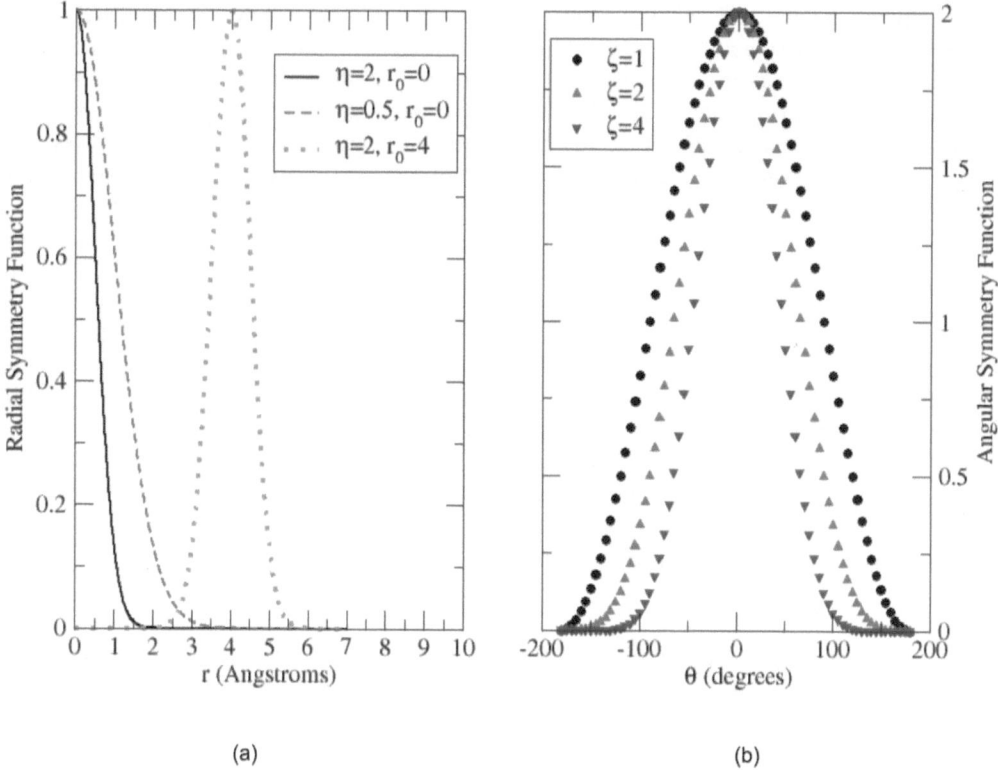

Figure 10.3 Examples of radial (a) and angular (b) symmetry functions used as input for atomic neural network potentials.

such as quantum DFT with the appropriate energy corrections. This is achieved by (i) entering the atomic coordinates in the symmetry functions and calculating the values taken by these functions, (ii) calculating the atomic energy with the current estimates for the weights of the neural network, and (iii) summing the atomic energies to obtain the total energy for the configuration. The training process then requires many iterations, or epochs, to optimize the weights of the neural networks. This is done by minimizing the difference between the value of the total energy predicted by the neural network and its actual value from quantum DFT calculations. This allows the NNPs to be of comparable accuracy to quantum DFT methods for a fraction of their cost. We discuss in the next section an application of an optimization algorithm on the example of the back-propagation algorithm.

10.2 LEARNING PARTITION FUNCTIONS

10.2.1 SINGLE-COMPONENT SYSTEMS

In the previous sections, we focused on using machine learning models to improve the accuracy with which potential energy is evaluated during simulations. We now examine an example of how machine learning can be used to post-process simulation results and predict partition functions. A question we set to address is, for instance, the following. If we are given sets of (N, V, T), can we predict the value of the canonical partition function $Q(N, V, T)$? Can we then use the predictions for different $Q(N, V, T)$ to predict the thermodynamic properties of a system over a wide range of conditions?

The network architecture shown in Figure 10.4 was recently used on this end [100]. There are four layers in the neural network, starting with an input layer of three neurons, two hidden layers of ten neurons each, and an output layer of two neurons. The three input neurons correspond to the three canonical variables N, V, and T, and the two output neurons are the $Q(N,V,T)$ functions and the total potential energy $U(N,V,T)$ for the system. The neural network thus provides analytical expressions for the machine-learned (ML) canonical partition functions (Q^{ML}) and machine-learned potential energies U^{ML}. For instance, the machine-learned partition function is given by

$$Q^{\mathrm{ML}} = f_4[b_3 + \sum_{l=1}^{10} W(3,4,l,1)f_3(b_2 + \sum_{j=1}^{10} W(2,3,j,l)f_2[(b_1 + \sum_{i=1}^{3} W(1,2,i,j)G_i])]) \qquad (10.12)$$

in which $W(i-1,i,l,j)$ denotes the weight parameter connecting neuron l from layer $i-1$ to neuron j from layer i. The f_i functions are activation functions. The tanh function for $i = 1, 2, 3$ and the linear function for $i = 4$ since numerical values must be returned as output by the network. b_i correspond to bias nodes that act as adjustable offsets and are used for numerical stability. Finally, the G_i are the input neurons that convey the required information, *i.e.*, the three numbers N, V, and T. Once the NN weights have been optimized, the partition functions can be calculated quickly and efficiently for any (N,V,T) through Eq. 10.12. This is much more efficient than running a new Wang-Landau simulation requiring generating the order of 10^9 configurations of the system to obtain this result.

An optimization algorithm is required to find NN weights that yield highly accurate partition functions and energy predictions. We explain how this can be achieved through the use of the

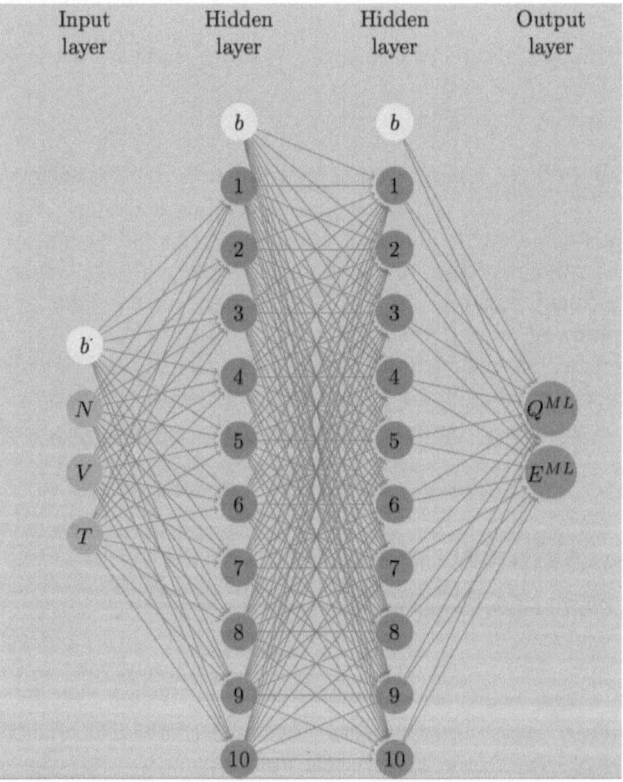

Figure 10.4 Network architecture for the prediction of partition functions (Q^{ML}) and of the energy (E^{ML}) of the system for any (N,V,T) set.

back-propagation algorithm [182,220] Alternative methods such as the Levenberg-Marquardt algorithm [297] and the global extended Kalman filter [181] are also often used. The optimization process consists of updating the weights to minimize the following error function.

$$\Delta = \frac{1}{2N_d} \sum_{i=1}^{N_d} \left[(Q_i(N,V,T) - Q_i^{\text{ML}})^2 + \alpha(E_i(N,V,T) - E_i^{\text{ML}})^2 \right] \tag{10.13}$$

in which N_d denotes the number of data points in the training set, and α is an adjustable parameter that balances the influence of Q and E during the training process. The adjustment of the weights takes place after each forward pass and is repeated until Δ becomes smaller than a threshold value. All the weights are updated once per iteration according to

$$W^{n+1}(i,j,k,l) = W^n(i,j,k,l) - \lambda \frac{\partial \Delta}{\partial W^n(i,j,k,l)} \tag{10.14}$$

in which λ is the learning rate, and W^n and W^{n+1} are the weights after n and $n+1$ iterations, respectively, and the partial derivatives of the error function Δ are calculated analytically.

One needs data for the partition functions and energies to train the NN weights. These are obtained from a few grand-canonical Wang-Landau simulations run for different (V,T) sets. The overall data (~ 4000 data points) is split between a training set and a hold-out set for validation purposes. Quantities such as the root mean squared error (RMSE) measure the model's performance. In the case of the ML prediction for $Q(N,V,T)$, the RMSE is given by

$$\text{RMSE}(Q) = \sqrt{\frac{1}{N_d} \sum_{i=1}^{N_d} \left(Q_i(N,V,T) - Q_i^{\text{ML}} \right)^2} \tag{10.15}$$

From a practical standpoint, we add that it is standard practice to shift the data to be centered on 0 and to scale it by the maximum value so that all the data fall within the $[-1,1]$ interval. We show in Figure 10.5 the results obtained with the machine learning model for the partition function. In particular, the plot on the right shows the excellent agreement between the predicted partition functions Q^{ML} when compared to unseen $Q(N,V,T)$ data from the validation dataset.

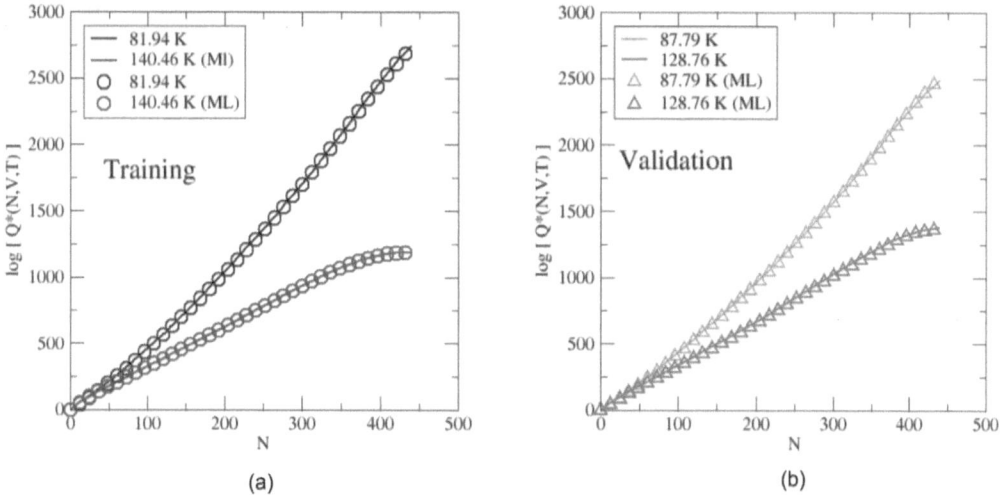

Figure 10.5 Panel (a) Q^{ML} predictions against Wang-Landau data from the training set. This provides a visual quality check of the training process. (b) Q^{ML} predictions against Wang-Landau data from the validation dataset. Since the validation data is not used to optimize the NN weights, the excellent agreement shows that the model can be extended to other (N,V,T) sets.

10.2.2   MULTICOMPONENT MIXTURES

We have seen from the previous example on the ML predicted partition functions for single-component systems that a neural network can provide an accurate and efficient way to interpolate and/or extrapolate Wang-Landau simulation results. The ability of ML to predict partition functions leads to a significant speed-up for single-component systems. This is because simulations, especially flat-histogram simulations, require sampling billions of configurations over which the interatomic distances and interaction energies must be evaluated. Such extensive sampling imposes, in turn, strict limitations on the size and complexity of systems tackled through simulations. This is even more so the case for multicomponent systems, in which the additional components increase the dimensions for the configurational space sampled during the simulations. However, the approach discussed in the previous section can be extended to the case of mixtures [103]. More specifically, for a binary mixture, the neural network will return a prediction for $Q^{\mathrm{ML}}(N_1, N_2, V, T)$, in which N_1 and N_2 stand for the number of molecules of type 1 and 2, respectively. Setting the volume V to a fixed value leads to an input layer with only three neurons, corresponding to the remaining variables N_1, N_2 and T. The rest of the architecture is as follows: two hidden neurons with five neurons each and an output neuron with a single neuron that provides the value of the canonical partition function $Q^{\mathrm{ML}}(N_1, N_2, V, T)$. After optimization of the NN weights, the ML prediction for the partition function takes the following form.

$$Q^{\mathrm{ML}}(N_1, N_2, V, T) = f_4[b_3 + \sum_{l=1}^{5} W(3,4,l,1) f_3(b_2 + \sum_{j=1}^{5} W(2,3,j,l) f_2[(b_1 + \sum_{i=1}^{3} W(1,2,i,j) G_i])])$$

(10.16)

in which $W(i-1, i, l, j)$ denote the NN weight connecting neuron l from layer $i-1$ to neuron j from layer i, f_i the activation functions (*tanh* for $i = 1, 2, 3$ and the linear function for $i = 4$), b_i bias nodes and G_i the input neurons (N_1, N_2 and T).

The NN is trained on a Wang-Landau dataset of the Argon-Krypton mixture simulation results, modeled with a many-body interatomic potential derived from *ab initio* calculations [268]. A question that often arises during the training process is the size of the training dataset that needs to be used. The minimum size necessary to train correctly the NN can be assessed by comparing the training RMSE to the validation RMSE. In this example [103], the NN is trained using different training dataset sizes. Specifically, the range covers a training dataset size between at most 3200 data points and 320 data points. The ratios of the two resulting RMSEs are plotted in Figure 10.6 against the training dataset size. The results show that, for the application considered here, the two RMSEs remain comparable as long as the training dataset contains ~ 1000 data points. Another consequence is the following. To cover the entire (N_1, N_2, T) space and predict accurate values for the canonical partition functions, one only needs, in this case, to carry out Wang-Landau simulations for over $\sim 17\%$ of the (N_1, N_2, T) space and use an ML model to predict the partition functions over the remaining space. In other words, using ML results in a ~ 6-fold gain in computing time. Once the NN weights have been optimized, the resulting ML model for the partition function can then be used to locate phase transitions in mixtures following the process outlined in the previous chapter.

10.2.3   ADSORBED PHASES

Let us now turn to the case of molecular systems adsorbed in a nanoporous material. If the focus is primarily on a single property such as, for instance, loading, it can be advantageous to train the NN on that property, which is, in that case, the adsorption isotherms [56,199,345]. However, having a machine learning model for partition functions of adsorbed fluids allows for calculating all thermodynamics simultaneously. In other words, the information on the loading and the free energy involved in the adsorption and desorption processes are immediately available. As we have seen, the partition function for an adsorbed fluid takes the same form as for the unconfined bulk.

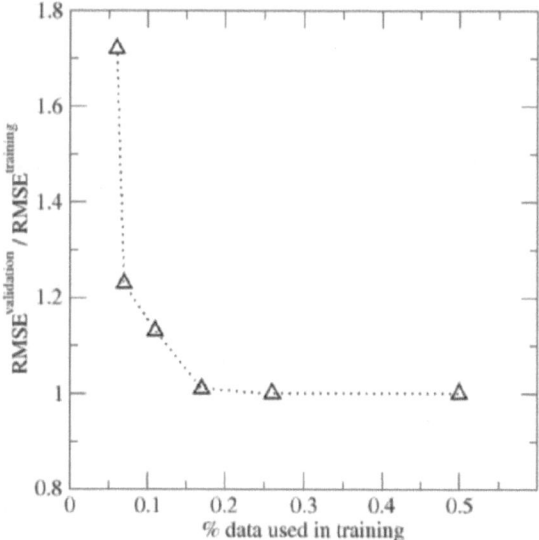

Figure 10.6 Ratio of the validation RMSE to the training RMSE as a function of the fraction of the overall dataset used for training.

The nanoporous material, or adsorbent, only impacts how the potential energy is calculated. To obtain the total interaction energy, guest-host interactions must be added to the intermolecular interactions between guest molecules. Similarly, the partition functions of adsorbed fluids and adsorbed mixtures can be learned with neural networks. Such an approach was tested recently [105] on a variety of systems, including on the example of the adsorption of CO_2 in IRMOF-1, to determine adsorption isotherms for H_2 in COF-102 and to determine the relative selectivity of COF-102 and COF-108 for the adsorption of ethane-methane mixtures. While the training and use of the ML model for the partition function can be carried out along the same lines as in the previous sections, we take advantage of this example to introduce a practical application of the concept of ensemble learning [109]. In ensemble learning, we recall that we average the predictions from k different NNs, which are the k elements that compose the ensemble. There are various strategies to obtain such an ensemble. In this example, the first strategy used is k-fold cross-validation, in which k NNs of the same architecture are trained on randomized subsets of the training dataset. The resulting k NN models as an ensemble [211]. In this example, the network included two hidden layers with $h_1 = 8$ and $h_2 = 5$ neurons, respectively, in the two hidden layers. The second strategy used was a bootstrap aggregation (bagging) approach with replacement [34] that allowed to generate k training subsets with different sample densities. The third strategy used was coined as a diversity approach, which varied the number of neurons in the hidden layers h_1 and h_2 trained on the k randomized subsets [166]. For the three strategies, the estimates provided by each of the k elements of the ensemble are then averaged to obtain the ML prediction. The results indicate that the diversity approach, specifically taking averages over predictions from 8 NN with different architectures, resulted in the lowest RMSE for Q^{ML} (see Figure 10.7) and led to high-accuracy predictions of thermodynamic properties when using Q^{ML} in statistical mechanical formula.

10.3 LEARNING TRANSITIONS

10.3.1 SPANNING PATHWAYS

We have seen how using simulations; we could collect data and information on the system that leads us to identify the location of a phase transition. However, the knowledge of the conditions

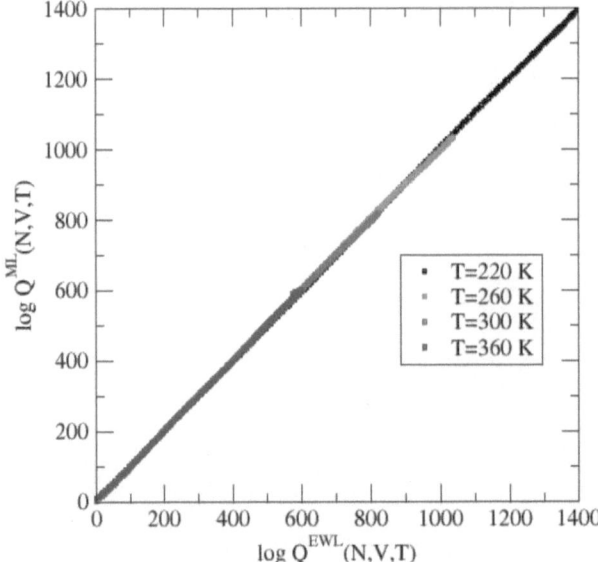

Figure 10.7 Plot of the machine learned partition functions $Q^{ML}(N,V,T)$ against the expected $Q^{EWL}(N,V,T)$ for CO_2 adsorbed in IRMOF-1. Results for both the training and validation datasets are shown together in this plot.

for coexistence does not tell us how the transition from one state of the system to another takes place. To understand the sequence of events that promotes this change in state or, in other words, the microscopic mechanism underlying the transition, we need to follow another approach and get a better sense of the pathway followed by the system during the transition [107]. This is especially important if the transition takes place slowly. In such a case, the system has to overcome a significant free energy barrier and may undergo a series of gradual changes involving several intermediate steps. To elucidate the phase transition mechanism, we will need to use a class of methods termed enhanced sampling simulations [143]. Flat-histogram simulations, such as Wang-Landau sampling [390] and transition matrix methods [118], are two examples of enhanced sampling methods. These are non-Boltzmann sampling methods because they allow for sampling high-energy configurations that would not be sampled in conventional molecular simulations. Indeed, these high-energy configurations are associated with a low Boltzmann weight and, thus, a low probability of occurrence. In other words, these are rare events that happen, for instance, when a molecular system goes through the formation of a nucleus of a new phase (nucleation process) along a phase transition pathway [7,43,84,101,102,135,356] or when a protein changes its conformation during folding. Methods that are specifically focused on the study of these rare events and the elucidation of the transition pathway include the umbrella sampling method [13,364] and metadynamics [40,67]. The idea behind the umbrella sampling method is to add an external potential energy Υ in the calculation of the total potential energy of the system that promotes the acceptance of high-energy configurations. In general, Υ is a harmonic function of a reaction coordinate ϕ that spans the transition pathway and takes the following form.

$$\Upsilon = \frac{1}{2a}k_\Upsilon \left(\phi - \phi_0\right)^2 \qquad (10.17)$$

In this equation, k_Υ denotes the force constant for the harmonic potential, and ϕ_0 is the target value we set for ϕ. During the simulation, a histogram $P(\phi)$ keeps track of the times a given interval of ϕ

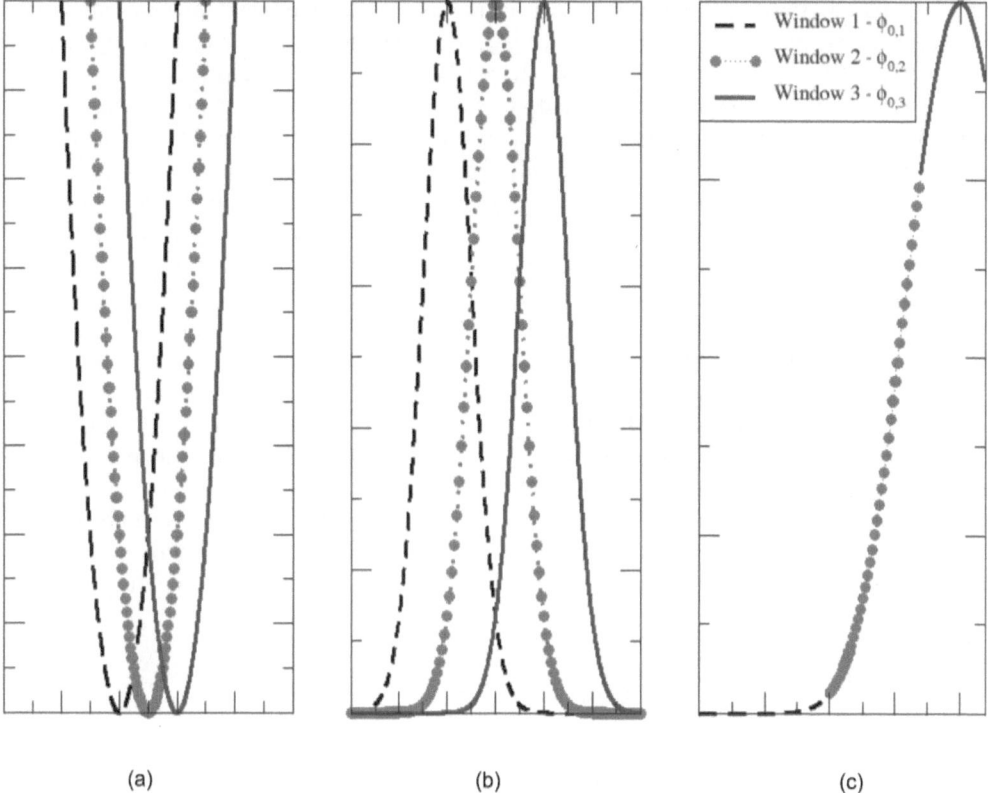

Figure 10.8 (a) Umbrella sampling potentials $\Upsilon(\phi)$ for increasing values of the target reaction coordinate ϕ_0. Here three successive umbrella sampling simulations (windows) are carried out with $\phi_{0,3} > \phi_{0,2} > \phi_{0,1}$. (b) Resulting histograms $P(\phi)$ obtained during the simulations. (c) Reconstructed free energy profile $\Delta G(\phi)$.

is visited. We show in Figure 10.8 an examples of umbrella sampling potentials Υ (left panel) and the resulting histograms $P(\phi)$

To obtain the free energy change along the transition pathway [204], we first subtract the external potential $\Upsilon(\phi)$ and then calculate the free energy change as $-k_B T \ln P(\phi)$. For a system with a given number of atoms N, pressure P and temperature T, the Gibbs free energy difference ΔG is given by

$$\Delta G(\phi) = -k_B T \ln P(\phi) - \Upsilon(\phi) \tag{10.18}$$

In this equation, the added energy term $\Upsilon(\phi)$ is removed to yield the free energy for the system we wanted to study, the system without $\Upsilon(\phi)$. Other methods sharing the same objective include transition path sampling [80] or transition interface sampling [30,375] approaches. These approaches generate an ensemble of paths connecting the initial and final states for the transition, and properties are averaged over this ensemble.

10.3.2 FROM PARTITION FUNCTIONS TO REACTION COORDINATES

We have introduced a reaction coordinate, or collective variable, as a milestone along a reaction pathway but have yet to discuss the mathematical definition of the reaction coordinate. The mathematical description for the reaction coordinate is expected to depend strongly on the phenomenon studied. During a conformational change, the molecule will change geometry. In this case, one may think of a reaction coordinate that maps the conformational change from the initial to the final state and is thus purely based on the geometric consideration. On the other hand, in a nucleation event, one expects many molecules to gather, cluster, and organize into a structure that closely resembles the structure of the new phase. An appropriate reaction coordinate should capture at least some of these aspects. Some approaches have indeed chosen to use the number of molecules belonging to the cluster, based on a distance criterion, as a reaction coordinate for the nucleation of the liquid from a supersaturated vapor [356]. Other approaches have used order parameters [339] that measure the amount of crystalline order in the system to quantify the crystal nucleation from the melt [355]. In recent work, entropic reaction coordinates have emerged as up-and-coming candidates to follow the onset of a phase transition on the macroscopic scale. Entropy is central to our understanding of the molecular association, pattern formation, and self-assembly processes [391]. Entropy, both in a thermodynamic sense [141] and information-theoretic definition [327], has been related to the onset of order, and conversely of disorder and to the amount of organization within a system. Entropy-driven organization processes have emerged as a novel paradigm for a wide range of phenomena including self-assembly in charged lock-key particles [273], segregation processes of polymer-grafted nanoparticles [404], crystallization of DNA grafted nanoparticles [359], polymers [203], proteins [82], droplet nucleation in single-component and binary systems [95,96], capillary condensation [98] and crystal nucleation [99,163,289]. We conclude this part by providing two examples of entropic reaction coordinates used in studies of crystal nucleation of atomic systems. The first is a two-body expression for entropy based on the radial distribution function $g(r)$ in a liquid of density ρ as [289]

$$S_2 = -2\pi\rho k_B \int_0^\infty [g(r)\ln g(r) - g(r) + 1]\, r^2 dr \tag{10.19}$$

This provides a connection between entropy and the amount of structural organization within the system via the use of $g(r)$. The second equation is directly related to thermodynamics since it is based on the machine-learned canonical partition function introduced in the previous section. In this case, entropy is given by Ref. [99]

$$S = \frac{E + k_B T \ln Q^{\mathrm{ML}}(N,V,T)}{T} \tag{10.20}$$

in which E denotes the internal energy of the system and $Q^{\mathrm{ML}}(N,V,T)$ the machine learned partition function. These expressions for entropy are then convenient to use in enhanced sampling simulations. For instance, Eq. 10.20 provides a way to track changes in entropy following a random move in a Monte Carlo simulation. This is because this term is the sum of a configuration-dependent term, the internal energy for a given configuration, and a term equal to the negative Helmholtz free energy. Umbrella sampling simulations can then be performed using this configuration-dependent term to drive the onset of the nucleation process.

10.3.3 ON-THE-FLY LEARNING OF COLLECTIVE VARIABLES

Recent work has also focused on informatics approaches to determine reaction coordinates. Early methods [233,287] relied on screening databases of reaction coordinates. They then extracted combinations of reaction coordinates from the databases using probabilistic likelihood and square errors. This prompted the beginning of using Machine Learning to go beyond interpolation and extrapolation tasks and enable the exploration of high-dimensional free energy landscapes [386]. Approaches combining machine learning and molecular simulation have increasingly been developed in recent years [269] to gain a deeper understanding of the role played by collective variables in the evolution of a molecular system [329]. These combined methods can also provide low-dimensional system representations, along which enhanced sampling simulations can be carried out. Autoencoders are recently used to learn nonlinear reaction coordinates [54]. Such reaction coordinates are defined as differentiable functions of atomic coordinates and, when used in enhanced sampling simulations, accelerate the exploration of folding free energy landscapes. Other approaches leverage variational inference [29] to determine collective variables via deep Bayesian models [28,322]. Machine learning is now often employed to guide free energy simulations for the study of solvation environments [14,403], protein folding [261,269], and ligand-protein systems [216]. When combined with molecular simulation, machine learning has recently emerged as a very efficient way to sample rugged energy landscape [172]. Among others, neural-network-based classification schemes can lead to the definition of path collective variables for phase transitions [310], and the recently introduced reinforced dynamics for enhanced sampling captures the protein structural changes [402] accurately.

Part III

Dynamic Trends: Motion Statistics

11 Molecular Evolution and Fluctuations

Time-Resolved Statistics

"A method is outlined by which it is possible to calculate exactly the behavior of several hundred interacting classical particles. The study (...) is carried out by an electronic computer which solves numerically the simultaneous equations of motion."

Berni J. Alder and Thomas Everett Wainwright,
Studies in molecular dynamics [2]

We have used the concept of ensemble averages so far to predict the properties of a group of molecules. In the case of an ensemble average, the idea is to collect statistics over an extensive collection of system replicas. Then, the measures of microscopic (mechanical) properties are ensemble-averaged, *i.e.*, calculated over the collection of replicas. In many ways, this approach is similar to polling. Polling involves selecting a representative sample for a group and averaging the responses from individuals selected in the sample. To gauge the opinion of the group in a meaningful manner, the poller needs to gather as significant a sample as possible, ensure that each individual polled belongs to the group and share with all other individuals the standard features that define the group, and finally, that the sample accounts for the entire group. If only a part of the group is over-represented in the sample, the poll results will be biased toward the opinion of this part of the group. This analogy emphasizes the role played by the set of replicas in a molecular group, most notably since, just like polls, we will, in practice, handle a finite, albeit significant, number of replicas in our samples. This, however, is unlike the "ideal" infinite number of replicas, covering all possible combinations of positions for the atoms in the system that the ensemble average should contain. As a result, the generation of replicas needs to follow a strict process. The Monte Carlo simulations we have discussed previously fulfill these conditions. First, they create a Markov chain of configurations for the system since each configuration is created via a random move attempted on the immediately preceding configuration in the chain. Second, each replica contributes according to the ensemble policy or, in other words, according to the probability density of the ensemble. Indeed, the Metropolis acceptance rules ensure that each replica appears several times proportional to its Boltzmann weight. Similarly, while flat-histogram and enhanced sampling methods perform a non-Boltzmann sampling, the contributions from the extensive set of configurations these methods generate are weighted according to the ensemble probability density. Are there, however, alternative ways to create representative samples? What happens if, instead of looking at multiple replicas of the system, we take snapshots of the system at regular time intervals to generate the sample of replicas? Will the average taken by "polling" a system at regular time intervals and averaging its mechanical properties over this set of replicas give the same result as an ensemble average? The answer is yes, in some instances. As we will see in the next section, if the policy does not change with time and the system reaches a stationary state, time averages and ensemble averages may become equivalent. In this case, as suggested by Alder and Wainwright, the numerical solving of the simultaneous equations of motion will provide access to time averages and, thus, ensemble averages directly.

DOI: 10.1201/9781003006411-14

11.1   COMPUTING MOLECULAR TRAJECTORIES

11.1.1   ENSEMBLE AND TIME AVERAGES EQUIVALENCY

The significance of stationary probability distributions can be traced back to Boltzmann [27] and Maxwell [248]. Boltzmann builds up on the concept of thermodynamic analogy introduced by Helmholtz [144,146] to show the role of stationary probability distributions. He defines "monodes" as systems with probability distributions invariant with time. He then makes the connection between "orthodes", which are collections of "monodes", and statistical ensembles to draw a thermodynamic analogy. Boltzmann then focuses on a specific "orthode" called an"ergode", which happens to be the two parameters (energy and volume) collection of microcanonical distributions. This leads us to the ergodic hypothesis, in which microcanonical distributions describe the statistics for a system modeled by the Hamiltonian microscopic equations.

Random processes provide another window into the ergodic hypothesis. We start by denoting as $x(t,i)$ the noisy output from a measuring device as a function of time (t), gathered during an experiment labeled α in a series of n experiments, $\alpha = 1, 2, \ldots, N$. Because of the noise, if we randomly pick two experiments or, in other words, two different values of α, we will obtain different results for $x(t, \alpha)$ even if we consider at the same time t. We will be able to tell, however, what the value of the average $< x >$ is when the noise averages out. This means that t does not define completely x. t only defines x in a probabilistic way, $i.e.$, by a set of properties (for instance, average and standard deviation) and by a set of probabilities. If we now think of α as a random continuous variable with a probability density $\rho(\alpha)d\alpha$, the functions $\{x(t,\alpha)\}$ now become a random or stochastic process. An example of a probability defining the random process x is the first probability distribution'$w_1(x,t)dx$, which characterizes the fraction of the total number of functions that have a value between x and $x + dx$ at a given time. Determining the probability distribution is, in general, an intensive task. It requires following the time evolution of an extensive collection of systems and counting, for all times t, the number of times a value between x and $x + dx$ is obtained. However, this task is much simpler if the random process is stationary. In this case, the probability distribution does not change with time and will be the same regardless of the origin chosen for the time $w_1(x,t) = w_1(x)$ or, equivalently, that $x(t, \alpha) = x(t + t_0, \alpha)$ for all values of t_0.

This means that if we record the time evolution for a single system over a very long period τ, we can cut the recording into N pieces (see Figure 11.1) and think of each piece as the time evolution of a different system over τ/N, provided that the period τ/N is longer than any characteristic time for the process. Thus, within this assumption, another statement of the ergodic hypothesis, making N observations at different times on a single system, is equivalent to considering N different systems at the same time t. Another consequence of the ergodic hypothesis is the equivalence between time averages and averages taken over a collection of systems, or ensemble averages. If we denote the former by $\bar{x}$, given by

$$\bar{x} = \lim_{T \to \infty} \frac{1}{T} \int_0^T x(t)dt \tag{11.1}$$

and the latter by $< x >$, defined as

$$< x >= \int_{-\infty}^{\infty} x w_1(x)dx \tag{11.2}$$

Both averages yield the same result for a stationary process, and thus

$$\bar{x} =< x > \tag{11.3}$$

11.1.2   MOLECULAR EQUATIONS OF MOTION

Now that we have established that, for stationary processes, the time average is equivalent to the ensemble average, we need to find a way to calculate the time evolution of groups composed of

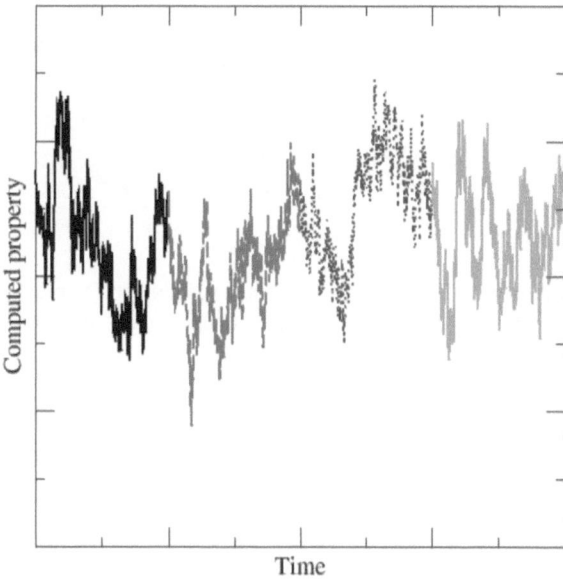

Figure 11.1 Time evolution of a phase space property over a long trajectory. The trajectory can be cut into several shorter trajectories as solid, dashed, and dotted lines. Each shorter trajectory exhibits similar features such as, for instance, the same mean.

many molecules. We will call this time evolution a trajectory. To determine such a trajectory, we need to integrate the equations of motion for each system component. Following the discussion from Chapter 1, we use Hamilton's formalism for the equations of motion. We recall that, in this case, we have for each position and momentum coordinates q_k and p_k, respectively,

$$\begin{aligned} \left(\frac{\partial \mathscr{H}}{\partial p_k}\right) &= \dot{q}_k \\ \left(\frac{\partial \mathscr{H}}{\partial q_k}\right) &= -\dot{p}_k \end{aligned} \tag{11.4}$$

in which the Hamiltonian $\mathscr{H}$ corresponds to the total energy of the system and is defined as $\mathscr{H} = U(\mathbf{q}) + K(\mathbf{p})$.

Recasting Hamilton's equations of motion for each atom i in cartesian coordinates, we obtain

$$\begin{aligned} \dot{\mathbf{r}}_\mathbf{i} &= \mathbf{p}_\mathbf{i}/m_i \\ \dot{\mathbf{p}}_\mathbf{i} &= -\nabla_{\mathbf{r}_\mathbf{i}} U = \mathbf{f}_\mathbf{i} \end{aligned} \tag{11.5}$$

In these equations, $\mathbf{r}_i$ and $\mathbf{p}_i$ denote the position and momentum vectors, respectively, $\dot{\mathbf{r}}_\mathbf{i}$ and $\dot{\mathbf{p}}_\mathbf{i}$ their respective time-derivatives, m_i the mass for particle i, and $\mathbf{f}_\mathbf{i}$ the force exerted on particle i obtained as the gradient $\nabla_{\mathbf{r}_\mathbf{i}}$ of the potential energy U with respect to $\mathbf{r}_i$. Computing the trajectory of N atoms will thus amount to integrating a set of $6N$ first-order differential equations.

What happens now if the system comprises molecules rather than atoms? To answer this question, we need more information on the force field used to model the molecules. In the first approach, we can think of the molecule as being held together by intramolecular forces that account for the existence of bonds and vibrations, such as stretching, bending, and torsion modes within the molecule. In this case, the force field includes a series of springs to model these vibrations, and each atom follows the equations of motion given above. For such a force field, U consists of intermolecular and intramolecular potential energies, and $\mathbf{f}_\mathbf{i}$ is a sum of intermolecular and intramolecular forces.

Alternatively, the force field may consider the molecule to be partly rigid (Figure 11.2). This choice arises from the greater strength of the intramolecular interactions and, as a result, the much

Figure 11.2   A given molecule may be modeled as flexible or rigid. In the flexible model, each double-arrowed line indicates that either a bond length (stretching mode), a bond angle (bending mode), or a dihedral angle (torsion mode) fluctuates with time.

faster timescale on which intramolecular vibrations occur. Similarly, when looking more closely at intramolecular interactions, stretching will, for instance, occur with a frequency at least an order of magnitude greater than bending and torsion. Furthermore, the amplitude of these vibrations will be minimal compared to other motions. Thus, it can be reasonable to assume that bond lengths are constant during the motion and freeze parts of the molecule only. We will see in the following sections how such constraints can be implemented when computing the trajectory of molecular systems. In that case, each atom follows the equations of motion given above, and the force exerted on each atom i includes an additional term due to the constraint forces (as we will see, there is no extra constraint potential energy in the equation for U).

Lastly, molecules may be considered completely rigid bodies. This can arise when considering, for instance, a diatomic molecule or systems of aromatic rings, such as polycyclic aromatic hydrocarbons. In such a case, the equations of motion for each molecule include two sets of equations. The first set is very similar to the equations of motion for atoms we have seen above and applies to the molecule's center of mass. More specifically, $\mathbf{r_i}$ now denotes the vector position for the center of mass of the molecule of mass m_i, obtained by summing the mass of all the atoms in the molecule, and $\mathbf{f_i}$ denotes the vector sum of all the forces acting on molecule i. The second set deals with the rotation of the rigid body of the molecule, which rotates due to the action of the torque Γ_i about its center of mass. If we denote by α an atom in molecule i, the torque is defined as

$$\Gamma_i = \sum_\alpha (\mathbf{r}_{i,\alpha} - \mathbf{r_i}) \times \mathbf{f}_{i,\alpha} \tag{11.6}$$

For instance, for a linear molecule, we obtain the following set of equations of motion for the rotation

$$\dot{\omega}_{\mathbf{i}} = \frac{\Gamma_i}{I}$$
$$\dot{\mathbf{u}}_{\mathbf{i}} = \omega_{\mathbf{i}} \times \mathbf{u}_{\mathbf{i}} \tag{11.7}$$

in which I denotes the moment of inertia, $\mathbf{u}_{\mathbf{i}}$ the unit molecular axis vector, and $\omega_{\mathbf{i}}$ the angular velocity.

For nonlinear molecules, it is advantageous to use quaternions rather than Euler angles (see Figure 11.3) to keep track of the orientation of the molecule [123]. A quaternion $\mathbf{Q}$ for a given molecule (we drop the index i in the rest of this section for the conciseness of the notation) can be thought of as a vector with four coordinates

$$\mathbf{Q} = \begin{bmatrix} q_0 \\ q_1 \\ q_2 \\ q_3 \end{bmatrix} \tag{11.8}$$

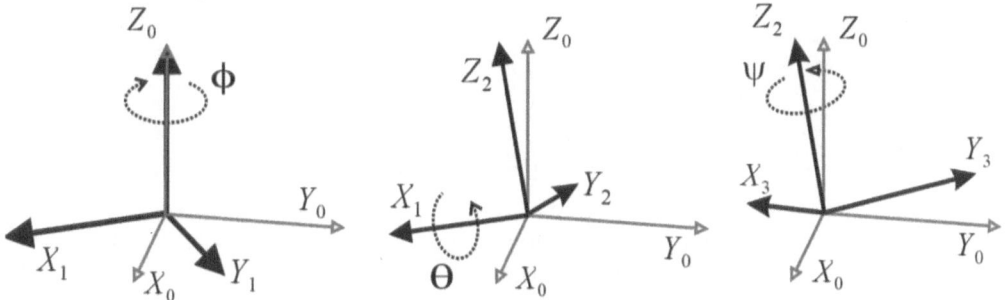

Figure 11.3 The orientation in space of a molecule can be tracked via the use of three Euler angles, denoted by ϕ, θ, and ψ as shown above.

with $q_0^2 + q_1^2 + q_2^2 + q_3^2 = 1$. The coordinates of $\mathbf{Q}$ are related to the Euler angles (θ, ϕ, ψ) through the following equations

$$q_0 = \cos\frac{\theta}{2}\cos\frac{1}{2}(\phi + \psi) \tag{11.9}$$

$$q_1 = \sin\frac{\theta}{2}\cos\frac{1}{2}(\phi - \psi) \tag{11.10}$$

$$q_0 = \sin\frac{\theta}{2}\sin\frac{1}{2}(\phi - \psi) \tag{11.11}$$

$$q_0 = \cos\frac{\theta}{2}\sin\frac{1}{2}(\phi + \psi) \tag{11.12}$$

The quaternions also specify the relation between vector coordinates in the referential whose axes are fixed in space (the "laboratory" x, y, and z axis) and vector coordinates in the body-fixed referential, defined by the axis of inertia of the molecule. A unit vector in the space-fixed laboratory $\mathbf{e}_s$ is related to the body-fixed laboratory $\mathbf{e}_b$ through

$$\mathbf{e}_b = \mathbf{A}\mathbf{e}_s \tag{11.13}$$

in which the rotation matrix $\mathbf{A}$ is defined as

$$\mathbf{A} = \begin{pmatrix} q_0^2 + q_1^2 - q_2^2 - q_3^2 & 2(q_1q_2 + q_0q_3) & 2(q_1q_3 - q_0q_2) \\ 2(q_1q_2 - q_0q_3) & q_0^2 - q_1^2 + q_2^2 - q_3^2 & 2(q_2q_3 + q_0q_1) \\ 2(q_1q_3 + q_0q_2) & 2(q_2q_3 - q_0q_1) & q_0^2 - q_1^2 - q_2^2 + q_3^2 \end{pmatrix} \tag{11.14}$$

This gives the two sets of equations of motion for the rotation of each molecule. First, the quaternions for each molecule follow the equations of motion below

$$\begin{pmatrix} \dot{q}_0 \\ \dot{q}_1 \\ \dot{q}_2 \\ \dot{q}_3 \end{pmatrix} = \frac{1}{2} \begin{pmatrix} q_0 & -q_1 & -q_2 & -q_3 \\ q_1 & q_0 & -q_3 & q_2 \\ q_2 & q_3 & q_0 & -q_1 \\ q_3 & -q_2 & q_1 & q_0 \end{pmatrix} \begin{pmatrix} 0 \\ \omega_{x,b} \\ \omega_{y,b} \\ \omega_{z,b} \end{pmatrix} \tag{11.15}$$

and the x, y, z-coordinates for the angular velocity in the body-fixed referential follow the second set of equations of motion given by

$$\dot{\omega}_{x,b} = \frac{\Gamma_x}{I_{xx}} + \left(\frac{I_{yy} - I_{zz}}{I_{xx}}\right)\omega_{y,b}\omega_{z,b}$$

$$\dot{\omega}_{y,b} = \frac{\Gamma_y}{I_{yy}} + \left(\frac{I_{zz} - I_{xx}}{I_{yy}}\right)\omega_{z,b}\omega_{x,b} \qquad (11.16)$$

$$\dot{\omega}_{z,b} = \frac{\Gamma_z}{I_{zz}} + \left(\frac{I_{xx} - I_{yy}}{I_{zz}}\right)\omega_{x,b}\omega_{y,b}$$

in which Γ_x, Γ_y, and Γ_z denote the coordinates for the torque and I_{xx}, I_{yy}, and I_{zz} are the three principal moments of inertia for the molecule [164].

11.1.3   INTEGRATION SCHEMES

We have defined in the previous section the systems of $6N$ equations that govern the time evolution of atomic and molecular systems. Although each term in the equations can usually be obtained analytically, solving these $6N$ equations is a different and much more complex task that cannot be performed analytically. This is because the equations are coupled since, for instance, the force exerted on an atom depends on the positions of many other atoms in the system. The idea is thus to solve numerically these equations using a finite difference approach. We start from the positions of the particles, the velocities, and other time derivatives of the positions at time t and integrate numerically and as accurately as possible, the positions and their time derivatives at time $t + \delta t$, in which δt is the integration time step. The value chosen for δt will impact the accuracy of the integration scheme and, in general, δt needs to be sufficiently small, *i.e.*, well below the time an atom takes to travel a length comparable to its dimension, to ensure reasonable accuracy. This can be done using an expansion, such as the Taylor expansion provides a simple way to predict the value of a function at time $t + \delta t$ given the value of the function at time t. Applied to the position $\mathbf{r_i}$ and velocity $\mathbf{v_i}$ of an atom i, the Taylor expansion, performed, for instance, to the third-order, gives

$$\mathbf{r_i}(t + \delta t) = \mathbf{r_i}(t) + \delta t \mathbf{v_i}(t) + \frac{\delta t^2}{2m_i}\mathbf{f_i}(t) + \frac{\delta t^3}{3!}\dddot{\mathbf{x}}_i(t) + \mathcal{O}(\delta t^4) \qquad (11.17)$$

$$\mathbf{v_i}(t + \delta t) = \mathbf{v_i}(t) + \frac{\delta t}{m_i}\mathbf{f_i}(t) + \frac{\delta t^2}{2}\ddot{\mathbf{v}}_i(t) + \frac{\delta t^3}{3!}\dddot{\mathbf{v}}_i(t) + \mathcal{O}(\delta t^4) \qquad (11.18)$$

The most straightforward approach then consists in using a Taylor expansion and propagating the equations of motion according to this expansion. Working solely with the terms that are either known (positions and velocities) or can be evaluated analytically (forces), we can use first-order Taylor expansions and integrate the equations according to the Euler algorithm as

$$\mathbf{r_i}(t + \delta t) = \mathbf{r_i}(t) + \delta t \mathbf{v_i}(t) + \mathcal{O}(\delta t^2) \qquad (11.19)$$

$$\mathbf{v_i}(t + \delta t) = \mathbf{v_i}(t) + \frac{\delta t}{m_i}\mathbf{f_i}(t) + O(\delta t^2) \qquad (11.20)$$

The Euler algorithm is a self-starting algorithm since one only needs the positions, velocities, and the forces at the beginning of the trajectory to provide accurate values (to the first-order) for entire trajectory. It is possible, however, to increase the accuracy of the integration scheme by including the terms of order 2, 3 and 4 in the Taylor expansion (Figure 11.4). To achieve this, we need to determine the successive time derivatives of the positions up to order 4. We use the Nordsieck representation, in which we keep track in a vector of the scaled time derivatives of the positions. Denoting by an index varying from 0 to 3 the order of differentiation, we have the position defined as $\mathbf{r_{i,0}}$ and the scaled time derivatives as $\mathbf{r_{i,1}} = \delta t(d\mathbf{r_{i,0}}/dt)$, $\mathbf{r_{i,2}} = \frac{1}{2}\delta t^2(d^2\mathbf{r_{i,0}}/dt^2)$, and $\mathbf{r_{i,3}} = \frac{1}{6}\delta t^3(d^3\mathbf{r_{i,0}}/dt^3)$. This provides the basis for the four-value Gear predictor-corrector algorithm, which, in matrix form, starts with the prediction of the time evolution of the following Nordsieck vector

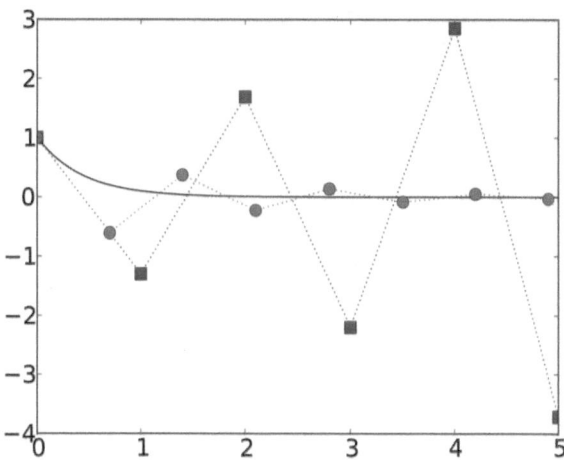

Figure 11.4   Integration with the Euler algorithm. The exact solution is a solid line, while results obtained with the Euler algorithm are shown as symbols for two different timesteps. The plot illustrates the gain in performance of the Euler algorithm as δt decreases from $\delta t = 1$ (shown as squares) to $\delta t = 0.5$ (shown as circles).

$$\begin{pmatrix} \mathbf{r}^{p}_{i,0}(t+\delta t) \\ \mathbf{r}^{p}_{i,1}(t+\delta t) \\ \mathbf{r}^{p}_{i,2}(t+\delta t) \\ \mathbf{r}^{p}_{i,3}(t+\delta t) \end{pmatrix} = \begin{pmatrix} 1 & 1 & 1 & 1 \\ 0 & 1 & 2 & 3 \\ 0 & 0 & 1 & 3 \\ 0 & 0 & 0 & 1 \end{pmatrix} \begin{pmatrix} \mathbf{r}_{i,0}(t) \\ \mathbf{r}_{i,1}(t) \\ \mathbf{r}_{i,2}(t) \\ \mathbf{r}_{i,3}(t) \end{pmatrix} \tag{11.21}$$

in which the superscript p denotes the predicted values, and the triangular matrix contains the Pascal triangle coefficients. The second stage in the Gear algorithm is the corrector, which takes the following form

$$\begin{pmatrix} \mathbf{r}_{i,0}(t+\delta t) \\ \mathbf{r}_{i,1}(t+\delta t) \\ \mathbf{r}_{i,2}(t+\delta t) \\ \mathbf{r}_{i,3}(t+\delta t) \end{pmatrix} = \begin{pmatrix} \mathbf{r}^{p}_{i,0}(t+\delta t) \\ \mathbf{r}^{p}_{i,1}(t+\delta t) \\ \mathbf{r}^{p}_{i,2}(t+\delta t) \\ \mathbf{r}^{p}_{i,3}(t+\delta t) \end{pmatrix} - \begin{pmatrix} c_0 \\ c_1 \\ c_2 \\ c_3 \end{pmatrix} \Delta \mathbf{r_i} \tag{11.22}$$

where the left-hand side gives the values for the time derivatives after applying the correction, the corrector coefficients c_0, c_1, c_2, and c_3 are constants that depend on the order of the differential equation, and $\Delta \mathbf{r_i} = \mathbf{r}^{p}_{i,1} - \mathbf{r}^{c}_{i,1}$ is the correction. $\mathbf{r}^{c}_{i,1}$ is the first derivative obtained by plugging $\mathbf{r}^{p}_{i,0}$ into the equation of motion. Later in Section 12.2.2, we provide a pseudo-code illustrating the implementation of a Gear predictor-corrector of the fourth order (or five-value algorithm).

Instead of considering the successive time derivatives, some integration schemes include intermediate position and velocity values along the time step, *e.g.*, at $t + \delta t/2$. This is the case of a very widely used integration scheme known as the velocity-Verlet algorithm [348,380]. The positions are advanced over the entire time step

$$\mathbf{r_i}(t + \delta t) = \mathbf{r_i}(t) + \delta t \mathbf{v_i}(t) + (1/2)\delta t^2 \mathbf{a_i}(t) \tag{11.23}$$

in which the acceleration $\mathbf{a_i}(t)$ is obtained from $\mathbf{f}_i(\mathbf{r}_i(t)/m_i)$, and the velocity is advanced in two successive stages, first to the midpoint of the timestep, and finally to the end of the timestep, according to

$$\mathbf{r}(t + \delta t) = \mathbf{r}(t) + \delta t \mathbf{v}(t) + (1/2)\delta t^2 \mathbf{a}(t) \tag{11.24}$$

$$\mathbf{v}(t + \delta t) = \mathbf{v}(t) + \frac{1}{2}\delta t \left(\mathbf{a}(t) + \mathbf{a}(t + \delta t) \right) \tag{11.25}$$

Algorithm 8 NVE integration with the velocity-Verlet algorithm

---

1: Start from an initial set of positions and velocities $(\mathbf{r}_i, \mathbf{v}_i)$

2: Compute the forces $\mathbf{F}_i$ on each atom i

3: **while** i_{step} is less than the number of time steps for the integration n_{step} **do**

4:     For each atom i, compute the velocity at the half-step $\mathbf{v}_i \leftarrow \mathbf{v}_i + (\delta t/2) \times \mathbf{F}_i$

5:     Using the half-step velocity, move each atom i to its position at the end of the step $\mathbf{r}_i \leftarrow$

   $\mathbf{r}_i + \delta t \times \mathbf{v}_i$

6:     Using the updated positions, compute the forces $\mathbf{F}_i$ on each atom i

7:     Using the updated forces, compute the velocity for each atom i at the end of the step

   $\mathbf{v}_i \leftarrow \mathbf{v}_i + (\delta t/2) \times \mathbf{F}_i$

8:     Accumulate the current value for the energy in the running total for the energy

9:     Accumulate running total for other quantities of interest

10: **end while**

11: Compute averages over the trajectory, *i.e.*, divide running totals by n_{steps}

Figure 11.5 Pseudo-code for a (N,V,E) integration with the velocity-Verlet integrator.

in which the acceleration $\mathbf{a}_i(t+\delta t)$ is obtained from $\mathbf{f}_i(\mathbf{r}_i(t+\delta t)/m_i)$. The implementation of this algorithm is illustrated in Figure 11.5.

Another method, known as Runge-Kutta, also utilizes evaluations of the functions along the timestep. The position and velocity at the end of the timestep are given by

$$\mathbf{r}_i(t+\delta t) = \mathbf{r}_i(t) + \frac{\delta t}{6}\left[\mathbf{k}_{1,i} + 2\mathbf{k}_{2,i} + 2\mathbf{k}_{3,i} + \mathbf{k}_{4,i}\right] \tag{11.26}$$

$$\mathbf{k}_{1,i} = \mathbf{v}_i(t,\mathbf{r}_i) \tag{11.27}$$

$$\mathbf{k}_{2,i} = \mathbf{v}_i(t+\delta t/2, \mathbf{r}_i + \delta t\mathbf{k}_{1,i}/2) \tag{11.28}$$

$$\mathbf{k}_{3,i} = \mathbf{v}_i(t+\delta t/2, \mathbf{r}_i + \delta t\mathbf{k}_{2,i}/2) \tag{11.29}$$

$$\mathbf{k}_{4,i} = \mathbf{v}_i(t+\delta t, \mathbf{r}_i + \delta t\mathbf{k}_{3,i}) \tag{11.30}$$

$$\tag{11.31}$$

and

$$\mathbf{v_i}(t+\delta t) = \mathbf{v_i}(t) + \frac{\delta t}{6}\left[\mathbf{k}_{1,i} + 2\mathbf{k}_{2,i} + 2\mathbf{k}_{3,i} + \mathbf{k}_{4,i}\right] \tag{11.32}$$

$$\mathbf{k}_{1,i} = \mathbf{a_i}(t,\mathbf{v_i}) \tag{11.33}$$

$$\mathbf{k}_{2,i} = \mathbf{a_i}(t+\delta t/2, \mathbf{v_i} + \delta t\mathbf{k}_{1,i}/2) \tag{11.34}$$

$$\mathbf{k}_{3,i} = \mathbf{a_i}(t+\delta t/2, \mathbf{v_i} + \delta t\mathbf{k}_{2,i}/2) \tag{11.35}$$

$$\mathbf{k}_{4,i} = \mathbf{a_i}(t+\delta t, \mathbf{v_i} + \delta t\mathbf{k}_{3,i}) \tag{11.36}$$

$$\tag{11.37}$$

respectively. As in the velocity-Verlet algorithm, we evaluate the acceleration $\mathbf{a}_i$ as the ratio $\mathbf{f}_i/m_i$.

11.2 MOLECULAR TRAJECTORIES

11.2.1 GAUSS' PRINCIPLE OF LEAST CONSTRAINT

So far, we have invoked Newton's equations of motion to provide the time evolution for a system with a Hamiltonian $\mathscr{H} = K(\mathbf{p}) + U(\mathbf{q})$. For such systems, the system's total energy is constant during the motion or, in other words, an inherent constraint of this dynamics. Since the system's energy is constant, the time average extracted from the computed trajectories will, therefore, according to the ergodic hypothesis, be equivalent to the microcanonical ensemble average. How do we obtain, however, a time average in a different ensemble, for which, for instance, the constraint consists of keeping the temperature constant during the dynamics? Similarly, if the molecule is treated as a rigid body, how do we factor these geometric constraints into the equations of motion? The answer was provided in 1829 by the famous physicist Gauss [154] in his principle of least constraint. This principle allows for a generalization of Newton's dynamics and is referred to as the most fundamental dynamical principle [198,285,392].

The curvature $\mathscr{C}$ plays a significant role in the application of Gauss' principle of least constraint. $\mathscr{C}$ is a function of the acceleration and is given by

$$\mathscr{C}(\ddot{\mathbf{r}}) = \frac{1}{2} \sum_{i=1}^{N} m_i \left(\ddot{r}_i - \frac{f_i}{m_i} \right)^2 \tag{11.38}$$

The principle of least constraint states that the actual acceleration of the system minimizes the curvature $\mathscr{C}$. In other words, to determine the constrained equations of motion, we need to differentiate $\mathscr{C}$, subject to the constraint. If we cast the constraint C in an acceleration-dependent form and denote by λ the corresponding multiplier, we obtain the following condition

$$\frac{\partial}{\partial \ddot{\mathbf{r}}} (\mathscr{C} - \lambda C) = 0 \tag{11.39}$$

Solving this equation yields the constrained equations of motion [198]. For instance, let us consider the example of a diatomic molecule with two atoms, a and b, and a bond length set to a constant d during the motion. This implies that, during the trajectory, we have

$$||\mathbf{r}_a - \mathbf{r}_b||^2 - d^2 = 0 \tag{11.40}$$

We now denote by $\mathbf{r}_{ab}$ the vector defined as $\mathbf{r}_{ab} = \mathbf{r}_b - \mathbf{r}_a$. To cast the constraint into an acceleration-dependent form, we differentiate this equation two times with respect to time and obtain

$$C = \mathbf{r}_{ab} \cdot \ddot{\mathbf{r}}_{ab} + (\dot{\mathbf{r}}_{ab})^2 = 0 \tag{11.41}$$

We then plug the expression for C into Eq. 11.39 and differentiate with respect to the acceleration of atom a. We obtain

$$\frac{\partial}{\partial \ddot{\mathbf{r}}_a} \left[\frac{m_a}{2} \left(\ddot{\mathbf{r}}_a - \frac{\mathbf{f}_a}{m_a} \right)^2 + \frac{m_b}{2} \left(\ddot{\mathbf{r}}_b - \frac{\mathbf{f}_b}{m_b} \right)^2 - \lambda \left(\mathbf{r}_{ab} \cdot \ddot{\mathbf{r}}_{ab} + (\dot{\mathbf{r}}_{ab})^2 \right) \right] = 0 \tag{11.42}$$

and thus the following equation of motion for atom a:

$$m_a \ddot{\mathbf{r}}_a = \mathbf{f}_a - \lambda \mathbf{r}_{ab} \tag{11.43}$$

Differentiating with respect to the acceleration of atom b provides the equation of motion for atom b

$$m_b \ddot{\mathbf{r}}_b = \mathbf{f}_b + \lambda \mathbf{r}_{ab} \tag{11.44}$$

The two constraint forces have opposite signs and, as a result, cancel out. This means that there is no net force arising from the constraint and that the momentum of the molecule is conserved. To determine the multiplier λ, we calculate the acceleration for the bond vector $\ddot{\mathbf{r}}_{ab} = \ddot{\mathbf{r}}_b - \ddot{\mathbf{r}}_a$ from the equations of motion for the two atoms a and b to obtain

$$\ddot{\mathbf{r}}_{ab} = \left(\frac{\mathbf{f}_b}{m_b} - \frac{\mathbf{f}_a}{m_a} \right) + \lambda \left(\frac{1}{m_b} + \frac{1}{m_a} \right) \mathbf{r}_{ab} \tag{11.45}$$

Multiplying both sides by $\mathbf{r}_{ab}$, we obtain

$$\mathbf{r}_{ab} \cdot \ddot{\mathbf{r}}_{ab} = \mathbf{r}_{ab} \cdot \left(\frac{\mathbf{f}_b}{m_b} - \frac{\mathbf{f}_a}{m_a} \right) + \lambda \left(\frac{1}{m_b} + \frac{1}{m_a} \right) \mathbf{r}_{ab} \tag{11.46}$$

From the equation defining C, we have $\mathbf{r}_{ab} \cdot \ddot{\mathbf{r}}_{ab} = -(\dot{\mathbf{r}}_{ab})^2$. Plugging this result into the previous equation yields the following expression for λ

$$\lambda = -\frac{\mathbf{r}_{ab} \cdot (m_a \mathbf{F}_b - m_b \mathbf{F}_a) + m_a m_b \dot{\mathbf{r}}_{ab}^2}{(m_a + m_b) \mathbf{r}_{ab}^2} \tag{11.47}$$

The example we covered here only involves a single geometric constraint. The approach can be readily extended to more complex molecules. Indeed, the generalization to multiple geometric constraints, and thus the determination of various multipliers, follows the same reasoning and leads to a set of coupled linear equations [116].

11.2.2   KEEPING THE TEMPERATURE IN CHECK

We now turn to the computation of molecular trajectories in ensembles that require a thermodynamic property, such as temperature, to be constant. We can apply another type of constraint that fixes the system's kinetic energy during the motion and, as a result, the temperature constant over the trajectories. To this end, we invoke Gauss' principle of least constraint to devise a "thermostat" for the system [127,198]. For a system of N atoms, we equate the kinetic energy to the contribution from the $3N$ degrees of freedom for the N atoms, with a contribution of $\frac{k_B}{2}T$ for each degree of freedom. This gives

$$\sum_{i=1}^{N} \frac{m_i}{2} \dot{\mathbf{r}}_i^2 - \frac{3}{2} N k_B T = 0 \tag{11.48}$$

Differentiating this equation with respect to time provides the acceleration-dependent form of the constraint equation as

$$C = \sum_{i=1}^{N} m_i \dot{\mathbf{r}}_i \cdot \ddot{\mathbf{r}}_i = 0 \tag{11.49}$$

Plugging the expression for C into Eq. 11.39 and differentiating with respect to the acceleration of atom i, we obtain

$$\frac{\partial}{\partial \ddot{\mathbf{r}}_i} \left[\frac{1}{2} \sum_{j=1}^{N} m_j \left(\ddot{\mathbf{r}}_j - \frac{\mathbf{f}_j}{m_j} \right)^2 - \lambda \sum_{j=1}^{N} m_j \dot{\mathbf{r}}_j \cdot \ddot{\mathbf{r}}_j \right] = 0 \tag{11.50}$$

and the following equation of motion

$$m_i \ddot{\mathbf{r}}_i = \mathbf{f}_i + \lambda m_i \dot{\mathbf{r}}_i \tag{11.51}$$

To derive the equation for the multiplier, we plug the equations of motion into the differential form for the constraint equation to obtain

$$\lambda = -\frac{\sum_{i=1}^{N} \mathbf{f}_i \cdot \dot{\mathbf{r}}_i}{\sum_{i=1}^{N} m_i \dot{\mathbf{r}}_i^2} \tag{11.52}$$

Written according to Hamilton's formalism, the equations of motion for a system with a Gaussian thermostat are thus given by

$$\dot{\mathbf{r}}_i = \frac{\mathbf{p}_i}{m_i} \tag{11.53}$$

$$\dot{\mathbf{p}}_i = \mathbf{f}_i + \lambda \mathbf{p}_i \tag{11.54}$$

with

$$\lambda = -\frac{\sum_{i=1}^N \mathbf{f}_i \cdot \mathbf{p}_i / m_i}{\sum_{i=1}^N \mathbf{p}_i^2 / m_i} \tag{11.55}$$

These equations of motion keep the temperature constant. The implementation of this algorithm is illustrated in Figure 11.6.

We need to ensure that they fulfill the conditions of the ergodic hypothesis and preserve the

---

Algorithm 9 NVT molecular dynamics (predictor-corrector & Gaussian thermostat)

1: Start from an initial set of positions and momenta $(\mathbf{r}_{i,0}, \mathbf{p}_{i,0})$

2: For each atom i, evaluate the force exerted on atom i to obtain $\mathbf{F}_i$

3: Set an initial value for the scaled time derivative $\mathbf{r}_{i,1} = \mathbf{p}_{i,0}/m_i \times \delta t$

4: Similarly, set an initial value for $\mathbf{p}_{i,1} = \mathbf{F}_i \times \delta t$

5: Initialize to 0 all successive scaled time derivatives $(\mathbf{r}_{i,2}, \mathbf{r}_{i,3}, \mathbf{r}_{i,4})$ and $(\mathbf{p}_{i,2}, \mathbf{p}_{i,3}, \mathbf{p}_{i,4})$

6: **while** i_{step} is less than the number of time steps for the integration n_{step} **do**

7:     For each atom i, predict the position $\mathbf{r}_{i,0}^p = \mathbf{r}_{i,0} + \mathbf{r}_{i,1} + \mathbf{r}_{i,2} + \mathbf{r}_{i,3} + \mathbf{r}_{i,4}$

8:     For each atom i, predict the position $\mathbf{r}_{i,1}^p = \mathbf{r}_{i,1} + 2\mathbf{r}_{i,2} + 3\mathbf{r}_{i,3} + 4\mathbf{r}_{i,4}$

9:     For each atom i, predict the position $\mathbf{r}_{i,2}^p = \mathbf{r}_{i,2} + 3\mathbf{r}_{i,3} + 6\mathbf{r}_{i,4}$

10:     For each atom i, predict the position $\mathbf{r}_{i,3}^p = \mathbf{r}_{i,3} + 4\mathbf{r}_{i,4}$

11:     Substitute $(\mathbf{r}_i \rightarrow \mathbf{p}_i)$ in steps 7-10 equations to predict $(\mathbf{p}_{i,0}^p, \mathbf{p}_{i,1}^p, \mathbf{p}_{i,2}^p, \mathbf{p}_{i,3}^p, \mathbf{p}_{i,4}^p)$

12:     Using the predicted positions, compute the forces $\mathbf{F}_i$ on each atom i

13:     Compute the thermostat multiplier $\lambda = (\sum_i \mathbf{p}_{i,0}^p \cdot \mathbf{F}_i)/(\sum_i (\mathbf{p}_{i,0}^p)^2$

14:     Add the thermostat term to the force on each atom i: $\mathbf{F}_i \leftarrow \mathbf{F}_i - \lambda \mathbf{p}_{i,0}^p$

15:     Compute the position correction for each atom i: $\Delta \mathbf{r}_i = \mathbf{r}_{i,1}^p - \mathbf{p}_{i,0}^p/m_i \times \delta t$

16:     Compute the momentum correction for each atom i: $\Delta \mathbf{p}_i = \mathbf{p}_{i,1}^p - \mathbf{F}_i \times \delta t$

17:     Apply the position correction for each atom i $(k = 0, .., 4)$: $\mathbf{r}_{i,k} = \mathbf{r}_{i,k}^p - \Delta \mathbf{r}_i \times c_k$

18:     Apply the momentum correction for each atom i $(k = 0, .., 4)$: $\mathbf{p}_{i,k} = \mathbf{p}_{i,k}^p - \Delta \mathbf{p}_i \times c_k$

19:     Accumulate running total for all quantities of interest

20: **end while**

21: Compute averages over the trajectory, *i.e.*, divide running totals by n_{steps}

Figure 11.6 Pseudo-code for a molecular dynamics simulation at constant (N, V, T) using a five-value (fourth order) predictor-corrector integration scheme and a Gaussian thermostat. The Gear coeeficients are $c_0 = 251/720$, $c_1 = 1$, $c_2 = 11/12$, $c_3 = 1/3$, and $c_4 = 1/24$.

canonical probability distribution during the trajectory. If we denote by f the probability distribution associated with the phase space point $\Gamma = (\mathbf{p}, \mathbf{q})$, the Liouville equation for the phase-space distribution function gives the rate of change in f as

$$\frac{\partial f}{\partial t} = -f \frac{\partial}{\partial \Gamma} \cdot \dot{\Gamma} - \dot{\Gamma} \cdot \frac{\partial f}{\partial \Gamma} = -\frac{\partial}{\partial \Gamma} \cdot (f\dot{\Gamma}) \tag{11.56}$$

The Liouville equation can also be written in terms of the streaming derivative or total time derivative of the distribution function as

$$\frac{df}{dt} = \frac{\partial f}{\partial t} + \dot{\Gamma} \cdot \frac{\partial f}{\partial \Gamma} = -f \frac{\partial}{\partial \Gamma} \cdot \dot{\Gamma} = -f\Lambda(\Gamma) \tag{11.57}$$

or, in a more compact form,

$$\frac{d \ln f(\Gamma, t)}{dt} = -\Lambda(\Gamma) \tag{11.58}$$

in which $\Lambda(\Gamma)$ is called the phase-space compression factor. When the equations of motion derive from a Hamiltonian, the phase-space compression factor vanishes

$$\Lambda(\Gamma) = \sum_{i=1}^{N} \left(\frac{\partial}{\partial \mathbf{q}_i} \cdot \dot{\mathbf{q}}_i + \frac{\partial}{\partial \mathbf{p}_i} \cdot \dot{\mathbf{p}}_i \right) = 0 \tag{11.59}$$

and the Liouville equation reduces to $\frac{df}{dt} = 0$ in this case.

The equations of motion supplemented with a Gaussian thermostat ensure that the kinetic energy, and, as a result, the system's temperature, is constant over the trajectory. We thus start by applying the Liouville equation to the isokinetic distribution function f_K

$$\frac{\partial f_K}{\partial t} = -f_K \frac{\partial}{\partial \Gamma} \cdot \dot{\Gamma} - \dot{\Gamma} \cdot \frac{\partial f_K}{\partial \Gamma} = (\beta \dot{\Phi} + 3N\lambda) f_K \tag{11.60}$$

in which $\dot{\Phi}$ denotes the time derivative of the potential energy, and the factor of $3N\lambda$ comes from the $3N$ identical terms $(\partial \dot{\mathbf{p}}_i)/(\partial \mathbf{p}_i) = (\partial \lambda \mathbf{p}_i)/(\partial \mathbf{p}_i) = \lambda$.

This equation also shows that the equations of motion preserve the isokinetic distribution since the right-hand side equals 0. We have

$$\beta \dot{\Phi} + 3N\lambda = -\beta \sum_i \frac{\mathbf{p}_i}{m_i} \cdot \mathbf{f}_i + 3N \frac{\Sigma_i(1/m_i)\mathbf{p}_i \cdot \mathbf{f}_i}{\Sigma_i \mathbf{p}_i^2/m_i} = 0 \tag{11.61}$$

given that $\beta = (3N)/(2K)$. Now let us examine if the dynamics preserve the canonical distribution f_C. In this case, we have

$$\frac{\partial f_c}{\partial t} = f_c(3N\lambda + \beta \dot{K} + \beta \dot{\Phi}) = f_c(\beta - \frac{3N}{2K(\Gamma)})\dot{\Phi}(\Gamma) = f_c \Delta(\beta)\dot{\Phi}(\Gamma) \tag{11.62}$$

From this equation, it appears that $\frac{\partial f_c}{\partial t}$ is not exactly equal to 0. Because we now consider an ensemble of trajectories with different initial conditions, and thus different kinetic energies, sampling the canonical distribution, $(3N)/(2K)$ is only equal to β on average. Similarly, $\dot{\Phi}$ is only equal to zero on average. It can be shown the equations of motion preserve the canonical distribution in the thermodynamic limit. Indeed, for an extensive phase variable B, we have the following time derivative for the ensemble average

$$\frac{d}{dt} \langle B(t) \rangle = \int d\Gamma B \frac{\partial f_c}{\partial t} = \int d\Gamma B \Delta(\beta) \dot{\Phi} f_c = \frac{\beta}{<K>} \langle B \Delta K \dot{\phi} \rangle + O(\Delta^2) \tag{11.63}$$

in which $\Delta K = K - \langle K \rangle$. Replacing B by $< B > +\Delta B$, we find

$$\frac{d}{dt} \langle B(t) \rangle = \frac{\beta}{<K>} [\langle B \rangle \langle \Delta K \dot{\Phi} \rangle + \langle \Delta B \Delta K \dot{\Phi} \rangle] = \frac{\beta}{<K>} \langle \Delta B \Delta K \dot{\phi} \rangle + O(1) \tag{11.64}$$

since $< \Delta K \dot{\Phi} >= 0$, and that the ensemble average of the product of three extensive, zero-mean phase variables is of order N, while $< K >$ is extensive. This shows that the change in $< B(t) >$ with time is of order one and is negligible when compared to $< B >$. Thus, the Gaussian isokinetic dynamics preserve the canonical distribution in the thermodynamic limit.

11.2.3 NOSÉ-HOOVER THERMOSTAT

Nosé was the first to introduce the concept of an integral feedback mechanism for the control of the temperature of the system [270,271], rather than a differential feedback mechanism as in the Gaussian thermostat. This results in an extended system with an additional degree of freedom or variable, whose time evolution is integrated along the trajectory. As explained by Nosé [272], "the extended-system method tries to mimic a situation which is written in a textbook on statistical mechanics. A physical system is connected to an external system. However, in the extended-system method, the external system consists of only one degree of freedom and is not as huge as we expect in a statistical mechanical treatment. Adding an external system allows the total energy of a physical system to fluctuate." Hoover rewrote the Nose equations of motion in a form that is similar to the Gaussian equations [186], with a multiplier which, in the Nosé-Hoover case, evolves with time according to an equation of motion. The equations of motion for the extended system take the following form [126]

$$\dot{\mathbf{q}}_i = \frac{\mathbf{p_i}}{m_i} \tag{11.65}$$

$$\dot{\mathbf{p}}_i = \mathbf{F}_i - \zeta \mathbf{p_i} \tag{11.66}$$

$$\dot{\zeta} = \frac{1}{Q} \left(\sum_{i=1}^{N} \frac{\mathbf{p}_i^2}{m_i} - 3Nk_BT \right) \tag{11.67}$$

$$\tag{11.68}$$

in which Q is a thermostat parameter. These equations contain a term akin to a friction term in the $\dot{\mathbf{p}}$ equation and a separate equation for the change in the thermostat, or heat bath, coefficient ζ. According to this scheme, the average of the kinetic energy rather than the instantaneous value of the kinetic energy is kept constant. If the kinetic energy of the system is greater than the average value for of $(3Nk_BT/2)$, then $\dot{\zeta} > 0$ and thus ζ increases, eventually becoming positive $\zeta > 0$ and thus acts as a friction that leads to a decrease in all $\dot{p}_i$ and the kinetic energy for the system. Conversely, if, during the trajectory, the kinetic energy of the system is smaller than the average value for of $(3Nk_BT/2)$, then $\dot{\zeta} < 0$ and thus ζ decreases, eventually becoming negative $\zeta < 0$ increasing all $\dot{p}_i$ and the kinetic energy for the system. The kinetic energy fluctuates around its average value, and the temperature remains constant and equal to T.

The distribution function $f(\Gamma, \zeta)$ generated by Eq. 11.68 is obtained by solving the Liouville equation as [198]

$$\frac{df}{dt} = -f \left(\frac{\partial}{\partial \Gamma} \cdot \dot{\Gamma} + \frac{\partial}{\partial \zeta} \dot{\zeta} \right) \tag{11.69}$$

The only nonzero contribution arises from the $\dot{\mathbf{p}}$ terms, providing the following equation

$$\frac{d}{dt} \ln f = 3N\zeta \tag{11.70}$$

If we now turn to the time derivative of the quantity $H_0 + \frac{1}{2}Q\zeta^2$, we obtain

$$\frac{d}{dt}\left(H_0 + \frac{Q\zeta^2}{2}\right) = \frac{d}{dt}H_0 + Q\zeta\dot{\zeta} = -3N\zeta k_B T \tag{11.71}$$

and find that the equilibrium distribution function is the extended canonical distribution f_c, given by

$$f_c(\Gamma, \zeta) = \frac{\exp[-\beta(H_0 + Q\zeta^2/2)]}{\int d\Gamma d\zeta \exp[-\beta(H_0 + Q\zeta^2/2)]} \tag{11.72}$$

11.3 MULTIPLE-TIME STEPS AND HYBRID SCHEMES

11.3.1 TIME-SPLITTING

Explicit reversible integrators can also be developed using operator factorization techniques [65, 371]. We discuss in this section the application of this approach to the propagation of equations of motion that preserve the canonical distribution. Here, we specifically examine the case of a system for which a chain of Nosé-Hoover thermostats (NHC) is used instead of the single feedback variable ζ as in the previous Section [244]. The equations of motion are, in this case, as follows [245]

$$\dot{\mathbf{r}}_i = \frac{\mathbf{p}_i}{m_i} \tag{11.73}$$

$$\dot{\mathbf{p}}_i = \mathbf{F}_i - \mathbf{p}_i \frac{p_{\zeta_i}}{Q_i} \tag{11.74}$$

$$\dot{\zeta}_i = \frac{p_{\zeta_i}}{Q_i} \tag{11.75}$$

$$\dot{p}_{\zeta_1} = \left[\sum_{i=1}^N \frac{\mathbf{p}_i^2}{m_i} - 3NkT\right] - p_{\zeta_1}\frac{p_{\zeta_2}}{Q_2} \tag{11.76}$$

$$\dot{p}_{\zeta_j} = \left[\sum_{i=1}^N \frac{\mathbf{p}_{\zeta_{j-1}}^2}{Q_{j-1}} - kT\right] - p_{\zeta_j}\frac{p_{\zeta_{j+1}}}{Q_{j+1}} \tag{11.77}$$

$$\dot{p}_{\zeta_M} = \left[\sum_{i=1}^N \frac{\mathbf{p}_{\zeta_{M-1}}^2}{Q_{M-1}} - kT\right] \tag{11.78}$$

in which the M additional equations correspond to M thermostats with extended variables ζ_j ($j = 1, .., N$), parameters Q_j, and momenta p_{ζ_j}. Here, the equations include the time evolution for the time derivative of the extended variables or, more precisely, their momenta and the time evolution of the extended variable. As shown in the equations, only the momenta of the extended variables act on the system through the friction terms, and the time evolution of the extended variables ζ_i does not impact the computation of the molecular trajectory. These equations of motion conserve the total energy of the extended system associated with the following Hamiltonian

$$\mathcal{H}' = \sum_{i=1}^N \frac{\mathbf{p}_i^2}{2m_i} + \sum_{i=1}^M \frac{p_{\zeta_i}^2}{2Q_i} + \phi(\mathbf{r}) + 3NkT\zeta_1 + kT\sum_{i=2}^M \zeta_i \tag{11.79}$$

We now turn to the derivation of the integrator. We start with the definition for the evolution operator, which evolves the equations of motion from time $t = 0$ to time t according to

$$\Gamma(t) = \exp(iLt)\Gamma(0) \tag{11.80}$$

$$iL = \dot{\Gamma} \cdot \nabla_\Gamma \tag{11.81}$$

In this equation, iL denotes the Liouville operator. In practice, a short-time approximation of the operator is applied P times over a timestep $\delta t = t/P$

$$\Gamma(t) = \prod_{i=1}^{P} \left(\prod_s \exp(iL_s \Delta t) \right) \Gamma(0) \tag{11.82}$$

Applying this approach to Eq. 11.78, we obtain the following expression for the Liouville operator

$$iL = \sum_{i=1}^{N} \mathbf{v}_i \cdot \nabla_{\mathbf{r}_i} + \sum_{i=1}^{N} \left[\frac{\mathbf{F}_i(\mathbf{r})}{m_i} \right] \cdot \nabla_{\mathbf{v}_i} \tag{11.83}$$

$$- \sum_{i=1}^{N} v_{\zeta_1} \mathbf{v}_i \cdot \nabla_{\mathbf{v}_i} + \sum_{i=1}^{M} v_{\zeta_i} \frac{\partial}{\partial \zeta_i} + \sum_{i=1}^{M-1} (G_i - v_{\zeta_i v_{\zeta_{i+1}}}) \frac{\partial}{\partial v_{\zeta_i}} + G_M \frac{\partial}{\partial v_{\zeta_M}} \tag{11.84}$$

with

$$G_1 = \frac{1}{Q_1} \left(\sum_{i=1}^{N} m_i \mathbf{v}_i^2 - 3NkT \right) \tag{11.85}$$

$$G_i = \frac{1}{Q_i} (Q_{i-1} v_{\zeta_{i-1}}^2 - kT) \quad i > 1 \tag{11.86}$$

Using the Trotter factorization formula, we can split the evolution operator into

$$\exp(iL\Delta t) = \exp\left(iL_{\text{NHC}} \frac{\Delta t}{2} \right) \exp\left(iL_1 \frac{\Delta t}{2} \right) \exp(iL_2 \Delta t) \exp\left(iL_1 \frac{\Delta t}{2} \right) \exp\left(iL_{\text{NHC}} \frac{\Delta t}{2} \right)$$
$$+ O(\Delta t^3) \tag{11.87}$$

where $iL_1 = \sum_i (\mathbf{f}_i/m_i) \nabla_{\mathbf{v}_i}$, $iL_2 = \sum_i \mathbf{v}_i \nabla_{\mathbf{r}_i}$, and $iL_{\text{NHC}} = iL - iL_1 - iL_2$. The Nosé-Hoover chain part of the evolution operator, $\exp(iL_{\text{NHC}} \Delta t/2)$, is given by

$$\exp\left(iL_{\text{NHC}} \frac{\Delta t}{2} \right) = \prod_{i=1}^{n_c} \exp\left(iL_{\text{NHC}} \frac{\Delta t}{2n_c} \right) \tag{11.88}$$

where

$$\exp\left(iL_{\text{NHC}} \frac{\Delta t}{2n_c} \right) = \exp\left(\frac{\Delta t}{4n_c} G_M \frac{\partial}{\partial v_{\zeta_M}} \right) \exp\left(-\frac{\Delta t}{8n_c} v_{\zeta_M} v_{\zeta_{M-1}} \frac{\partial}{\partial v_{\zeta_{M-1}}} \right)$$

$$\times \exp\left(\frac{\Delta t}{4n_c} G_{M-1} \frac{\partial}{\partial v_{\zeta_{M-1}}} \right) \exp\left(-\frac{\Delta t}{8n_c} v_{\zeta_M} v_{\zeta_{M-1}} \frac{\partial}{\partial v_{\zeta_{M-1}}} \right)$$

$$\times \exp\left(-\frac{\Delta t}{2n_c} \sum_{i=1}^{N} v_{\zeta_1} \mathbf{v}_i \cdot \nabla_{\mathbf{v}_i} \right) \exp\left(\frac{\Delta t}{2n_c} \sum_{i=1}^{M} v_{\zeta_i} \frac{\partial}{\partial \zeta_i} \right) \tag{11.89}$$

$$\times \ldots \exp\left(-\frac{\Delta t}{8n_c} v_{\zeta_M} v_{\zeta_{M-1}} \frac{\partial}{\partial v_{\zeta_{M-1}}} \right) \exp\left(\frac{\Delta t}{4n_c} G_{M-1} \frac{\partial}{\partial v_{\zeta_{M-1}}} \right)$$

$$\times \exp\left(-\frac{\Delta t}{8n_c} v_{\zeta_M} v_{\zeta_{M-1}} \frac{\partial}{\partial v_{\zeta_{M-1}}} \right) \exp\left(\frac{\Delta t}{4n_c} G_M \frac{\partial}{\partial v_{\zeta_M}} \right)$$

in which a multiple-time step approach is used ($n_c > 1$). From a practical standpoint, the integration involves first applying the operator $\exp(iL_{\text{NHC}} \Delta t/2)$ to update the $\{\zeta, \mathbf{v}_\zeta, \mathbf{v}\}$. Then, the updated velocities are used as input to the usual velocity-Verlet step before $\exp(iL_{\text{NHC}} \Delta t/2)$ is applied to the output of this velocity-Verlet step.

11.3.2 CONTROLLING PRESSURE

The multiple-time step approach can also be applied to propagate the equations of motion for a system subjected to a constant pressure and obtain a dynamics that preserves the isothermal-isobaric distribution. In this case, the equations of motion include additional integral feedback for the volume fluctuations and thus the control of pressure [187]. In the case of an uniform (*i.e.*, in the three spatial directions) constraint on pressure [245], the equations are as follows

$$\dot{\mathbf{r}}_i = \frac{\mathbf{p}_i}{m_i} + \frac{p_\varepsilon}{W}\mathbf{r}_i \tag{11.90}$$

$$\dot{\mathbf{p}}_i = \mathbf{F}_i - \left(1 + \frac{d}{3N}\right)\frac{p_\varepsilon}{W}\mathbf{p}_i - \frac{p_\zeta}{Q}\mathbf{p}_i \tag{11.91}$$

$$\dot{V} = \frac{dV p_\varepsilon}{W} \tag{11.92}$$

$$\dot{p}_\varepsilon = dV(P_{\text{system}} - P_0) + \frac{d}{3N}\sum_{i=1}^{N}\frac{\mathbf{p}_i^2}{m_i} - \frac{p_\zeta}{Q}p_\varepsilon \tag{11.93}$$

$$\dot{\zeta} = \frac{p_\zeta}{Q} \tag{11.94}$$

$$\dot{p}_\zeta = \sum_{i=1}^{N}\frac{\mathbf{p}_i^2}{m_i} + \frac{p_\varepsilon^2}{W} - (3N+1)kT \tag{11.95}$$

in which P_0 denotes the target value for the pressure and the pressure inside the system, P_{system}, is given by

$$P_{\text{system}} = \frac{1}{dV}\left[\sum_{i=1}^{N}\frac{\mathbf{p}_i^2}{m_i} + \sum_{i=1}^{N}\mathbf{r}_i \cdot \mathbf{F}_i - (dV)\frac{\partial\phi(\mathbf{r},V)}{\partial V}\right] \tag{11.96}$$

Here, p_ε denotes the momentum associated with the logarithm of the volume, and W is the "barostat" parameter. This leads to the following energy being conserved, with the corresponding Hamiltonian below

$$\mathcal{H}' = \sum_{i=1}^{N}\frac{\mathbf{p}_i^2}{2m_i} + \frac{p_\varepsilon^2}{2W} + \frac{p_\zeta^2}{2Q} + \phi(\mathbf{r},V) + (3N+1)kT\zeta + P_{\text{ext}}V \tag{11.97}$$

Alternatively, if the shape of the cell is also allowed to fluctuate [258,284] so that the non-diagonal elements of the pressure tensor are constrained to remain equal to 0 and the diagonal elements are kept fixed at the target pressure P_0, the equations of motions become

$$\dot{\mathbf{r}}_i = \frac{\mathbf{p}_i}{m_i} + \frac{\overleftrightarrow{\mathbf{p}}_g}{W_g}\mathbf{r}_i \tag{11.98}$$

$$\dot{\mathbf{p}}_i = \mathbf{F}_i - \frac{\overleftrightarrow{\mathbf{p}}_g}{W_g}\mathbf{p}_i - \frac{1}{N_f}\frac{Tr[\overleftrightarrow{\mathbf{p}}_g]}{W_g}\mathbf{p}_i - \frac{p_\zeta}{Q}\mathbf{p}_i \tag{11.99}$$

$$\dot{\overleftrightarrow{\mathbf{h}}} = \frac{\overleftrightarrow{\mathbf{p}}_g\overleftrightarrow{\mathbf{h}}}{W_g} \tag{11.100}$$

$$\dot{\overleftrightarrow{\mathbf{p}}}_g = V(\overleftrightarrow{\mathbf{P}}_{\text{system}} - \overleftrightarrow{\mathbf{I}}P_{\text{ext}}) + \left[\frac{1}{N_f}\sum_{i=1}^{N}\frac{\mathbf{p}_i^2}{m_i}\right]\overleftrightarrow{\mathbf{I}} - \frac{p_\zeta}{Q}\overleftrightarrow{\mathbf{p}}_g \tag{11.101}$$

$$\dot{\zeta} = \frac{p_\zeta}{Q} \tag{11.102}$$

$$\dot{p}_\zeta = \sum_{i=1}^{N}\frac{\mathbf{p}_i^2}{m_i} + \frac{1}{W_g}Tr[\overleftrightarrow{\mathbf{p}}_g^t\overleftrightarrow{\mathbf{p}}_g] - (3N+d^2)kT \tag{11.103}$$

where $\overset{\leftrightarrow}{\mathbf{h}}$ denotes the cell matrix, V the volume $V = \det[\overset{\leftrightarrow}{\mathbf{h}}]$, $\overset{\leftrightarrow}{\mathbf{I}}$ the identity matrix, and $Tr[\overset{\leftrightarrow}{\mathbf{p}}{}_g^t \overset{\leftrightarrow}{\mathbf{p}_g}]$ the sum of the squares of all the elements of $\overset{\leftrightarrow}{\mathbf{p}}_g$, the cell variable momentum matrix. The pressure tensor is defined as

$$(P_{\text{system}})_{\alpha\beta} = \frac{1}{V}\left[\sum_{i=1}^{N} \frac{(\mathbf{p}_i)_\alpha (\mathbf{p}_i)_\beta}{m_i} + (\mathbf{F}_i)_\alpha (\mathbf{r_i})_\beta - (\overset{\leftrightarrow}{\phi'}\,\overset{\leftrightarrow}{\mathbf{h}}{}^t)_{\alpha\beta}\right] \tag{11.104}$$

$$(\phi')_{\alpha\beta} = \frac{\partial \phi(\mathbf{r}, \overset{\leftrightarrow}{\mathbf{h}})}{\partial (h)_{\alpha\beta}} \tag{11.105}$$

and the energy conserved during the dynamics is given by the following Hamiltonian

$$\mathcal{H}' = \sum_{i=1}^{N} \frac{\mathbf{p}_i^2}{2m_i} + \frac{1}{2W_g} Tr[\overset{\leftrightarrow}{\mathbf{p}}{}_g^t \overset{\leftrightarrow}{\mathbf{p}}_g] + \frac{p_\zeta^2}{2Q} + \phi(\mathbf{r}, \overset{\leftrightarrow}{\mathbf{h}}) + P_{\text{ext}} \det[\overset{\leftrightarrow}{\mathbf{h}}] + (3N + d^2)kT\zeta \tag{11.106}$$

For both dynamics, the equations of motion can then be supplemented by chains of thermostats instead of a single thermostat and integrated with a multiple-time step algorithm similar to that outlined in the canonical case [245].

11.3.3 HYBRID SCHEMES

Integrating the equations of motion can also be used in a hybrid simulation approach known as Hybrid Monte Carlo (HMC) simulations. In this case, the Monte Carlo random moves corresponding to molecules' translation, rotation, or conformational changes are replaced by short molecular dynamics trajectories performed in the microcanonical ensemble. This often allows for very efficient sampling of the configurational space. For instance, in a dense molecular liquid, the acceptance probability of random translations or rotations of large molecules is very close to zero. Similarly, the acceptance probability of random changes in the molecular conformation of chain molecules also becomes extremely small in condensed phases. This hybrid approach was initially proposed by Mehlig et al. [255]. If a "molecular dynamics trajectory" move is attempted from an initial configuration Γ to a new configuration Γ', the Metropolis criterion for this type of move is given by

$$\text{acc}(\Gamma \to \Gamma') = \min\left[1, \exp\left(-\frac{(E(\Gamma') - E(\Gamma))}{k_B T}\right)\right] \tag{11.107}$$

However, the integration scheme to compute the short molecular dynamics trajectory must satisfy two criteria. Integrating the equations of motion must be carried out with a molecular dynamics algorithm that is both time-reversible and area-preserving. Examples of such integrators include, for instance, the case in atomic fluids [85], the velocity-Verlet integrator. For rigid molecules [90], the integrator may consist of a velocity-Verlet integrator for the equations of motion for the center of mass and an integration scheme, such as those proposed by Matubayasi and Nakahara [247], for the equations of motion for the rotation. Such schemes rely on a Trotter factorization of the evolution operator. This, in turn, ensures that the integrator is time reversible in addition to being area preserving. For flexible and chain molecules [90], the multiple time step RESPA-NVE algorithm [245] provides an integration scheme that is both time-reversible and area-preserving. In practice, forces exerted on each atom are split between two terms. The first term $\mathbf{F}_1$ contains the forces due to the high-frequency intramolecular vibrations, which include stretching, bending, and torsion. The second term $\mathbf{F}_2$ encompasses the intermolecular forces. Following the Trotter factorization, the integration is carried out in three successive steps. First, velocities are advanced over a half-time step $\delta t/2$ by applying only the force $\mathbf{F}_2$. Second, velocities and positions are advanced n times over a small time step $\delta t' = \delta t/n$ by applying the force $\mathbf{F}_1$. Third, velocities are again advanced over a half-time step $\delta t/2$ using the force $\mathbf{F}_2$.

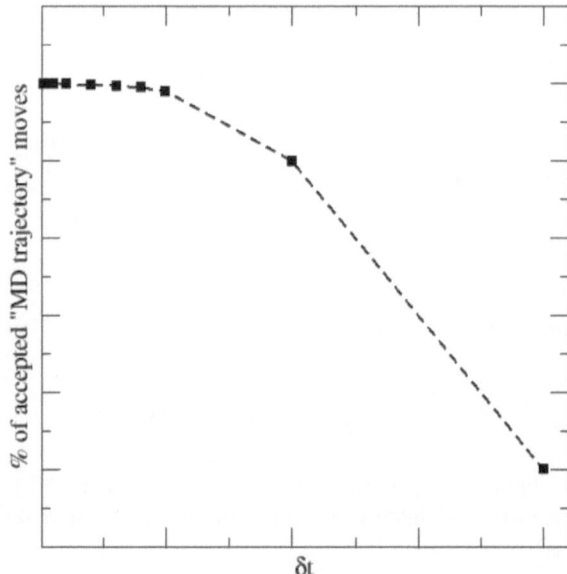

Figure 11.7 Dependence of the acceptance rate for MD moves as a function of the timestep δt. Shorter timesteps ensure excellent energy conservation and thus high acceptance rates, while longer timesteps lead to a less accurate integration and lower acceptance rates for the MD moves.

Once one of these integrators has been implemented, the acceptance rule in Eq. 11.107 ensures that the HMC simulations sample the canonical ensemble. Generally, shorter timesteps will ensure better energy conservation and, thus, higher acceptance rates (see Figure 11.7). If the HMC simulation needs to sample the isothermal-isobaric ensemble instead, a second type of random move must be performed to allow the volume to change and pressure to be controlled. In this case, random changes in the system's volume can be implemented with the same Metropolis acceptance rule as the one employed in conventional MC simulations in the (N, P, T) ensemble.

12 Noise and Information
Correlation Functions

"A general type of fluctuation-dissipation theorem is discussed to show that the physical quantities such as (...) complex conductivity for electric conduction are rigorously expressed in terms of time-fluctuation of dynamical variables associated with such irreversible processes."

Ryogo Kubo,
Statistical-mechanical theory of irreversible processes [212]

In Chapter 11, we saw how the ergodic hypothesis allows us to connect ensemble averages, taken over an extensive collection of molecular systems, to time averages obtained by computing molecular trajectories over long times. If determining average properties is the main goal when drawing statistics on a molecular system, then both approaches are essentially equivalent and will be a matter of efficiency. On the one hand, Monte Carlo is ideally suited to design random moves not bound by the equations of motion. It thus can significantly accelerate the exploration of the configurational space. We have discussed, for instance, the insertion of a molecule at a random location in the system in grand-canonical Monte Carlo simulations or identifying exchange moves in semi-grand Monte Carlo simulations. On the other hand, integrating the equations of motion performed during molecular simulations can be incredibly efficient in simulating changes in the conformation of complex molecules, biomolecules, and polymers or concerted motions involving different molecules in condensed phases. While both approaches can be used interchangeably to calculate thermodynamic or structural properties, molecular dynamics simulations also provide information on the system's dynamics. Such results come with additional work through the computation of forces and time-evolved molecular trajectories. However, the insights provided by gathering statistics on the dynamical features of the molecular system are incredibly potent. For instance, quantifying the time evolution of collective properties, such as the overall motion for one type of ions, or the overall shear stress in the system, provides access to transport coefficients and, in these two examples, conductivity and viscosity. This is the case even for a system under stationary conditions and thus at equilibrium. While the time-dependent behavior of properties and their fluctuations around the average value may appear unusable noise, it contains precious information when unraveled through the computation of time-correlation functions, as discussed in this chapter.

12.1 MOTION AND TRANSPORT

12.1.1 BROWNIAN MOTION

The botanist Robert Brown initially observed the existence of constant and irregular motion in suspensions or fluids during the first part of the 19th century. In his 1828 paper [37], Brown describes the straightforward setup used for the experiments: "The observations, of which it is my object to give a summary in the following pages, have all been made with a simple microscope, and indeed with the same lens, the focal length of which is about 1/32nd of an inch". He then reports taking pollen grains from the plant *Clarckia pulchella* and immersing them in water. Brown observes that "many of them very evidently in motion" and that "in a few instances the particle was seen to turn on its longer axis". He checks that these motions "arose neither from currents in the fluid, nor from its gradual evaporation, but belonged to the particle itself." This motion became later known as Brownian motion, and a typical trajectory is shown in Figure 12.1. Brown also verified that the same behavior was observed in aqueous solutions of various biological and mineral extracts.

DOI: 10.1201/9781003006411-15

Figure 12.1   Example of Brownian motion, with the trajectory for the Brownian particle shown here as a line.

As discussed by Zwanzig [407], the theory of Brownian motion applies to many other systems, such as *e.g.*, to the motion of ions in water, but also abstract functions and collective properties of molecular system, such as *e.g.* concentrations.

Despite the apparent random nature of the trajectory shown in Figure 12.1, Hamilton's and Newton's equations of motion apply to a Brownian particle or function, just as they apply to any purely deterministic object. Focusing on the example of a one-dimensional (1D) for a particle or radius a and mass m, Newton's equation for the Brownian particle is as follows

$$m\frac{dv}{dt} = F_{\text{tot}}(t) \tag{12.1}$$

where x denotes the position of the Brownian particle, v its velocity, η the viscosity of the fluid in which the particle is immersed, and $F_{\text{tot}}(t)$ is the total instantaneous force on the particle at time t. $F_{\text{tot}}(t)$ arises from the interactions between the Brownian particle and the molecules composing the fluid in which the particle is immersed. If the positions of the molecules are known for all values of t, F_{tot} can be computed and is thus not truly a "random force". As a first approximation, we can interpret F_{tot} as a frictional force $-\zeta v$ proportional to the velocity v. From Stokes' law, we have $\zeta = 6\pi\eta a$ and Eq. 12.1 can now be written as

$$m\frac{dv}{dt} = -\zeta v \tag{12.2}$$

Solving this equation provides the following solution for $v(t)$

$$v(t) = e^{-\zeta t/m} v(0) \tag{12.3}$$

which implies that the velocity of the Brownian particle exhibits an exponential decay to 0 when $t \to \infty$. However, this is incompatible with the experimental observation that the mean squared velocity of the particle is such that $< v^2 >_{eq} = k_B T/m$ at thermal equilibrium. This calls into question the approximation that $F_{tot}(t)$ is only a frictional force.

12.1.2 LANGEVIN EQUATION & FLUCTUATION-DISSIPATION

The next step consists of supplementing the frictional term with a fluctuating or noise term corresponding to the Brownian particle's interaction with the environment or heat bath. The resulting equation is called the Langevin equation, given by

$$m \frac{dv}{dt} = -\zeta v + F_f(t) \tag{12.4}$$

We can interpret the fluctuating force $F_f(t)$ as arising from the successive collisions between the Brownian particle and the molecules of the fluid in which the particle is immersed. The fluctuating force has a Gaussian distribution determined by its first and second moments, which, when averaged over an infinitesimal time interval, are equal to

$$\langle F_f(t) \rangle = 0 \tag{12.5}$$

$$\langle F_f(t).F_f(t') \rangle = 2B\delta(t - t') \tag{12.6}$$

where B is the characteristic strength of the fluctuating force, the delta function indicates no correlation between impacts in the infinitesimal time intervals around two distinct times t and t'. In other words, we have $\delta(t' - t) = 0$ if $t' \neq t$.

Solving the Langevin equation provides the following expression for $v(t)$

$$v(t) = e^{-\zeta t/m} v(0) + \int_0^t \frac{F_f(t')}{m} e^{-\zeta(t-t')/m} dt' \tag{12.7}$$

On the right-hand side of the equation, the first term, *i.e.* the exponential decay from the initial velocity, corresponds to the effect of the frictional force. In contrast, the second term corresponds to the contribution of the fluctuating force or random noise. We now examine how adding the fluctuating force satisfies the condition imposed by thermal equilibrium for the mean-squared velocity $< v^2 > = k_B T/m$. Let us now see the effect of the fluctuating force on the mean-squared velocity. We start from $v^2(t)$ using Eq. 12.7

$$< v^2(t) > = e^{-2\zeta t/m} v^2(0) + 2v(0)e^{-\zeta t/m} < \int_0^t \frac{F_f(t')}{m} e^{-\zeta(t-t')/m} dt' >$$
$$+ \frac{1}{m^2} < \int_0^t F_f(t') m e^{-\zeta(t-t')/m} dt' \int_0^t F_f(t'') e^{-\zeta(t-t'')/m} dt'' > \tag{12.8}$$

Since the first moment for the fluctuating force is 0, the second term in this equation vanishes. The third term can be rearranged into

$$\frac{1}{m^2} \int_0^t e^{-\zeta(t-t')/m} \int_0^t < F_f(t')F_f(t'') > e^{-\zeta(t-t'')/m} dt'' dt' \tag{12.9}$$

which, using the second moment for the fluctuating force, leads to

$$\frac{1}{m^2} \int_0^t e^{-\zeta(t-t')/m} \int_0^t e^{-\zeta(t-t'')/m} 2B\delta(t' - t'') dt'' dt' \tag{12.10}$$

and thus to

$$\frac{2B}{m^2} \int_0^t e^{-2\zeta(t-t')/m} dt' \tag{12.11}$$

Overall, we obtain the mean-squared velocity

$$\langle v(t)^2 \rangle = e^{-2\zeta t/m} v(0)^2 + \frac{B}{\zeta m}(1 - e^{-2\zeta t/m}) \tag{12.12}$$

When $t \to \infty$, we obtain

$$\langle v(t)^2 \rangle = \frac{B}{\zeta m} \tag{12.13}$$

and thus, since for very large t, the system has reached thermal equilibrium, resulting in the following condition for B

$$B = \zeta kT \tag{12.14}$$

Equation 12.14 is the fluctuation-dissipation theorem. The theorem captures the relation between friction, through ζ, and the fluctuating force or random noise, through B. As emphasized by Zwanzig [407], friction results in an exponential decay of velocity and eventually drives the system to "a completely dead state", while noise maintains a non-zero mean-squared velocity and thus "keeps the system alive".

12.1.3 EINSTEIN DIFFUSION EQUATION

After focusing on the mean squared velocity, we now examine the mean squared displacement for the Brownian particle. To this end, we start with the solution to the Langevin equation. We note as $\Delta x(t)$ the 1D-displacement of the particle along the axis x as a function of time and calculate it by integrating $v(t)$ with respect to time

$$\Delta x(t) = \int_0^t dt' v(t') \tag{12.15}$$

where

$$v(t) = e^{-\zeta t/m} v(0) + \int_0^t dt' e^{-\zeta(t-t')/m} F_f(t')/m \tag{12.16}$$

We follow the same reasoning as for the mean-squared velocity and obtain the mean-squared displacement $\langle \Delta x(t)^2 \rangle$

$$\langle \Delta x(t)^2 \rangle = 2\frac{k_B T}{\zeta} \left[t - \frac{m}{\zeta} + \frac{m}{\zeta} e^{-\zeta t/m} \right] \tag{12.17}$$

For $t \to 0$, we can expand the exponential around 0 to obtain

$$e^{-\zeta t/m} = 1 - \frac{\zeta t}{m} + \frac{\zeta^2 t^2}{2m^2} + \mathcal{O}(t^3) \tag{12.18}$$

and thus, at short times, the $\langle \Delta x(t)^2 \rangle$ exhibits a ballistic behavior with a t^2 time-dependence

$$\langle \Delta x(t)^2 \rangle = \frac{k_B T \zeta t^2}{m^2} + \mathcal{O}(t^3) \tag{12.19}$$

When $t \to \infty$, the noise term becomes predominant, and $\langle \Delta x(t)^2 \rangle$ is a linear function of t

$$\langle \Delta x(t)^2 \rangle \to 2\frac{kT}{\zeta} t + \mathcal{O}(1) \tag{12.20}$$

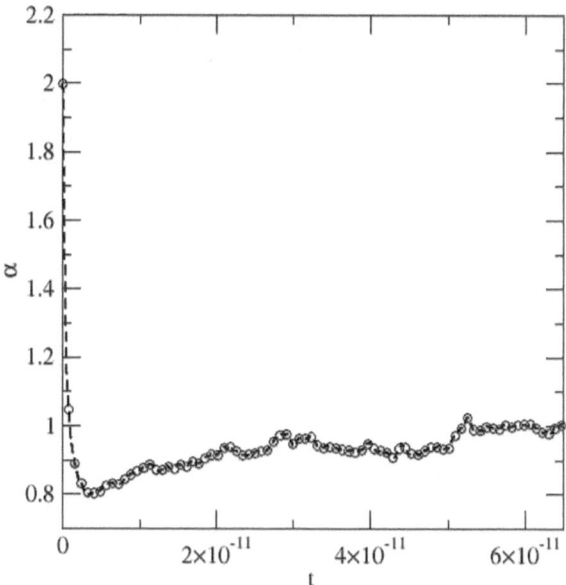

Figure 12.2 Behavior of the mean-squared displacement $\langle \Delta x(t)^2 \rangle$. If $\langle \Delta x(t)^2 \rangle \propto t^\alpha$, the quantity $\frac{d\ln\langle \Delta x(t)^2\rangle}{d\ln t}$ provides directly the value for the exponent α. The plot shows the two limiting behaviors observed for $t \to 0$ (ballistic) and $t \to \infty$ (diffusive). When $t \to 0$, the plot yields $\alpha \to 2$. On the other hand, when $t \to \infty$, $\langle \Delta x(t)^2 \rangle$ varies linearly with time.

Figure 12.2 shows the time evolution of the mean-squared displacement, with the two limiting behaviors observed for $t \to 0$ and $t \to \infty$. Einstein defines D, the self-diffusion coefficient of the Brownian particle, through

$$\langle \Delta x(t)^2 \rangle = 2Dt \tag{12.21}$$

which, in turn, yields Einstein's formula for the self-diffusion coefficient, given by

$$D = \frac{kT}{\zeta} \tag{12.22}$$

This equation is used in conjunction with Stokes' law for ζ, which gives rise to the well-known Stokes-Einstein formula.

12.2 TRANSPORT FROM CORRELATION

12.2.1 *D* FROM A CORRELATION FUNCTION

Correlation functions are powerful tools for analyzing the behavior of molecular systems. Correlation functions can quantify spatial correlations, which are measures of the spatial interdependence between atoms and molecules, and provide microscopic insight into the structure of atomic and molecular systems. The pair correlation function is the most straightforward spatial correlation function, corresponding to the probability of finding two atoms at a given distance. For instance, if we consider two atoms, place the first atom at the origin, and denote by $\mathbf{r}$ the vector joining the two atoms, the two-body (pair) correlation function $g(r)$ corresponds to the probability of finding the second atom within a small interval $d\mathbf{r}$ around a set distance $r = ||\mathbf{r}||$. $g(r)$ can be calculated over a series of configurations of the system, generated either through Monte Carlo simulations or through the computation of molecular trajectories, by counting how often the distance between two atoms is equal to r

$$g(r) = \frac{V}{N^2} < \sum_i \sum_{j \neq i} \delta(\mathbf{r} - \mathbf{r}_{ij}) > \qquad (12.23)$$

An example of a pair correlation function, or pair distribution function, is shown in Figure 12.3. We can also analyze the behavior of the system through time correlation functions. In this case, the idea is to analyze the statistical behavior of a time-dependent quantity $B(t)$ over the computed molecular trajectory. The quantity $B(t)$ can be the velocity of a single atom or a collective property of the system, such as, for instance, the shear stress. We first define the fluctuation $\delta B(t)$ as

$$\delta B(t) = B(t) - \langle B \rangle \qquad (12.24)$$

in which $\langle B \rangle$ denotes the time average for $B(t)$, obtained as follows over a molecular trajectory of duration τ

$$\langle B \rangle = \frac{1}{\tau} \int_0^\tau dt B(t) \qquad (12.25)$$

The time-correlation function is defined as the time average of the product of fluctuations at different times. It is given by

$$C(t) = \frac{1}{\tau} \int_0^\tau ds \delta B(s).\delta B(t+s) \qquad (12.26)$$

The time-correlation function measures how fluctuations at different times are correlated (see Figure 12.4). This is the equivalent of time-dependent information to how spatial correlation functions like $g(r)$ measure correlations between atoms at different positions. Since the system is at equilibrium, choosing a specific time for the origin has no impact on the results. Thus, from a practical standpoint, the time-correlation function is calculated as an ensemble average of the product of two fluctuations rather than through its long-time average. This is achieved by taking many different time origins and computing the correlation product for a set time difference between the two fluctuations. We add that experiments tend to provide the Fourier transform of time correlations rather than the time correlation itself. The relation between the two is defined below, with a Fourier transform of $C(t)$ taking the following form

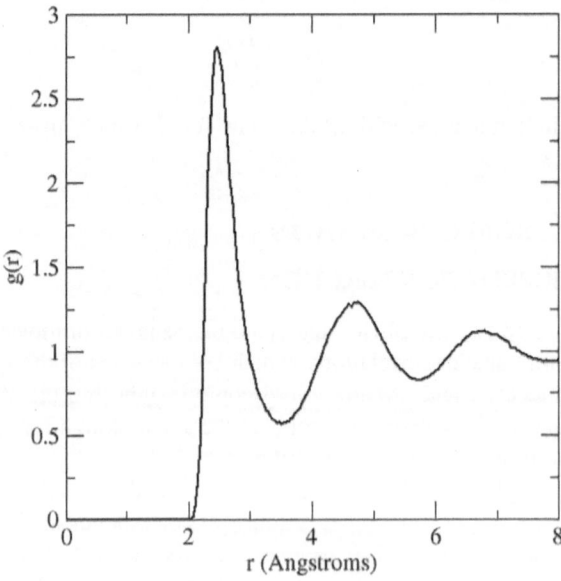

Figure 12.3 Example of a radial distribution function. The peaks correspond to distances for which the probability of finding another atom is especially high.

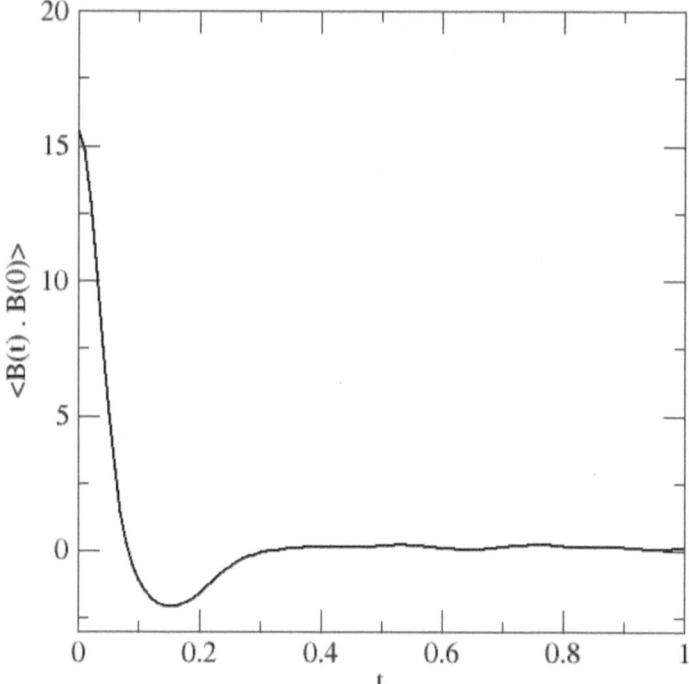

Figure 12.4 Example of a time-correlation function $< B(t) \cdot B(0) >$ for a phase variable $B(t)$. Fluctuations are highly correlated for short times, resulting in large correlation products and large values for the time-correlation function. Fluctuations become uncorrelated at long times, and the time-correlation function converges toward 0.

$$C_\omega = \int_{-\infty}^{\infty} dt\, e^{-i\omega t} C(t) \tag{12.27}$$

where ω denotes the frequency and the Fourier transform C_ω the spectral density.

Time-correlation functions provide a direct access to transport coefficients. For instance, computing the time-correlation function for the velocity of a particle yields the self-diffusion coefficient D. To show this, we start from the equation for net displacement of the position of the particle over the time interval $(0,t)$

$$x(t) = \int_0^t ds\, v(s) \tag{12.28}$$

By taking the square of the displacement and performing an ensemble average, we obtain

$$\langle x^2 \rangle = \left\langle \int_0^t ds_1 v(s_1) \int_0^t ds_2 v(s_2) \right\rangle = \int_0^t ds_1 \int_0^t ds_2 \langle v(s_1)v(s_2) \rangle \tag{12.29}$$

Differentiating this equation with respect to time and combining the two terms on the right-hand side, we have

$$\frac{\partial}{\partial t} \langle x^2 \rangle = 2 \int_0^t ds \langle v(t)v(s) \rangle \tag{12.30}$$

Making the change in variable $u = t - s$ gives

$$\frac{\partial}{\partial t} \langle x^2 \rangle = 2 \int_0^t ds \langle v(t-s)v(0) \rangle = 2 \int_0^t du \langle v(u)v(0) \rangle \tag{12.31}$$

Since for $t \to \infty$, $\frac{\partial}{\partial t} \langle x^2 \rangle = 2D$, we obtain the following relation between the self-diffusion coefficient and the velocity correlation function for a one-dimensional system

$$D = \int_0^\infty dt \, \langle v(t)v(0) \rangle \tag{12.32}$$

For a three-dimensional system, this equation generalizes to

$$D = \frac{1}{3} \int_0^\infty dt \, \langle \mathbf{v}(t) \cdot \mathbf{v}(0) \rangle \tag{12.33}$$

where $\mathbf{v}$ is the vector velocity.

12.2.2 THE MORI-ZWANZIG APPROACH

The correspondence between time correlation functions and transport coefficients goes well beyond the example of diffusion. Mori and Zwanzig introduced a formalism, based on the use of projection operators, that enabled this generalization to a wide range of transport phenomena [262,406]. To provide a brief account of the formalism, we use the operator algebra for phase variables we started to introduce in Chapter 11 with the Liouville operator [22,194,198]. The Mori-Zwanzig approach consists of introducing projection operators that apply to phase variables. The idea is to split the phase variable B_2 into two parts. The first part contains the fraction of B_2 that is correlated with, or colinear with, B_1. The second part contains the fraction that is normal to or has no correlation with B_1. We define the projection operator $\mathbf{P}$ that projects B_2 onto a vector colinear with B_1 as

$$\mathbf{P}B_2 = \frac{(B_2, B_1)}{(B_1, B_1)} \cdot B_1 \tag{12.34}$$

and the projector $\mathbf{Q} = 1 - P$ that projects B_2 onto a vector normal to, and thus random with respect to, B_1 as shown by

$$\begin{aligned}
(\mathbf{Q}B_2, B_1) &= \left(B_2 - \frac{(B_2, B_1)}{(B_1, B_1)} \cdot (B_1, B_1) \right) \\
&= (B_2, B_1) - \frac{(B_2, B_1)}{(B_1, B_1)} \cdot (B_1, B_1) \\
&= 0
\end{aligned} \tag{12.35}$$

The idea is to apply projections operators and partition the Liouville operator into

$$L = \mathbf{P}L + (1 - \mathbf{P})L \tag{12.36}$$

and introduce the following operator identity [407], or Dyson decomposition [198],

$$e^{tL} = e^{t(1-\mathbf{P})L} + \int_0^t ds \, e^{(t-s)L} \mathbf{P}L e^{s(1-\mathbf{P})L} \tag{12.37}$$

Here we have dropped the imaginary factor i in the notation of the Liouville operator since both notations can be used interchangeably [407]. Applying the operator identity to a $\mathbf{Q}$-projection of phase variable B, or $(1 - \mathbf{P})LB$, we obtain for the left-hand side

$$\begin{aligned}
e^{tL(1-\mathbf{P})}LB &= e^{tL}LB - e^{tL}\mathbf{P}LB \\
&= \frac{\partial}{\partial t} e^{tL} B - e^{tL} \cdot \frac{(LB, B)}{(B, B)} \cdot B \\
&= \frac{\partial}{\partial t} B(t) - \frac{(LB, B)}{(B, B)} \cdot B(t) \\
&= \frac{\partial}{\partial t} B(t) - i\Omega \cdot B(t)
\end{aligned} \tag{12.38}$$

where we define the frequency Ω such that $i\Omega = \frac{(LB,B)}{(B,B)}$. Introducing $F(t) = e^{t(1-\mathbf{P})L}(1-\mathbf{P})LB$ as the uncorrelated or "noise" term, the right-hand side of the operator identity applied to $(1-\mathbf{P})LB$ becomes

$$F(t) + \int_0^t ds\, e^{(t-s)L} \frac{(LF(s),B)}{(B,B)} \cdot B = F(t) + \int_0^t ds \frac{(LF(s),B)}{(B,B)} \cdot B(t-s) \qquad (12.39)$$

We finally define the kernel $\mathbf{K}(t)$ as

$$\begin{aligned}
\mathbf{K}(t) &= -\frac{(LF(t),B)}{(B,B)} \\
&= \frac{(F(t),LB)}{(B,B)} \\
&= \frac{(e^{t(1-\mathbf{P})L}(1-\mathbf{P})LB,LB)}{(B,B)}
\end{aligned} \qquad (12.40)$$

where we have used that the operator L is anti-Hermitian. Finally, after equating the two sides of the equation and rearranging the terms, we obtain

$$\frac{\partial}{\partial t}B(t) = i\Omega \cdot B(t) - \int_0^t ds\, \mathbf{K}(s) \cdot B(t-s) + F(t) \qquad (12.41)$$

This equation can be interpreted as a generalization of the Langevin equation. We add that, in this generalized Langevin equation, the integral is non-Markovian since it involves a correlation product of two functions taken at two different times, *i.e.*, s and $(t-s)$ for the kernel $\mathbf{K}$ and the phase variable B, respectively. As a result, the kernel $\mathbf{K}$ is often called a memory kernel.

12.2.3   EVALUATION OF TRANSPORT COEFFICIENTS

As discussed in this section, for each phase variable B, the memory kernel $\mathbf{K}$ can be written as a time-correlation function and provides a formula for the transport coefficient associated with B. Since many phase variables vary slowly with time, we can note that

$$\frac{\partial}{\partial t}B(t) = LB(t) = \mathscr{O}(\lambda) \qquad (12.42)$$

where λ is a parameter that takes small values. We then analyze the generalization of the Langevin equation, provided in Eq. 12.41. For $\lambda \to 0$, we can replace the non-Markovian integral with its Markovian equivalent to obtain

$$\frac{\partial}{\partial t}B(t) = i\Omega \cdot B(t) - \int_0^t ds\, \mathbf{K}(s) \cdot B(s) + F(t) + \mathscr{O}(\lambda^3 B) \qquad (12.43)$$

since the term $i\Omega$ includes LB and thus is first-order in λ and $\mathbf{K}$ is second-order in λ. We return to the equation for the kernel $\mathbf{K}$ and rearrange it into a time correlation in two steps

$$\begin{aligned}
\mathbf{K}(t) &= \frac{(e^{t(1-\mathbf{P})L}(1-\mathbf{P})LB,LB)}{(B,B)} \\
&= \frac{(e^{t(1-\mathbf{P})L}(1-\mathbf{P})LB,(1-\mathbf{P})LB)}{(B,B)} \\
&= \frac{(e^{tL}(1-\mathbf{P})LB,(1-\mathbf{P})LB)}{(B,B)} + \mathscr{O}(\lambda^3)
\end{aligned} \qquad (12.44)$$

Since $1-\mathbf{P}$ appears on the left of both factors in the left-hand side of the product, we can apply, during the first step, the $1-\mathbf{P}$ operator to the right-hand side of the product without changing the result. In the second step, we utilize the fact that the difference between the projected dynamics $e^{t(1-P)L}$ and the ordinary dynamics e^{tL} is of the order of $\mathcal{O}(\lambda)$. The final result in Eq. 12.44 shows that the time-correlation function for $(1-P)LB)$, generated using the ordinary dynamics, is at the center of the kernel definition.

We now examine how the diffusion coefficient D can be obtained from a memory kernel. D is the transport coefficient associated with a change in the tagged particle's concentration $C(x,t)$ along the x-axis. Next, we take the spatial Fourier transform of the concentration to obtain

$$C(x,t) = \sum_q A_q(t) e^{iqx} \tag{12.45}$$

where $A_q = e^{-iqR}$ in which R denotes the position of the tagged particle. Calculating the rate of change of the Fourier components, we obtain

$$LA_q = -iq\dot{R}e^{iqR} = -iqV + \mathcal{O}(q^2) \tag{12.46}$$

where V is the velocity of the tagged particle. The $i\Omega$ term is the average of an odd function of the velocity and, as a result, vanishes, while the memory kernel involves the time-correlation function for the velocity through

$$K(t) = -q^2 <Ve^{tL}V> + \mathcal{O}(q^3) \tag{12.47}$$

In the small q limit, A_q varies slowly, and we obtain the following equation, following a Markovian approximation,

$$\frac{d}{dt}A_q(t) = -q^2 \int_0^\infty ds <V(s)V> A_q(t) + \mathcal{O}(q^3) \tag{12.48}$$

This equation defines the Fourier components obtained from the Fourier transform of the diffusion equation

$$\frac{d}{dt}C(x,t) = D\nabla^2 C(x,t) \tag{12.49}$$

leading to the equation relating D to the velocity time-correlation function. For instance, the same approach can be applied to the x-component of the momentum density corresponding to the product ρv_x, where ρ denotes the density. In this case, the kernel is obtained through a Fourier transform of the Navier-Stokes equation defining the viscosity η

$$\rho\frac{\partial}{\partial t}v_x = \eta\frac{\partial^2}{\partial v^2}v_x \tag{12.50}$$

leading to the following relation between shear viscosity and the xy-components of the molecular stress tensor $\mathbf{P}_{xy}$ according to

$$\eta = \frac{1}{Vk_BT}\int_0^\infty \langle \mathbf{P}_{xy}(t) \cdot \mathbf{P}_{xy}(0) \rangle \tag{12.51}$$

From a practical standpoint, the integral of the shear stress time-correlation function (see left panel of Figure 12.5) is calculated using a finite time as the upper bound in the integral, according to

$$\eta(\tau) = \frac{1}{Vk_BT}\int_0^\tau \langle \mathbf{P}_{xy}(t) \cdot \mathbf{P}_{xy}(0) \rangle \tag{12.52}$$

As shown in the right panel of Figure 12.5, we vary the time τ, which corresponds to the upper bound in the integral, and obtain the shear viscosity when the integral reaches a plateau.

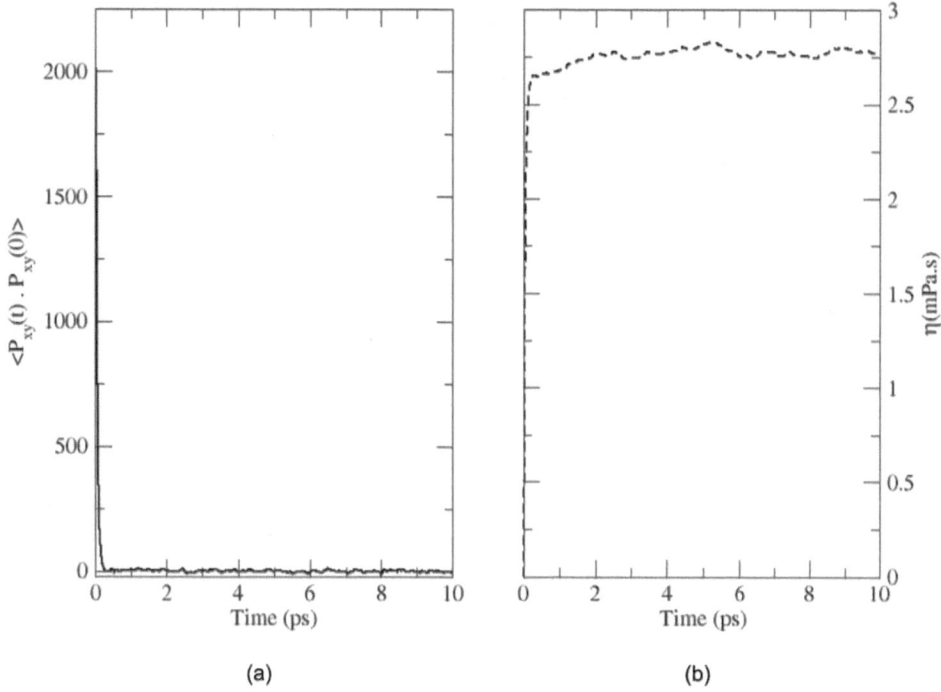

Figure 12.5 (a) Time-correlation function $< P_{xy}(t) \cdot P_{xy}(0) >$ for the shear stress in an atomic liquid [87] (*Pt* at 3375 K and 16.2 g/cm^3). (b) Viscosity obtained after integration of the time-correlation function for the shear stress (Eq. 12.52).

12.3   RESPONSE THEORY

12.3.1   LINEAR RESPONSE THEORY

Onsager prefigured the identification between memory kernels and transport coefficients in his well-known regression hypothesis [275,276]. Onsager hypothesized that equilibrium fluctuations for a phase variable and changes in the same phase variable during its relaxation after a perturbation relied on the same transport coefficient. In other words, when looking at the time-dependent behavior of a phase variable over a short molecular trajectory, it is impossible to say if the system is already at equilibrium or if it returns to equilibrium after a small perturbation. This means that we can also think of extracting transport coefficients from equilibrium time-correlation functions or measuring how the system responds to a perturbation such as, for instance, an external field.

We first examine the response of a system to a small perturbation, which does not vary with time. When an external field of intensity E acts on a molecular system, it changes the Hamiltonian of the system, which becomes $\mathcal{H} = \mathcal{H}_0 - M \cdot E$. Here, $\mathcal{H}_0$ denotes the unperturbed Hamiltonian for the system in the absence of any external field, M is a characteristic function of the system, *e.g.*, an electric dipole or magnetic moment, involved in the coupling with the external field through the extra term $-M \cdot E$. When the external field acts on the system, the average of a phase variable B is given by

$$< B >_E = \int d\Gamma B f_E(\Gamma) = \int d\Gamma B \frac{e^{-\beta \mathcal{H}_0 + \beta M \cdot E}}{Q_E} \tag{12.53}$$

where $f_E(\Gamma)$ is the perturbed distribution function, and $Q_E = \int d\Gamma e^{-\beta \mathcal{H}_0 + \beta M \cdot E}$. To obtain the linear response to the field, we carry out a first-order expansion of the numerator and denominator of $f_E(\Gamma)$ to obtain the numerator

$$e^{-\beta \mathscr{H}_0 + \beta M \cdot E} = \left(1 + \beta M \cdot E + \mathscr{O}(E^2)\right) e^{-\beta \mathscr{H}_0} \tag{12.54}$$

and for the denominator

$$Q_E = \left(1 + \beta < M >_0 \cdot E + \mathscr{O}(E^2)\right) Q \tag{12.55}$$

$< .. >_0$ indicates that the average is calculated using the unperturbed distribution. Thus, the perturbed distribution function is given by

$$f_E(\Gamma) = (1 + \beta [M - < M >_0] \cdot E) f + \mathscr{O}(E^2) \tag{12.56}$$

Assuming that $< M >_0 = 0$, we obtain for the phase variable average $< B >_E$

$$< B >_E = < B >_0 + \beta < B \cdot M > E + \mathscr{O}(E^2) \tag{12.57}$$

The average linear response of the system to the field is thus given by $\beta < B \cdot M >$. $< B \cdot M >$ measures the system's susceptibility to the perturbation. We can use this equation to determine transport coefficients through choices for the phase variable B and the perturbation E. Under this type of steady-state nonequilibrium setup, transport coefficients will be provided by computing the ratio of the perturbed average $< B >_E$ over the intensity of the perturbation. We return to this point in Chapter 13.

12.3.2   TIME-DEPENDENT LINEAR RESPONSE

Let us examine what happens if the external perturbation is a time-dependent function. In this case, the perturbed Hamiltonian is now defined as $\mathscr{H}(t) = \mathscr{H}_0 - M \cdot E(t)$. Invoking the Liouville equation, we have the following equation for the time-dependent distribution function

$$\frac{\partial}{\partial t} f = -L_0 f - L_1 E(t) f \tag{12.58}$$

L_0 refers to the unperturbed Liouville operator, and L_1 to the Liouville operator for the perturbation. We expand the distribution function to include the linear response terms into $f = f_0 + f_1 + \mathscr{O}(E^2)$. The time evolution of the f_0 and f_1 distributions are then, to first order,

$$\begin{aligned}
\frac{\partial}{\partial t} f_0 &= -L_0 f_0 \\
\frac{\partial}{\partial t} f_1 &= -L_0 f_1 - L_1 E(t) f_0
\end{aligned} \tag{12.59}$$

If we start from a system that is initially at equilibrium and switch the field on from $t = 0^+$ on, the initial conditions are $f_0(0) = f_{eq}$ and $f_1(0) = 0$. The first-order distribution functions are given by $f_0(t) = f_{eq} = \frac{e^{-\beta \mathscr{H}}}{Q}$ and

$$f_1(t) = -\int_0^t ds\, e^{-(t-s)L_0} L_1 E(s) f_0(s) \tag{12.60}$$

Since $f_0(s) = f_{eq}$ for all times s, we can evaluate $L_1 E(s) f_0(s)$ as [407]

$$\begin{aligned}
L_1 E(s) f_0(s) &= -E(s)[\frac{\partial M}{\partial \mathbf{p}} \cdot \frac{\partial f_{eq}}{\partial \mathbf{q}} - \frac{\partial M}{\partial \mathbf{q}} \cdot \frac{\partial f_{eq}}{\partial \mathbf{p}}] \\
&= \beta f_{eq} E(s)[\frac{\partial M}{\partial \mathbf{p}} \cdot \frac{\partial \mathscr{H}}{\partial \mathbf{q}} - \frac{\partial M}{\partial \mathbf{q}} \cdot \frac{\partial \mathscr{H}}{\partial \mathbf{p}}] \\
&= -\beta f_{eq} E(s) \dot{M}
\end{aligned} \tag{12.61}$$

For simplicity, we assume that the equilibrium average of phase variable $B(t)$ is 0 and continue to make the same assumption for M. The first-order time-dependent average for $B(t)$ can thus be obtained through the following integral [212]

$$
\begin{aligned}
\langle B(t) \rangle_E &= \beta \int_0^t ds E(s) \int d\Gamma B(\Gamma) e^{-(t-s)L_0} \dot{M} f_{\text{eq}} \\
&= \beta \int_0^t ds E(s) < B(t-s) \dot{M}(0) >_0 \\
&= \beta \int_0^t ds E(t-s) < B(s) \dot{M}(0) >_0
\end{aligned}
\tag{12.62}
$$

While the previous section gave access to the system's response in the steady state for a constant field, Eq. 12.62 provides the average for any phase variable for a time-dependent field and in the transient regime.

12.3.3 NONLINEAR RESPONSE, DYNAMICAL STABILITY, AND CHAOS

In this last section, we highlight some caveats and active research areas that extend nonequilibrium statistical mechanics beyond the results from static and dynamic linear response theory. The first caveat concerns the underlying assumption we have made systematically so far. Indeed, we have discussed the linear response of the system to an external field so far. As a result, the expressions found so far are thus valid to the first-order and neglect terms of the order of $\mathscr{O}(E^2)$ and beyond. While this is perfectly acceptable for small perturbations in the $E \to 0$ limit, the neglected terms increase quadratically as the intensity of the field increases and become of great significance for large perturbations. Very early on, Kubo [212] realized this and built an extension to linear response theory, known as nonlinear response theory. Kubo's initial approach calculated the perturbed distribution function as an expansion (power-series) or the perturbation, *i.e.*, in powers of the field intensity. The perturbed distribution so obtained allows the computation of any time-dependent average of phase variables. However, studies quickly revealed the nonanalyticity of some transport coefficients, such as the shear viscosity [205], and prompted Yamada and Kawasaki to propose a different approach [396]. Yamada and Kawasaki introduced a novel expression for the perturbed distribution function, which, in its renormalized version, is given by

$$
f(t) = \frac{\exp[-\beta H_0(-t)]}{\int d\Gamma \exp[-\beta H_0(-t)]}
\tag{12.63}
$$

Evans and Morriss [198] explain how this formalism applies to adiabatic systems only and leads to several difficulties in practice. Since there is no mechanism for heat dissipation, the system constantly heats up and cannot reach a steady state. In Chapter 13, we will examine a thermostatted approach, known as the transient-time correlation function, that provides access to the nonlinear response in the steady state.

The knowledge of the distribution function also provides a direct route to determining entropy. At equilibrium, Gibbs had already proposed [159] that the entropy S be defined as

$$
S(t) = -k_B \int d\Gamma f(\Gamma) \ln f(\Gamma)
\tag{12.64}
$$

Extending this expression to evaluate entropy production in nonequilibrium systems leads to

$$
\dot{S}(t) = -k_B \int d\Gamma [1 + \ln f(\Gamma)] \frac{\partial f}{\partial t}
\tag{12.65}
$$

This equation leads to several apparent paradoxes [124]. These can be resolved by: (i) considering a coarse-grained entropy [312], obtained by decreasing the resolution for the distribution function

Figure 12.6   Example of a chaotic system: a Lorenz attractor.

instead, and (ii) taking into account the chaotic behavior of the molecular dynamical systems. The second point stems from the reduction in the dimension of the accessible phase space that takes place when a system reaches a nonequilibrium steady state. In other words, phase space trajectories are chaotic (see Figure 12.6 for a well-known example of a chaotic system), separate exponentially with time at a rate characterized by an exponent known as the "largest Lyapunov exponent", and the accessible phase space is termed as a "strange attractor" of reduced dimension when compared to the equilibrium phase space. We refer the interested reader to the work of Bennetin [19] and Hoover and Posch [190] for the calculation of Lyapunov exponents and to the following studies by Hurewicz and Wallman [193], Mandelbrot [240], Balatoni and Renyi [10] for the characterization of the phase space dimension in terms of topology, capacity, and information dimension, respectively.

In his paper on *The Laws of Chaos*, Prigogine summarizes the connection between the complex behavior of dynamical systems, distributions, and entropy: "We emphasize the following sequence: instability (chaos) → probability → irreversibility. Our approach from certain points of view follows Boltzmann's brilliant insights. Today we know that his approach applies to unstable dynamical systems." Prigogine concludes, "Instead of thinking about trajectories or wave functions, we can think in terms of probabilities and properties of the evolutionary operator associated with the Liouville-von Neumann equation. By way of the evolutionary operators, we can unify dynamics and thermodynamics and better grasp the lesson of the second principle of thermodynamics."

13 External Fields and Agents
New Communication Paradigms

"Contrary to what happens at equilibrium, or near equilibrium, systems far from equilibrium do not conform to any minimum principle that is valid for functions of free energy or entropy production."

Ilya Prigogine and Isabelle Stengers,
The End of Certainty [298]

While we extracted thermodynamic properties from equilibrium averages in the book's second part; we focused on time-dependent information, especially fluctuations around equilibrium averages, to obtain dynamical properties. As previously discussed, this idea is central to Onsager's regression hypothesis [275]. Onsager drew a parallel between equilibrium fluctuations and time-dependent changes during a molecular system's relaxation after applying a perturbation or external field. This led us to obtain nonequilibrium properties, such as transport coefficients, from equilibrium time-correlation functions through the Green-Kubo relations introduced in the previous chapter. In other words, through equilibrium time-correlation functions, we can thus gain insight into the nonequilibrium response of the system to an external field. There is, however, a wealth of information that can only be accessed when we directly study the response and, for instance, compute the molecular trajectories for a molecular system subjected to an external field. For example, the external field may change the structure of the molecular system and, thus, the spatial arrangement of molecules. This may also result in the onset of nonlinear phenomena for a high intensity of the field, for which linear response theory no longer applies. To access this information, we need a protocol to generate the nonequilibrium trajectories and mirror the statistical treatment we have discussed for equilibrium systems. Some of these extensions are intuitive. For instance, steady-state averages can be considered the analogs of equilibrium averages and provide access to both thermodynamic and dynamical properties out of equilibrium. For example, as we will see in this chapter, transport coefficients can be calculated from the ratio of steady-state averages of specific phase variables to the field intensity via a direct application of linear response theory. Some of the extensions are more involved and rely on nonlinear response theory. As we will see shortly, transient-time correlation functions can be interpreted as nonequilibrium analogs to equilibrium partition functions and allow for calculating time-dependent averages of phase variables both in the transient regime and the steady state. There are, however, several practical - how do we model perturbation and dissipation? - and conceptual challenges - which nonequilibrium criteria replace the equilibrium minimum principles invoked by Stengers and Priogine? These are some of the key questions we address in the following sections.

13.1 NONEQUILIBRIUM MOLECULAR TRAJECTORIES

13.1.1 BOUNDARY-DRIVEN AND SYNTHETIC SETUPS

We start in this section with the first practical challenge one comes across when setting up the computation of nonequilibrium molecular trajectories. How do we account for the perturbation? Different phenomena or origins may cause a system to be out of equilibrium. The first type of transport coefficient is termed mechanical transport. In this case, a system may be subjected to an external field, and the molecules may simply respond to that field (see the left panel in Figure 13.1). For instance, charged species, such as sodium or chloride ions, are set in motion along the direction of an

DOI: 10.1201/9781003006411-16

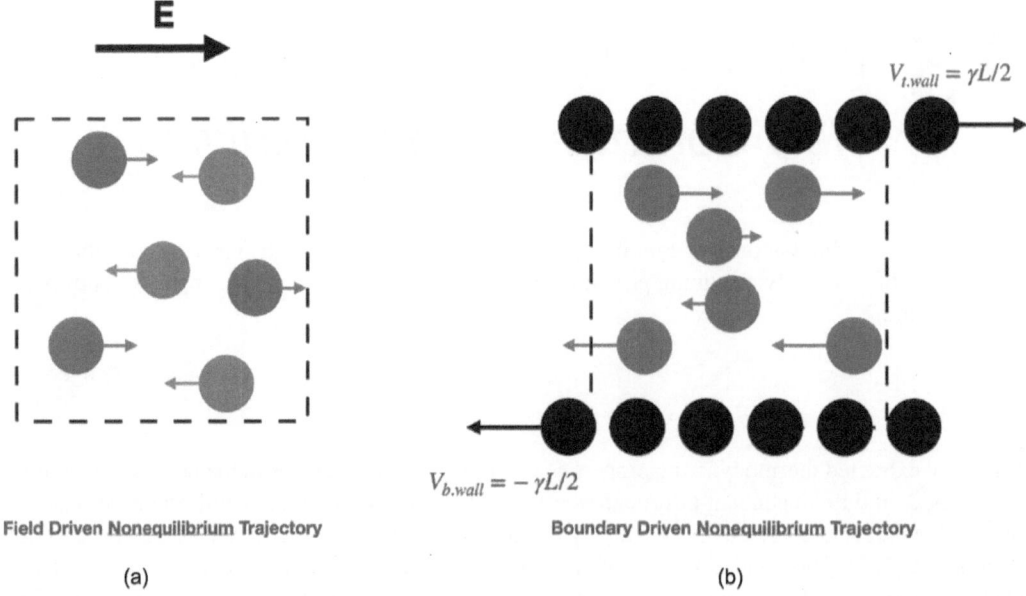

Figure 13.1 Field-driven & boundary-driven systems. The panel (a) shows an example of a field-driven transport, with ions traveling in opposite directions in response to an electric field (the vectors for the particles' velocities are shown as arrows on the plots). The right panel (b) the onset of a linear velocity profile due to the two boundary walls moving in opposite directions. In both plots, the arrows on the particles indicate their velocity.

imposed electric field. In that case, the perturbation is directly included in the equations of motion as an additional, field-induced force. We referred to this first type of perturbation in the previous chapter when we discussed how linear response theory allowed for determining transport coefficients. Here, we can calculate the system's evolution when subjected to an external perturbation by adding the field-induced additional force in the equations of motion and computing the nonequilibrium molecular trajectories. Once the system has reached a steady state, we need to perform a steady-state average of a time-dependent phase variable $B(t)$ and determine the transport coefficient L through a linear constitutive relation

$$L = \frac{\langle B(t) \rangle_{\text{st}}}{E} \tag{13.1}$$

where E denotes the intensity of the external field and $< .. >_{\text{st}}$ indicates that the average is taken over the steady state, *i.e.* after the molecular trajectories have run a time long enough to exceed any relaxation time of the molecular system. For instance, the conductivity may be calculated with this approach by computing the ratio of the current's steady-state average over the electric field's intensity. In the $E \to 0$ limit, the result obtained for L in the steady-state using Eq. 13.1 coincides with the equilibrium approach outlined in the previous chapter, which relies on the computation of time-correlation functions. We add that this approach hinges on using some mechanism to account for heat dissipation, such as, for instance, a thermostat. In the absence of such a mechanism, the application of the external field results in a constant influx of energy, *e.g.*, through the work resulting from the motion of ions along the direction of the external field. If there is no heat dissipation, then the system's energy constantly increases. This, in turn, prevents the system from reaching a steady state. In the next section, we will return to the implementation of thermostats and other mechanisms for heat dissipation.

The second type of transport coefficient is termed a thermal transport coefficient. In nature, thermal transport processes are boundary-driven (see the right panel in Figure 13.1) rather than driven by an external field. An example of a thermal transport process is heat conduction. Suppose a molecular system is confined between two walls, or boundaries, of different temperatures. In that case, the molecular system is subjected to a temperature gradient, and a heat flux will arise. Fourier's law states that the heat flux J_q is related to the temperature gradient through the following relation

$$J_q = -\kappa \frac{dT(x)}{dx} \tag{13.2}$$

where κ is the thermal conductivity and, for simplicity, the temperature exhibits a spatial dependence along a single dimension, here the x-axis. This means there is a striking similarity between the constitutive relations defining mechanical and thermal transport coefficients. Similarly, the Green-Kubo formula can evaluate thermal transport coefficients from equilibrium time-correlation functions, and linear response theory can also be invoked to determine thermal transport coefficients. As we will see shortly, this can be achieved by framing the computation of thermal transport coefficients into the same formalism as mechanical transport coefficients. In other words, we will introduce synthetic, rather than boundary-driven, methods that include an artificial force term in the equations of motion that mimic the effect of gradients. Supplemented with a thermostat, this synthetic setup and synthetic equations of motion enable the computation of steady-stage averages for fluxes. Thus, when divided by the intensity of the perturbation, it provides direct access to thermal transport coefficients.

13.1.2 ACCOUNTING FOR HEAT DISSIPATION

We have already mentioned several times the need to include a thermostat when a molecular system is subjected to an external field that constantly brings energy to the system in the form of work. Thus, without a thermostat, the system is said to be "adiabatic", and its energy and temperature steadily increase. On the other hand, using a thermostat removes energy from the system in the form of heat, balances the constant influx of energy due to the external field, and allows the system to reach a steady state. In analogy with what we have seen at equilibrium in Part II, such systems are termed isothermal. Both cases exist in nature, with nonequilibrium systems responding adiabatically or isothermally. Strictly speaking, these are idealized limits. There is always some heat dissipated through the boundaries, implying that no nonequilibrium system is perfectly adiabatic. Similarly, heat transfer is slow on the microscopic scale, leading to non-zero, albeit small, temperature gradients, and thus only quasi-thermal, rather than isothermal, nonequilibrium systems exist. With this caveat in mind, we assume quasi-thermal systems are adequately modeled as isothermal systems.

To illustrate how a thermostat acts on a non-equilibrium system, we consider the example of a Gaussian feedback mechanism applied to a generic set of (nonequilibrium) equations of motion

$$\dot{\mathbf{r}}_i = \frac{\mathbf{p}_i}{m_i} + \mathbf{C}_i E(t) \tag{13.3}$$

$$c_i = \mathbf{f}_i + \mathbf{D}_i E(t) + \lambda \dot{\mathbf{p}}_i \tag{13.4}$$

where the last term in the second equation corresponds to the Gaussian thermostat, with the multiplier λ defined according to Gauss' principle of least constraint, $E(t)$ denotes the intensity of the time-dependent field and the additional terms, involving $\mathbf{C}_i$ and $\mathbf{D}_i$, model the effect of the field on the system. The following sections will provide specific examples for determining ionic conduction and shear viscosity.

This set of equations relies on several assumptions. First, the momenta are defined with respect to the streaming velocity and are called "peculiar momenta". If the external field results in the overall flow of the system, only the fluctuations in the peculiar momenta and peculiar velocities should be interpreted as thermal and thus as contributing to the temperature. For instance, in a moving train, a glass of water does not get warmer because of the motion of the train - hence the water temperature can be calculated from the peculiar velocities by subtracting the streaming velocity, here $\mathbf{u}$ the velocity of the train, from the velocity of all water molecules, which leads to the following nonequilibrium definition for temperature

$$T = \sum_{i=1}^{N} \frac{\mathbf{p}_i^2}{2m_i} = \sum_{i=1}^{N} \frac{m_i}{2} (\mathbf{v}_i - \mathbf{u})^2 \tag{13.5}$$

Second, the sum of the peculiar momenta over the N molecules of the system should equal 0. This can be ensured by simply setting $\sum_{i=1}^{N} \mathbf{p}_i = 0$ initially. After the application of Gauss' principle of least constraint, the expression for the thermostat multiplier is as follows

$$\alpha = \frac{\sum_i \frac{\mathbf{F}_i \cdot \mathbf{p}_i}{m_i}}{\sum_i \frac{\mathbf{p}_i^2}{m_i}} + \frac{\sum_i \frac{\mathbf{D}_i \cdot \mathbf{p}_i}{m_i}}{\sum_i \frac{\mathbf{p}_i^2}{m_i}} E(t) \tag{13.6}$$

This Gaussian thermostat provides a way for the system to reach a steady, while the external, potentially synthetic, field is on through the inclusion of the $\mathbf{C}_i$ and $\mathbf{D}_i$ terms in the equations of motion. The thermostat relies on a nonequilibrium extension of the equipartition principle applied to the streaming kinetic energy. This is one of the many ways to define a nonequilibrium temperature, and this streaming kinetic temperature might not be well suited to all cases and situations, especially when the streaming velocity profile is ill-defined such as, *e.g.*, for complex flows or in complex and crowded environments. We will return to this point with a practical definition later in this chapter.

13.1.3 EXTRACTING TRANSPORT COEFFICIENTS

We consider in this section the example of conductivity. Here we focus on the motion of ions in a fluid when an electric field $\mathbf{E}$ is applied. In this section, we take the example of a molten salt with N ions at equilibrium. As with all transport coefficients, we can find an appropriate time-correlation function to obtain the conductivity. Here, the fundamental quantity is the current density $\mathbf{j}$ defined as

$$\mathbf{j} = \frac{e}{V} \sum_{i=1}^{N} \frac{z_i \mathbf{p}_i}{m_i} \tag{13.7}$$

where e is the unit charge ($e = 1.6 \times 10^{-19}$ C), z_i the formal charge of an ion i, m_i its mass, and $\mathbf{p}_i$ its momentum. We can evaluate the conductivity from the current density time-correlation function as

$$\langle \sigma(t) \rangle = \frac{V}{3k_B T} \int_0^t ds \, \langle \mathbf{j}(s) \cdot \mathbf{j}(0) \rangle \tag{13.8}$$

If we are now interested in the system's response to an electric field, we can extract the conductivity from a nonequilibrium trajectory using linear response theory. Let us subject the system to an external field $\mathbf{E} = E\mathbf{e}_x$, with E the intensity of the field constant with time and $\mathbf{e}_x$ the unit vector for the x-axis. During the nonequilibrium trajectory, the equations of motion for an ion i are as follows:

$$\dot{\mathbf{r}}_i = \frac{\mathbf{p}_i}{m_i} \tag{13.9}$$

$$\dot{\mathbf{p}}_i = \mathbf{F}_i + z_i e E \mathbf{e}_x - \alpha \mathbf{p}_i \tag{13.10}$$

where $\mathbf{F}_i$ denotes the force on ion i resulting from the inter-ions potential, and α the Gaussian multiplier, which is temperature constant. For simplicity, in this section, we will take the usual definition for temperature rather than a more involved definition based on the peculiar kinetic energy.

We refer the reader to prior work [235,288] for the temperature and multiplier full expressions when subtracting the streaming velocity for each type of ion. Provided that a low intensity of the electric field is used [288], the two approaches give results for the conductivity in good agreement with each other and with the Green-Kubo equilibrium estimate of Eq. 13.8. Defining the temperature as

$$T = \frac{1}{3Nk_B} \left\langle \sum_{i=1}^{N} \mathbf{p}_i^2 m_i \right\rangle \tag{13.11}$$

we obtain the following equation for the multiplier α using Gauss' principle of least constraint

$$\alpha = \frac{\sum_{i=1}^{N} \mathbf{p}_i \cdot \mathbf{F}_i/m_i + eE \sum_{i=1}^{N} z_i p_{xi}/m_i}{\sum_{i=1}^{N} \mathbf{p}_i^2/m_i} \tag{13.12}$$

The molecular trajectories are computed by integrating the nonequilibrium equations of motion. Once the nonequilibrium system has reached a steady state, we collect the average of the current density - more specifically, $< j_x >$ here since the field is along the x-axis. Linear response theory then provides the value for the field-dependent conductivity directly as

$$\sigma(E) = \frac{< j_x >_{st}}{E} \tag{13.13}$$

where $< .. >_{st}$ denotes a steady-state average. The low-field intensity response, or asymptotic limit when $E \to 0$, for $\sigma(E)$ coincides with the Green-Kubo estimate for the conductivity through

$$\lim_{E \to 0} \sigma(E) = \lim_{E \to 0} \frac{< j_x >_{st}}{E} = \frac{V}{3k_B T} \int_0^{\infty} ds < \mathbf{j}(s) \cdot \mathbf{j}(0) >_{eq} \tag{13.14}$$

13.2 COMPUTING NONEQUILIBRIUM TRAJECTORIES

13.2.1 PHYSICAL BOUNDARIES *VS* PERIODIC BOUNDARIES

Subjecting a system to a mechanical perturbation involves applying a field and adding an extra term corresponding to the field-dependent force to the equations of motion. On the other hand, thermal perturbations are more complex since these are boundary-driven in Nature. Let's consider the example of a fluid subjected to shear. The most direct way to impose such a perturbation is through two moving boundaries, *i.e.*, two "physical" walls moving in opposite directions. While this strategy seems *a priori* straightforward, it leads to questions and challenges. For instance, what model and structure should we use for the wall? How do the walls impact the solid-liquid interface? Will the fluid molecules stick or slip? Since the structure of the fluid will be affected by the presence of the walls over several molecular layers on both sides, how large should the system be to yield reliable results for the corresponding transport coefficient, *i.e.*, shear viscosity? Since the motion of the walls will be slowly imparted, on a "layer-by-layer" basis, from the boundary to the center of the "channel", how long will the simulations need to be run before the onset of a steady-state linear streaming velocity profile across the fluid? The answer to this question will be a complex interplay between wall velocity, shear rate γ, system size and cost of computing interactions, and the timescale that can be accessed through simulations. How do we account for heat dissipation? Should we include a thermostat in the wall to be closest to the actual experiment? While such a setup allows the system to reach a steady state, it also implies that the temperature of the fluid will not be controlled. Indeed, the temperature will increase as we get further away from the walls and exhibit, in the case of a fluid undergoing shear flow, the quadratic profile predicted by the Navier-Stokes equations. The less realistic alternative consists in applying a thermostat to the fluid and including an additional term in the equations of motion for the fluid molecules. In other words, to alleviate the unwanted boundary effects on the structure and gain control over the thermodynamic conditions, such as temperature, one needs to use a synthetic field to mimic a boundary-driven perturbation.

We will see in the next section how to replace a boundary-driven streaming velocity profile with a synthetically generated profile. However, this is only part of the story, as we also need to replace the actual boundaries with a set of periodic boundary conditions that match the boundary-driven perturbation. We have seen in Chapter 6 how periodic boundary conditions allow us to simulate quasi-infinite systems at equilibrium by replication of the central simulation box. However, the equilibrium periodic conditions are inappropriate for a system undergoing shear flow. For instance, if the flow is along the x-axis, a particle i with coordinates (x_i, y_i, z_i) and velocity $\mathbf{v}_i = v_i \mathbf{e}_x$ exits the "bottom" of the simulation cell when its y-coordinate becomes less than, for instance, 0, where L is the edge of the simulation cell. This particle is reinserted at the "top" of the simulation cell with a new y-coordinate $y_i \leftarrow y_i + L$, or in modulo notation $(y_i)_{\text{mod } L}$, and will keep the same velocity v_i. This conflicts with a flow profile of slope γ. For instance, one would typically expect the velocity of a particle at the "top" of the simulation cell, and hence steered in an experiment by the top "wall" moving at a $\gamma L/2$, to be greater than the velocity of a particle at the "bottom" of the simulation cell, and hence steered in the opposite direction by the bottom "wall" moving at a $-\gamma L/2$. So the periodic boundary conditions should act on positions and velocities in a system undergoing some flow. This example calls for introducing new sets of periodic boundary conditions suitable for a given flow geometry. In the rest of this section, we continue our analysis of a system subjected to a shear flow and provide the complete expression for the Lees-Edwards periodic boundary conditions [222], designed for this type of flow geometry (Figure 13.2). The idea behind the Lees-Edwards periodic boundary conditions is to drive the shear flow through the motion of the periodic images of the simulation cell. Since the particles sense the motion of neighboring particles through the interactions, the motion of image particles above and below the simulation cell results in a linear velocity profile

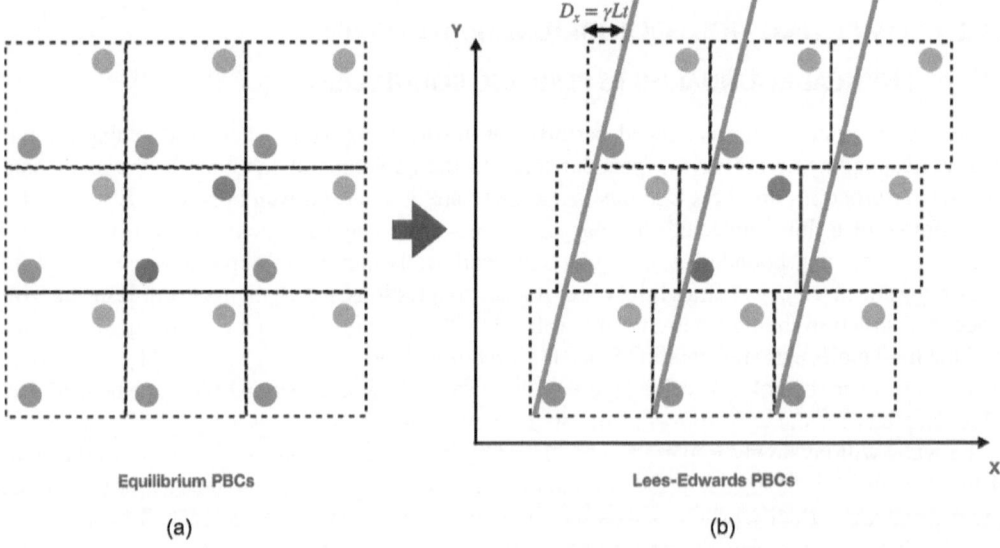

Equilibrium PBCs   Lees-Edwards PBCs

(a)   (b)

Figure 13.2 (a) Conventional periodic boundary conditions (PBCs) and (b) Lees-Edwards PBCs or "sliding brick" PBCs. In the conventional PBCs used for equilibrium systems (see also Chapter 6), a particle from the central simulation cell and its images in the cell replica all have the same velocity. In the Lees-Edwards PBCs, the top row of cells moves at a higher velocity than the middle row (by γL) and the bottom row (by $2\gamma L$). For instance, after a time t, the top row will have "traveled" further along the x-axis than the central row by a distance of $D_x = \gamma L t$. In other words, a particle from the central cell exiting by the bottom of the cell will reenter the cell by its top with the following updated velocity $(v_x)_i \rightarrow (v_x)_i + \gamma L$.

$\mathbf{u}(\mathbf{r}) = \gamma y \mathbf{e_x}$ across the system. The Lees-Edwards periodic boundary conditions are often called a sliding brick representation. Let us now examine the motion of particles under such periodic boundary conditions. If we consider a simulation cell of side L and set to 0 the streaming velocity at the origin. Particles within the simulation cell have coordinates (x_i, y_i, z_i) such that $0 < \{x_i, y_i, z_i\} < L)$, and we define the laboratory velocity of particle i as the sum of a peculiar (thermal) velocity $\mathbf{c}_i$ and of a streaming velocity $\mathbf{u}(\mathbf{r}_i)$. This results in the following equation for the velocity laboratory $\mathbf{v}_i$

$$\dot{\mathbf{r}}_i = \mathbf{c}_i + \mathbf{u}(\mathbf{r}_i) \tag{13.15}$$

We now focus on boundary crossings in the y direction and consider the two nearest images along that direction. Initially ($t = 0$), the two nearest images for a particle in the simulation with position $\mathbf{r}_i$ are located at $\mathbf{r}'_i = \mathbf{r}_i + \hat{y}L$ and $\mathbf{r}''_i = \mathbf{r}_i - \hat{y}L$, respectively. The positions of particle i and its two images evolve as a function of time according to

$$\mathbf{r}_i(t) = \mathbf{r}_i(0) + \int_0^t ds\dot{\mathbf{r}}_i(s) = \mathbf{r}_i(0) + \int_0^t ds(\mathbf{c}_i + \mathbf{e_x}\gamma y_i) \tag{13.16}$$

$$\mathbf{r}'_i(t) = \mathbf{r}_i(0) + \mathbf{e_y}L + \int_0^t ds(\mathbf{c}'_i + \mathbf{e_x}\gamma(y_i + L)) = \mathbf{r}_i(0)$$
$$+ \mathbf{e_x}\gamma t L + \mathbf{e_y}L + \int_0^t ds(\mathbf{c}'_i + \mathbf{e_x}\gamma y_i) \tag{13.17}$$

$$\mathbf{r}''_i(t) = \mathbf{r}_i(0) - \mathbf{e_y}L + \int_0^t ds(\mathbf{c}''_i + \mathbf{e_x}\gamma(y_i + L)) = \mathbf{r}_i(0)$$
$$- \mathbf{e_x}\gamma t L - \mathbf{e_y}L + \int_0^t ds(\mathbf{c}''_i + \mathbf{e_x}\gamma y_i) \tag{13.18}$$

The peculiar velocities of a particle and its periodic images are the same, which means that $\mathbf{c}_i = \mathbf{c}'_i = \mathbf{c}''_i$. We thus have

$$\mathbf{r}_i(t) = \mathbf{r}_i(0) + \int_0^t ds(\mathbf{c}_i + \mathbf{e_x}\gamma y_i) \tag{13.19}$$

$$\mathbf{r}'_i(t) = \mathbf{r}_i(t) + \mathbf{e_x}\gamma t L + \mathbf{e_y}L \tag{13.20}$$

$$\mathbf{r}''_i(t) = \mathbf{r}_i(t) - \mathbf{e_x}\gamma t L - \mathbf{e_y}L \tag{13.21}$$

If the particle exits the simulation via the "bottom" of the cell, $\mathbf{r}_i(t)$ is replaced by the image particle at $\mathbf{r}'_i(t)$ and $\mathbf{r}_i^{new} = (\mathbf{r}'_i)_{\bmod L} = (\mathbf{r}_i + \mathbf{e_x}\gamma L t)_{\bmod L}$. Alternatively, if the particle exits the simulation via the "top" of the cell, $\mathbf{r}_i(t)$ is replaced by the image at $\mathbf{r}''_i(t)$ and $\mathbf{r}_i^{new} = (\mathbf{r}''_i)_{\bmod L} = (\mathbf{r}_i - \mathbf{e_x}\gamma L t)_{\bmod L}$. We summarize the Lees-Edwards periodic boundary conditions in Figure.

13.2.2   NONEQUILIBRIUM DEFINITIONS FOR TEMPERATURE

In the previous section, we set periodic boundary conditions corresponding to the sliding brick representation of Lees and Edwards. These periodic boundary conditions model, in a setup free from explicit boundaries, the dynamical effect induced by two moving walls, with velocities differing by γL. The next step in setting up a homogeneous nonequilibrium simulation is to mimic the mechanisms by which heat is dissipated through a thermostat. We have previously seen how to build a Gaussian or Nosé-Hoover thermostat that uses the peculiar kinetic energy as temperature and interprets fluctuations in the peculiar velocities as thermal fluctuations. Any departure from the assumed streaming velocity profile will be dampened and constrained to match the target temperature. This approach is appropriate if the modeled system is, for instance, a homogeneous atomic fluid subjected to a low to moderate shear rate. In this case, the streaming velocity profile will remain linear, and a thermostat that keeps the peculiar kinetic energy and temperature constant works properly.

However, many possibilities exist when the assumption of a specific streaming velocity or flow profile does not apply [367]. For instance, the system under shear may become inhomogeneous if it undergoes shear-induced ordering or a shear-induced phase transition [72,343] and the assumption of a linear flow profile will not be appropriate any longer. If the fluid flows past an obstacle or a wall, the molecules close to the surface could either stick or slip depending on the surface-fluid interaction and the surface corrugation, resulting in local departures from the assumed flow profile. Another example occurs when very high shear rates are applied. In such a case, hydrodynamic stability theory predicts a nonlinear flow profile, with a sinusoidal secondary flow developing on top of the linear flow profile [79,254]. In such cases, using an alternate definition for temperature, known as the configurational temperature, avoids making assumptions about the flow profile and provides a versatile way to thermostat a non-equilibrium system. To define the configurational temperature, we invoke the microcanonical ensemble and start from the expression for the entropy $S(N,V,E)$ given by

$$S(E) = k_B \ln(\Omega(N,V,E)) = k_B \int_{\mu_C(E)} d\Gamma \tag{13.22}$$

where $\Gamma = (\mathbf{q}_1, \ldots, \mathbf{q}_N, \mathbf{p}_1, \ldots, \mathbf{p}_N)$ denotes a phase space point, and $\mu_C(E)$ represents the set of points for which the energy is between E and $E + \delta E$ ($|\delta E| \ll E$) with Boltzmann's equal a priori probability assumption. Temperature is defined as

$$\frac{1}{T} = \left(\frac{\partial S}{\partial E} \right)_V \tag{13.23}$$

Rugh [318] proposes to evaluate this derivative to obtain temperature. To achieve this, he translates Γ to Γ' along the phase space gradient of the Hamiltonian,

$$\Gamma' = \Gamma + \mathbf{n}(\Gamma)\Delta E = \Gamma + \frac{\mathbf{G} \cdot \nabla H(\Gamma)}{\nabla H(\Gamma) \cdot \mathbf{G} \cdot \nabla H(\Gamma)} \Delta E \tag{13.24}$$

where ΔE is a change in the system's energy. The matrix $\mathbf{G}$ can take many forms. However, it needs to satisfy the following condition

$$\nabla H(\Gamma) \cdot \mathbf{G} \cdot \nabla H(\Gamma) \neq 0 \tag{13.25}$$

Considering the $(N, V, (E + \Delta E))$ microcanonical ensemble, the entropy is given by

$$S(E + \Delta E) = k_B \ln \int_{\mu_C(E+\Delta E)} d\Gamma = k_B \ln \int_{\mu_C(E)} J(\Gamma) d\Gamma \tag{13.26}$$

where $J(\Gamma)$ is the Jacobian of the transformation $\Gamma \to \Gamma'$. To first order in ΔE, the Jacobian is equal to

$$J(\Gamma) = \left| \frac{\partial \Gamma'(\Gamma)}{\partial \Gamma} \right| = 1 + \nabla \cdot \mathbf{n}(\Gamma)\Delta E \tag{13.27}$$

Since $\left(\frac{\partial S}{\partial E} \right)_V$ can be evaluated as $\frac{S(E+\Delta E)-S(E)}{\Delta E}$, we can plug in the results we obtained for $S(E)$ and $S(E + \Delta E)$ to obtain the following expression for the temperature [76]

$$\frac{1}{T} = \left(\frac{\partial S}{\partial E} \right)_V = k_B \langle \nabla \cdot n \rangle \tag{13.28}$$

If the matrix $\mathbf{G}$ is such that $g_{ab} = \delta_{ab}$ when both a and b refer to position coordinates and 0 otherwise, we have

$$\frac{1}{T} = \left\langle \frac{-\sum_i \nabla_i \cdot \mathbf{F}_i}{\sum_i \mathbf{F}_i^2} - \frac{\sum_{ij} \mathbf{F}_i \mathbf{F}_j : \nabla_i \mathbf{F}_j}{(\sum_i \mathbf{F}_i^2)^2} \right\rangle \tag{13.29}$$

where the indices i and j refer to atoms of the system, $\mathbf{F}_i$ the force exerted on i, ∇_i the gradient vector with respect to Cartesian coordinates of i, and: is the dyadic operator. The dyadic operator is such that, for vectors $\mathbf{a}, \mathbf{b}$ and matrix $\mathbf{M}$, we have $\mathbf{ab} : \mathbf{M} = \sum_{\alpha,\beta} a_\alpha b_\beta M_{\beta\alpha}$. For short-ranged interatomic interactions, the second term on the right-hand side is $\mathcal{O}(N^{-1})$ and vanishes in the thermodynamic limit. We can thus define the following order one expression for the temperature:

$$\frac{1}{k_B T_{\mathrm{con1}}} = \left\langle \frac{-\sum_i \nabla_i \cdot \mathbf{F}_i}{\sum_i \mathbf{F}_i^2} \right\rangle \tag{13.30}$$

A more general expression of Rugh's theorem is given by

$$\frac{1}{T} = k_B \frac{\langle \nabla \cdot \mathbf{B}(\Gamma) \rangle}{\langle \nabla H \cdot \mathbf{B}(\Gamma) \rangle} \tag{13.31}$$

where $\mathbf{B}(\Gamma)$ is an arbitrary phase space vector field. If we now substitute $\mathbf{G} \cdot (\nabla H)$ for $\mathbf{B}(\Gamma)$, we obtain a second configurational expression for the temperature as a fraction of two averages [42, 202]. The resulting T_{conF} is given by

$$\frac{1}{k_B T_{\mathrm{conF}}} = \frac{\langle -\sum_i \nabla_i \cdot \mathbf{F}_i \rangle}{\langle \sum_i \mathbf{F}_i^2 \rangle} \tag{13.32}$$

This configurational temperature can then define a configurational thermostat [75,230] in the same spirit as the Nosé-Hoover thermostat. Supplementing the equations of motion for a system at equilibrium with a configurational thermostat yields the following set [33]

$$\dot{\mathbf{r}}_i = \frac{\mathbf{p}_i}{m_i} - \eta \frac{\partial U}{\partial \mathbf{r}_i} \tag{13.33}$$

$$\dot{\mathbf{p}}_i = -\frac{\partial U}{\partial \mathbf{r}_i} \tag{13.34}$$

$$\dot{\eta} = \frac{1}{Q_\eta} \left(\sum_{i=1}^N \left(\frac{\partial U}{\partial \mathbf{r}_i} \right)^2 - k_B T \sum_{i=1}^N \frac{\partial^2 U}{\partial \mathbf{r}_i^2} \right) \tag{13.35}$$

where U is the potential energy of the system. As shown by Braga and Travis [33], implementing a configurational temperature-based thermostat generates the canonical distribution. Such a thermostat, added to a set of equations of motion for an equilibrium system, can be readily applied to a system subjected to an external field [86].

13.2.3 TRANSPORT IN THE STEADY-STATE

We now examine an example of a synthetic method to determine a thermal transport coefficient. In this section, we focus on the example of shear viscosity. As previously discussed, we may carry out equilibrium molecular dynamics simulations and calculate the shear viscosity through the following Green-Kubo expression [198]

$$\eta = \frac{V}{k_B T} \int_0^\infty \langle P_{xy}(s) \cdot P_{xy}(0) \rangle \, ds \tag{13.36}$$

where P_{xy} is the shear stress [198], corresponding to an off-diagonal pressure tensor element. The microscopic representation for the pressure tensor is due to Irving and Kirkwood [197] given by Ref. [198,362]

$$\mathbf{P}(\mathbf{r},t) = \frac{1}{V} \left[\sum_i m_i (\mathbf{v}_i - \mathbf{u}(\mathbf{r},t))(\mathbf{v}_i - \mathbf{u}(\mathbf{r},t)) - \frac{1}{2} \sum_{i,j} \mathbf{r}_{ij}(t) \mathcal{O}_{ij} \mathbf{F}_{ij}(t) \right] \tag{13.37}$$

where $\mathbf{u}$ denotes the streaming velocity, $(\mathbf{v}_i - \mathbf{u})$ the peculiar velocity, $\mathscr{O}_{ij}$ is an operator equal to 1 for systems involving pair interactions [362].

To simulate a fluid undergoing shear flow using a synthetic method, we need to supplement the equations of motion with field-dependent terms. The approach we discuss is the nonequilibrium equations of motion introduced by Evans and Morriss [198], following earlier developments by Hoover, Ladd, Moran, Ashurst, Hickman, and Evans [188,189]. These equations are named the SLLOD equations of motion and take the following form:

$$\dot{\mathbf{r}}_i = \frac{\mathbf{p}_i}{m_i} + \gamma y_i \mathbf{e}_x$$
$$\dot{\mathbf{p}}_i = \mathbf{F}_i - \gamma p_{y_i} \mathbf{e}_x$$

(13.38)

where $\mathbf{r}_i$, $\mathbf{p}_i$ and m_i denote the position, peculiar momentum and mass of atom i, $\mathbf{F}_i$ the force exerted on i, N the total number of atoms, $\mathbf{e}_x$ a unit vector along the x axis, and γ the applied shear rate. These equations of motion establish a linear velocity profile with a slope equal to the shear rate γ at time $t = 0^+$ right after the equations have been applied to the system. To sustain the linear flow profile beyond $t = 0^+$, the SLLOD equations of motion need to be implemented with the Lees-Edwards boundary conditions to retain the dynamical effect that would be induced by moving boundaries [78]. To dissipate heat in the system under shear, we need to supplement these equations of motion with a thermostat. Using a thermostat that assumes a linear flow profile will enforce a linear flow profile throughout the system, which, as we discussed, can be incompatible with the onset of local inhomogeneities. In such cases, it is preferable to use a profile unbiased thermostat, either by subtracting the local average velocity from the laboratory velocity to obtain a more accurate estimate for the peculiar velocity and, thus, for the temperature [367] or through the use of a configurational thermostat. With the addition of a configurational thermostat, the SLLOD equations of motion become

$$\dot{\mathbf{r}}_i = \frac{\mathbf{p}_i}{m_i} + \dot{\gamma} y_i \mathbf{e}_x - \zeta \nabla_i U$$
$$\dot{\mathbf{p}}_i = \mathbf{F}_i - \dot{\gamma} p_{y_i} \mathbf{e}_x$$
$$\zeta = \frac{1}{Q_\zeta} \left(\sum_{i=1}^N (\nabla_i U)^2 - k_B T \sum_{i=1}^N \nabla_i^2 U \right)$$

(13.39)

where U denotes the system's potential energy, T is the target value for the temperature, and ζ is an additional dynamical variable akin to a friction coefficient. Q_η is generally interpreted as the mass associated with the heat bath. This thermostat fixes the configurational temperature $\left\langle \sum_{i=1}^N (\nabla_i U)^2 \right\rangle / \left(k_B \left\langle \sum_{i=1}^N \nabla_i^2 U \right\rangle \right)$ to the target value.

To determine the shear viscosity during a nonequilibrium molecular dynamics simulation run, we apply the equations of motion given in Eq. 13.39 to the system until a steady state has been reached. In the steady state, we may apply linear response theory to obtain the shear viscosity η as

$$\eta = -\frac{\langle P_{xy} \rangle_{\text{st}}}{\gamma}$$

(13.40)

From a practical standpoint, the successful application of linear response theory hinges on two assumptions. First, the signal, or shear rate γ we impose through the equations of motion, must be low enough to observe a linear response. For very high γ, nonlinear effects will arise. The fluid will undergo significant structural changes, either through an "alignment" of the fluid structure with the direction of shear and a decrease in viscosity as a result, or shear-thinning, or through the formation of densely packed clusters and a steep increase in viscosity as a result of shear-thickening. Second, to obtain a reliable estimate for the shear viscosity, the signal must be large enough to overcome the

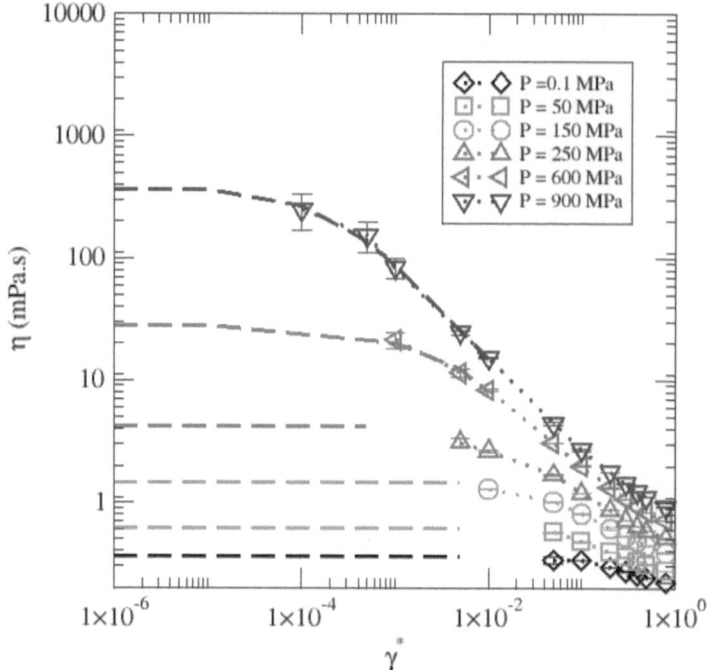

Figure 13.3 Shear viscosity against shear rate for 2,2,4-tri- methylhexane at 293 K and for pressures ranging from 0.1 to 900 MPa. The dashed line indicates the extrapolated value at zero shear rate for the shear viscosity [373].

noise and yield an estimate for $\langle P_{xy} \rangle$ with a reasonably low standard deviation. This means there can be a large gap between equilibrium, *i.e.*, the zero-shear rate limit, and the lowest shear rates that yield reliable estimates for shear viscosity (Figure 13.3). We discuss in the next section an approach known as the transient-time correlation function method, which allows nonequilibrium molecular dynamics to deliver accurate results even for low signal-to-noise ratios.

13.3 TRANSIENT-TIME CORRELATION FUNCTION

13.3.1 FORMALISM

The transient-time correlation function (TTCF) formalism consists of a nonlinear generalization of the Green-Kubo relations. The theory underlying the TTCF approach directly connects the averages of phase variables in the steady state and the functions we will define in this section as transient-time correlation functions. The TTCF theory was initially developed for adiabatic systems by Visscher [384], Dufty and Lindenfeld [112], and Cohen [59], and later extended to isothermal (isokinetic) systems by Morriss and Evans [263]. Recent applications of the approach have enabled the study of the rheology of a wide range of fluids at low, experimentally accessible shear rates [31,87–89,250,280,361], the response of fluids subjected to an elongational flow [179], ionic conductivity [74], transport in nanoconfined systems [73,86] and tribology [237].

We consider a phase variable $B(\Gamma, t)$. In the Heisenberg representation, the average of B at time t is given by

$$\langle B(t) \rangle = \int d\Gamma f(0) B(\Gamma, t) \tag{13.41}$$

where $f(0)$ is the initial distribution function. For time-independent external fields, differentiating this expression with respect to time gives

$$\frac{d\langle B(t)\rangle}{dt} = \int d\Gamma f(0)\frac{d\Gamma}{dt} \cdot \frac{\partial B(t)}{\partial\Gamma} \tag{13.42}$$

After an integration by parts, we obtain

$$\frac{d\langle B(t)\rangle}{dt} = \int d\Gamma B(t)\frac{\partial}{\partial\Gamma} \cdot \frac{d\Gamma}{dt}f(0) \tag{13.43}$$

where we have used that the boundary term vanishes in periodic systems. Integrating with respect to time, we obtain the nonlinear nonequilibrium response as

$$\langle B(t)\rangle = \langle B(0)\rangle - \int_0^t ds \int d\Gamma B(s)\frac{\partial}{\partial\Gamma} \cdot \frac{d\Gamma}{dt}f(0) \tag{13.44}$$

This equation is the generic expression for the transient time-correlation function, which evaluates the time-correlation of phase variable B with phase space derivative $\frac{\partial(\dot{\Gamma}f(0))}{\partial\Gamma}$. Choosing B to match the system's response, such as, e.g., $B = P_{xy}$ for a fluid subjected to a flow profile such that $\mathbf{u} = \gamma y\mathbf{e}_x$, yields the shear viscosity η through the usual ratio $\eta = -P_{xy}/\gamma$.

If the initial distribution is canonical, then,

$$\frac{\partial}{\partial\Gamma} \cdot \left(f(0)\frac{d\Gamma}{dt}\right) = f(0)\left[\frac{\partial\dot{\Gamma}}{\partial\Gamma} - \dot{\Gamma}\frac{\partial}{\partial\Gamma}\left(\frac{H_0(\Gamma)}{k_BT}\right)\right] \tag{13.45}$$

where $H_0(\Gamma) = \sum_i \mathbf{P}_i^2/(2m) + U(\mathbf{q}_i)$ is the internal energy. In the case of an isokinetic flow where the temperature is maintained constant via a Gaussian thermostat with multiplier α, we can write that

$$\frac{\partial}{\partial\Gamma} \cdot \left(f(0)\frac{d\Gamma}{dt}\right) = f(0)\left[-3\alpha N - \sum_{i=1}^N \frac{\dot{\mathbf{P}}_i\mathbf{P}_i}{k_BTm_i} + \sum_{i=1}^N \frac{\dot{\mathbf{q}}_i\mathbf{F}_i}{k_BT}\right] \tag{13.46}$$

The exact expression for the transient-time correlation function then depends on the equations of motion and, thus, on the type of external field applied to the system. We examine in the following two sections the applications of the TTCF approach for a realistic (electric) field and a synthetic (shear rate) field.

13.3.2   BRIDGING BETWEEN EQUILIBRIUM AND NONEQUILIBRIUM

We first focus on the case of an ionic fluid subjected to an electric field with low intensity. In this case, the nonequilibrium trajectory is obtained from the nonequilibrium equations of motion defined as follows for ion i

$$\dot{\mathbf{r}}_i = \frac{\mathbf{p}_i}{m_i} \tag{13.47}$$

$$\dot{\mathbf{p}}_i = \mathbf{F}_i + z_i e E\mathbf{e}_x - \alpha\mathbf{p}_i \tag{13.48}$$

where E denotes the intensity of the electric field, m_i and z_i the mass and the formal charge of the ion i, $\mathbf{F}_i$ the force on ion i resulting, $\mathbf{e}_x$ the unit vector for the direction of the field (here aligned with the x-axis), and α the Gaussian thermostat multiplier. Since the electric field is of low intensity, we expect the kinetic and configurational expressions for the temperature to give similar results and elect to use the usual kinetic expression for the temperature for simplicity

$$T_{\text{kin}} = \frac{\langle\sum_{i=1}^N \mathbf{p}_i^2/m_i\rangle}{3Nk_B} \tag{13.49}$$

where N is the total number of positive and negative ions. The multiplier α is given by

$$\alpha = \frac{\sum_{i=1}^N \mathbf{P}_i \cdot \mathbf{F}_i/m_i + eE\sum_{i=1}^N z_ip_{xi}/m_i}{\sum_{i=1}^N \mathbf{P}_i^2/m_i} \tag{13.50}$$

Using these equations of motion, $\frac{\partial}{\partial \Gamma} \cdot \left(f(0) \frac{d\Gamma}{dt} \right)$ becomes

$$\frac{\partial}{\partial \Gamma} \cdot \left[f(0) \frac{d\Gamma}{dt} \right] = -\frac{f(0)VE j_x}{k_B T} \tag{13.51}$$

where $j_x = (e/V) \sum_{i=1}^{N} z_i p_{xi}/m_i$ is the current density. Plugging this result into Eq. 13.44, we obtain the average of any phase variable B at time t as

$$\langle B(t) \rangle = \langle B(0) \rangle + \frac{VE}{k_B T} \int_0^t \langle B(s) \cdot j_x(0) \rangle \, ds \tag{13.52}$$

Choosing B such that $B(t) = j_x(t)$ allows us to obtain the conductivity. Since the equilibrium average $\langle j_x(0) \rangle$ is equal to 0, we obtain the following expression for $\langle j_x(t) \rangle$:

$$\langle j_x(t) \rangle = \frac{VE}{k_B T} \int_0^t \langle j_x(s) \cdot j_x(0) \rangle \, ds \tag{13.53}$$

Dividing by the field intensity, we obtain the conductivity as

$$\langle \sigma(t) \rangle = \frac{V}{k_B T} \int_0^t \langle j_x(s) \cdot j_x(0) \rangle \, ds \tag{13.54}$$

From a practical point of view, the TTCF approach consists of collecting the system's response over a large number of nonequilibrium trajectories. A scheme illustrating the TTCF approach is shown in Figure 13.4. To this end, we generate many equilibrium configurations during a long equilibrium trajectory (*i.e.*, with the field switched off). Each of these configurations is the starting point for a nonequilibrium trajectory. We show here some results obtained on molten sodium chloride for a density of 1500 kg.m^{-3} and at 1500 K for a system of 512 ions, with 256 ions of each type [74]. After an initial equilibration of the system, we generate a second simulation run over which we select 20,000 equilibrium configurations. For each configuration, we apply three different phase mappings to obtain three additional equilibrium configurations, either by reversing the sign of the momenta of

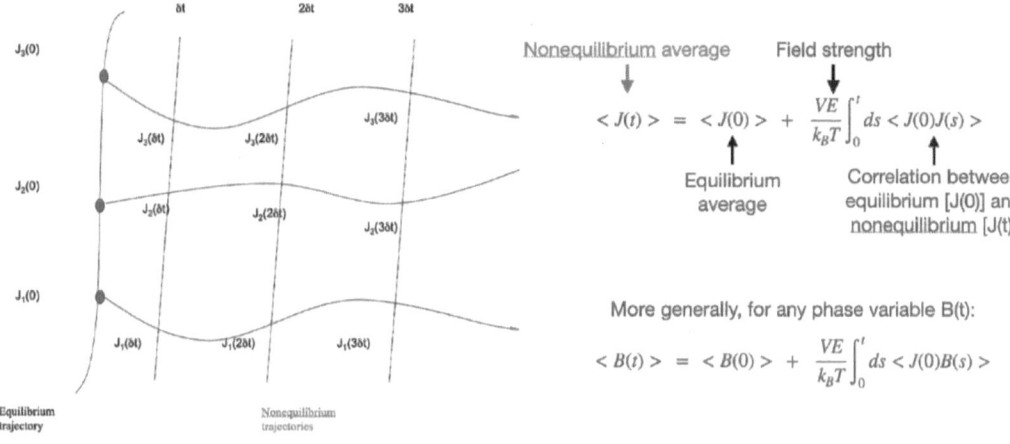

Figure 13.4 Schematic setup for a simulation leveraging the TTCF formalism. Many phase space points, shown as red dots on the left of the plot, are generated during an equilibrium trajectory (vertical line on the left). Each phase space point j, with a value for the nonequilibrium response function (or flux) $J_j(0)$, is the start of a nonequilibrium trajectory (horizontal lines) over which the time-dependent functions $J_j(t)$ and/or $B_j(t)$ are collected to calculate the average correlation products on the right. The integrals for the phase variables are also given on the right of the plot.

all ions (time-reversal mapping) or by changing the sign of x and p_x for all ions (mirror symmetry), or by applying both the time-reversal mapping and the mirror symmetry. The 80,000 configurations are then used as starting points for nonequilibrium trajectories generated using Eq. 13.47.

We show the current density as a function of time in Figure. The behavior is qualitatively the same for all fields, with a transient regime associated with a current density overshoot, followed by a steady state. For very strong fields, we have a vast signal-to-noise ratio. This means that nonequilibrium simulations reach a steady state rapidly, and the average for the current in the steady state exhibits a reliable estimate for the conductivity with a slight standard deviation. Figure shows the excellent agreement obtained between the TTCF average and a direct average, carried out over 80,000 nonequilibrium simulations according to

$$\langle j_x(t) \rangle = \frac{\sum_{i=1}^{N_t} j_x^i(t)}{N_t} \tag{13.55}$$

For a low intensity of the field and, thus, a low signal-to-noise ratio, the direct average for the current becomes dominated by the noise. No reliable estimate for the conductivity can be extracted via a simple average from the nonequilibrium simulations. On the other hand, the TTCF approach yields reliable results, even for the lowest intensity. We then focus on the comparison between the TTCF conductivity in the linear regime and the Green-Kubo estimate given by

$$\sigma = \frac{V}{3k_B T} \int_0^\infty \langle \mathbf{j}(t) \cdot \mathbf{j}(0) \rangle \, dt \tag{13.56}$$

As shown in Figure, the TTCF approach bridges the Green-Kubo estimate and the far-from-equilibrium conductivity obtained at large fields. Using the TTCF approach, it becomes possible to locate the onset of nonlinearity and changes in the properties of the ionic fluid for strong fields. The plot, shown in Figure, reveals a transition from a linear regime, with conductivity in excellent agreement with the Green-Kubo estimate, to a nonlinear behavior for field intensities beyond 10^3 V m^{-1}.

13.3.3 TRANSPORT CLOSE(R) TO EQUILIBRIUM

We now apply the TTCF approach to determine the shear viscosity in a system subjected to shear. From a practical point of view, we will follow the same approach as previously and collect the response of the system over a large number of nonequilibrium trajectories. More specifically, we select many equilibrium configurations over a long equilibrium trajectory, with the synthetic field switched off, and use each of these configurations as a starting point for a nonequilibrium trajectory. As discussed in the previous section, we also use phase space-mappings to augment the number of equilibrium configurations and thus of nonequilibrium trajectories over which we compute the system's response (we refer the reader to prior work [88] for details on the appropriate phase space-mappings in this case). Unlike the TTCF approach to conductivity which relied on the use of a real (electric) field, we generate the nonequilibrium trajectories with a synthetic shear field. To this end, we use the SLLOD equations of motion with the Lees-Edwards periodic boundary conditions. For a fluid undergoing a Couette flow in the x-direction with a velocity gradient along the y-direction, the SLLOD equations for atom i are given by

$$\dot{\mathbf{r}}_i = \frac{\mathbf{p}_i}{m} + \gamma y_i \mathbf{e}_x \tag{13.57}$$

$$\dot{\mathbf{p}}_i = \mathbf{F}_i - \gamma p_{yi} \mathbf{e}_x \tag{13.58}$$

We recall that applying the SLLOD equations of motion amounts to (i) initially superimposing the appropriate linear velocity profile to the actual velocities of the molecules of the fluid and (ii) afterward applying Newton's equations of motion to this fluid. The internal energy is defined as

$H_0(\Gamma) = \sum_i \mathbf{p}_i^2/(2m) + \phi(\mathbf{q}_i)$. The rate of change in internal energy due to shear is equal to $\dot{H}_0 = -\dot{\gamma} V P_{xy}$, where $P_{xy} = \sum_i (p_{xi}p_{yi}/m + F_{xi}y_i)/V$ is the opposite of the shear stress and V is the volume of the system. To account for the dissipation of viscous heat, we supplement the equations of motion with a Gaussian thermostat to obtain the following equations of motion:

$$\dot{\mathbf{r}}_i = \frac{\mathbf{p}_i}{m} + \dot{\gamma} y_i \mathbf{e}_x \tag{13.59}$$

$$\dot{\mathbf{p}}_i = \mathbf{F}_i - \dot{\gamma} p_{yi} \mathbf{e}_x - \alpha \mathbf{p}_i \tag{13.60}$$

where α is the thermostat multiplier, defined as

$$\alpha = \frac{\sum_{i=1}^{N}(\mathbf{p}_i \cdot \mathbf{F}_i - \gamma p_{xi}p_{yi})/m_i}{\sum_{i=1}^{N}\mathbf{p}_i^2/m_i} \tag{13.61}$$

Since we are interested in the shear rate response for a low shear rate, we do not anticipate any departure from the linear flow profile induced by the SLLOD equations of motion and Lees-Edwards periodic boundary conditions. In other words, the Gaussian thermostat is such that it keeps the usual expression of the temperature $\left(T = \frac{\sum_{i=1}^{N}\mathbf{p}_i^2}{3Nk_Bm_i} \right)$, based on the peculiar kinetic energy, constant.

For a system with a canonical distribution $f(0)$ at $t = 0$ and with a time evolution that follows the dynamics modeled by the isokinetic SLLOD equations, we have, to the first order in the number of particles,

$$(\partial/\partial\Gamma) \cdot (f(0)d\Gamma/dt) = \dot{\gamma} V P_{xy} f(0)/(k_B T) \tag{13.62}$$

The average of a phase variable B at time t is equal to

$$\langle B(t) \rangle = \langle B(0) \rangle - \frac{V\dot{\gamma}}{k_B T} \int_0^t \langle B(s)P_{xy}(0) \rangle ds \tag{13.63}$$

If we choose $B(t) = P_{xy}(t)$, the equilibrium average $\langle P_{xy}(0) \rangle$ is equal to 0, and we obtain the following expression for $\langle P_{xy}(t) \rangle$:

$$\langle P_{xy}(t) \rangle = -\frac{V\dot{\gamma}}{k_B T} \int_0^t \langle P_{xy}(s)P_{xy}(0) \rangle ds \tag{13.64}$$

It is then straightforward to evaluate the shear viscosity according to

$$\langle \eta(t) \rangle = \frac{V}{k_B T} \int_0^t \langle P_{xy}(s)P_{xy}(0) \rangle ds \tag{13.65}$$

The conventional nonequilibrium approach collects averages of the steady state's shear stress. This works very well at a very high shear rate where the signal-to-noise ratio and the steady-state average are reached very quickly, and shear stress fluctuations are very small compared to the shear stress average. This can be confirmed by simply taking the direct average of the shear stress over a set of nonequilibrium trajectories (Figure 13.5). If we denote by N_t the number of trajectories in the set, the direct average for the shear stress is defined as

$$\langle P_{xy}(t) \rangle = \frac{1}{N_t} \sum_{i=1}^{N_t} P_{xy,i}(t) \tag{13.66}$$

and the direct average for the shear viscosity is

$$\langle \eta(t) \rangle = \frac{-1}{\gamma N_t} \sum_{i=1}^{N_t} P_{xy,i}(t) \tag{13.67}$$

Such averages become dominated by the noise at low shear rates when the signal-to-noise ratio becomes too weak. As shown in Figure 13.6, the TTCF approach is particularly advantageous over conventional nonequilibrium simulation methods since it allows access to the fluid's response at low shear rates, *i.e.*, for shear rates that can be achieved in experiments.

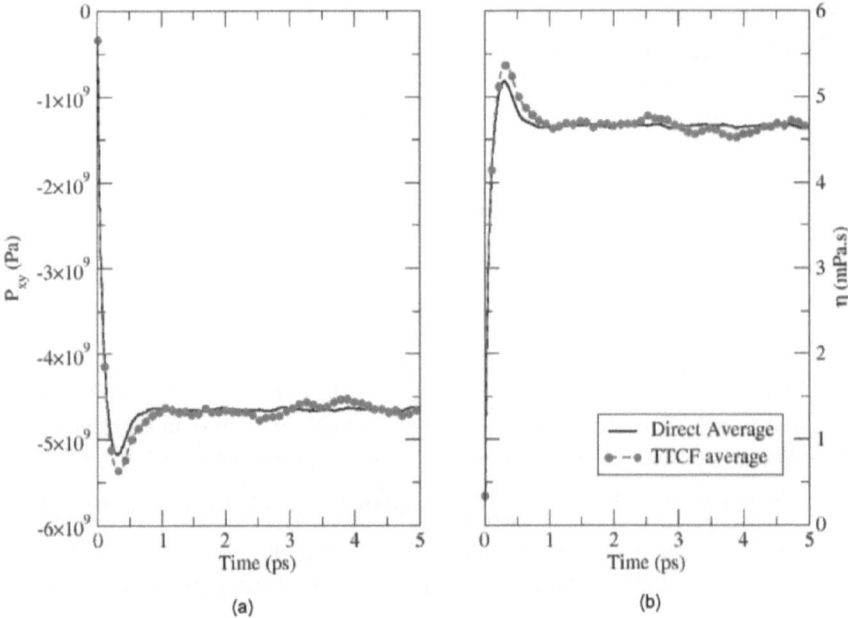

Figure 13.5 Comparison between TTCF averages and direct averages for the shear stress (a) and viscosity (b). The results here are shown for liquid lead at 1925 K and a density of 18.4 g/cm³. The shear rate is extremely large $\gamma = 1 \times 10^{12}\ s^{-1}$. The signal-to-noise ratio is huge in this case, and the TTCF and direct averages coincide.

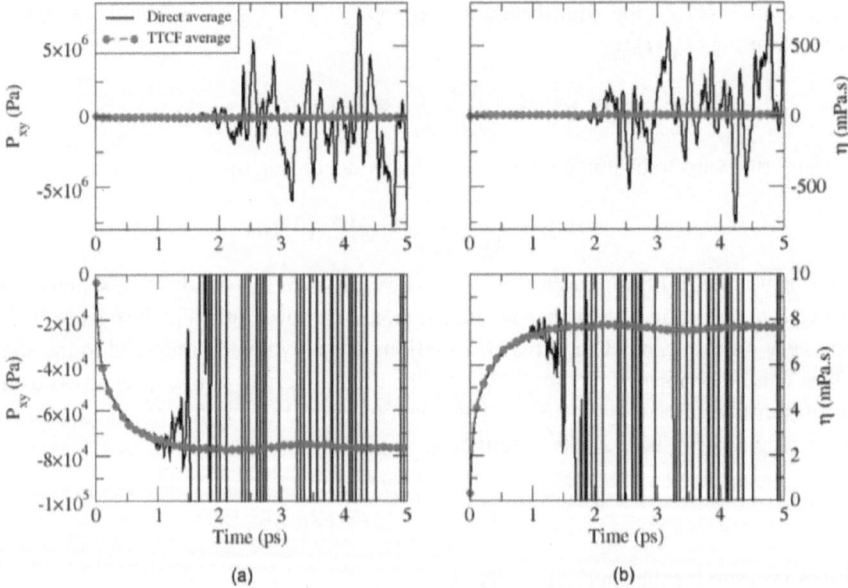

Figure 13.6 Comparison between TTCF averages and direct averages for the shear stress (a) and for the viscosity (b). The results are shown for liquid lead at 1925 K and a density of 18.4 g/cm³ as in Figure 13.5, but the shear rate is much lower by five orders of magnitude. As a result, the signal-to-noise ratio is very low, and the direct average becomes unreliable. On the other hand, the TTCF approach continues to provide reliable and accurate results regardless of how low the shear rate is. In both panels, the top plots show the full range of values taken by the direct average, while the bottom plots are scaled to show the TTCF average).

14 Fluctuation Theorems, Molecular Machines and Emergent Behavior in Active Matter

"There is a lack of unifying principles in nonequilibrium statistical mechanics, compared to the equilibrium case. So it is not surprising that the fluctuation theorem has received a lot of attention."

Ramses van Zon and Ezechiel Godert David Cohen,
Extension of the fluctuation theorem [377]

This quote from van Zon and Cohen echoes in two respects the statement from Stengers and Prigogine we placed at the beginning of Chapter 13. First, as we pivot towards nonequilibrium systems, the fundamental difference between an equilibrium and a non-equilibrium system is the absence of the well-defined criteria of evolution based on the pursuit of an extremum for a thermodynamic potential [317]. Examples discussed in Parts I & II of the book included the definition of an equilibrium state as corresponding to a maximum of entropy at constant (N, V, E) in the microcanonical case, a minimum of Helmholtz free energy at constant (N, V, T) in the canonical case, and a minimum of Gibbs free energy at constant (N, P, T) in the isothermal-isobaric ensemble. Second, the term "fluctuation" stands out in the quotes, and both statements point to a characterization of nonequilibrium states, and most notably, nonequilibrium steady states via the behavior of fluctuations rather than simply in terms of steady-state averages. They also highlight the role played by entropy production, which is a hallmark of time-reversal symmetry breaking or, in other words, irreversibility and provides a signature of the system's departure from equilibrium [308]. As we will see in this chapter, applying the concepts of fluctuation and entropy production to living systems has led to a new understanding of the far-from-equilibrium physics that governs these systems. For instance, by introducing an invariant measure for nonequilibrium steady states [316,333], fluctuation theorems [41,125,128,129,147,148] quantify the non-zero probability that, over finite time intervals, entropy production becomes negative. This, in turn, results in observations contrary to the second law of thermodynamics. Furthermore, extensions of these theorems show that nonequilibrium work measurements can yield free energy differences between equilibrium states [66,201], opening the door to an experimental confirmation of these theoretical developments [57,62,173] and the design of synthetic molecular motors and nanomachines [55,153,180,215,290,302,326,365,366,381].

14.1 FLUCTUATION THEOREMS

14.1.1 FORMALISM

To introduce the fluctuation theorem (FT), we focus on a specific nonequilibrium system consisting of a fluid subjected to shear. To this end, we generate nonequilibrium trajectories using the set of nonequilibrium equations we have introduced in Chapter 13, *i.e.*, the SLLOD equations. Using these equations of motion, we generate nonequilibrium trajectory segments of finite duration and evaluate the probability of observing a given shear stress fluctuation in the steady state. The FT predicts a non-zero probability that, over limited time intervals, the shear stress reverses its sign, which violates the second law of thermodynamics. A fluid undergoing shear flow produces heat typically dissipated to a heat bath. However, the FT predicts that the fluid under shear can absorb heat from the

DOI: 10.1201/9781003006411-17

heat bath, leading to trajectories over which there is a negative entropy production. The FT provides a way to quantify the entropy production fluctuations in a dissipative nonequilibrium system. We use the following set of nonequilibrium equations of motion

$$
\begin{aligned}
\dot{\mathbf{q}}_i &= \frac{\mathbf{p}_i}{m} + \hat{\mathbf{x}}\gamma y_i \\
\dot{\mathbf{p}}_i &= \mathbf{F}_i - \hat{\mathbf{x}}\gamma p_{yi} - \alpha\mathbf{p}_i
\end{aligned}
\tag{14.1}
$$

where shear is applied along the x direction ($\hat{x}$ is the unit vector for the x axis) and the velocity gradient is along the y direction. The shear rate is denoted by γ, the force exerted on particle i as $\mathbf{F}_i$, and heat is dissipated to the heat bath through the $\alpha\mathbf{p}_i$ term. Here, we use an ergostat [106] that always keeps the rate of change in internal energy equal to zero. Writing the rate of change in internal energy as

$$
\dot{H} = -\gamma\sum_{i=1}^{N}\left(\frac{p_{x_i}p_{y_i}}{m} + F_{x_i}y_i\right) - \alpha\sum_{i=1}^{N}\frac{\mathbf{p}_i^2}{m}
\tag{14.2}
$$

where the first term on the right-hand side corresponds to the time derivative for the work function $\dot{W} = -\gamma P_{xy}V$, while the second term corresponds to the time derivative for the heat transferred to the system $\dot{Q} = -\alpha\sum_{i=1}^{N}\frac{\mathbf{p}_i^2}{m}$. To keep the constant internal energy constant ($\dot{H} = 0$), the instantaneous value for α has to satisfy the following relation

$$
\alpha = -\frac{-\gamma P_{xy}V}{\sum_{i=1}^{N}\frac{\mathbf{p}_i^2}{m}}
\tag{14.3}
$$

where $P_{xy}V = \sum_{i=1}^{N}\left(\frac{p_{x_i}p_{y_i}}{m} + F_{x_i}y_i\right)$ denotes the shear stress.

We now focus on fluctuations of the average amount of heat dissipated over trajectory segments of length τ in the steady state [125,128,147,148,377]. For a dissipative system, the FT specifies the properties of the probability distribution $p(\bar{\Omega}_\tau)$, that is the probability of observing a given value of $\bar{\Omega}_\tau$ for the average dissipation over a time interval τ in the steady state. The FT characterizes the ratio of observing the average dissipation of opposite signs over a trajectory of length τ through the following relation:

$$
\frac{p(\bar{\Omega}_\tau)}{p(-\bar{\Omega}_\tau)} \sim e^{\tau\bar{\Omega}_\tau}
\tag{14.4}
$$

where the symbol $\sim$ indicates that this result becomes exact for a large enough τ and $\tau\bar{\Omega}_\tau$ corresponds to a generalized entropy production over a time interval τ. FT is thus a theorem for entropy production, which applies to systems arbitrarily far from equilibrium regardless of how large $\bar{\Omega}_\tau$ and its fluctuations. This relation is also called a large deviation theorem [158,213,219,236].

Cohen and Gallavotti showed that this generalized entropy production is associated with dissipation and phase space compression. For the nonequilibrium equations of motion given in Eq. 14.1, the dissipation function is provided by

$$
\Omega(t) = \frac{\partial}{\Gamma}\cdot\dot{\Gamma} = \sum_{i=1}^{N}\left(\frac{\partial}{\partial\mathbf{q}_i}\cdot\dot{\mathbf{q}}_i + \frac{\partial}{\partial\mathbf{p}_i}\cdot\dot{\mathbf{p}}_i\right) = 2N\alpha(\Gamma(t))
\tag{14.5}
$$

The average over a time interval of length τ is given by

$$
\bar{\Omega}_\tau = \frac{1}{\tau}\int_0^\tau\Omega(s)ds = \frac{1}{\tau}\int_0^\tau 2N\alpha(s)ds = 2N<\alpha>_\tau
\tag{14.6}
$$

This result provides further insight on why $\bar{\Omega}_\tau$ can be interpreted as the entropy production. Using the equipartition principle for the kinetic energy, we find that $\dot{Q} = -\frac{2N\alpha}{\beta}$, in which $\beta = 1/(k_BT)$. This yields

$$
\bar{\Omega}_\tau = \frac{1}{\tau}\int_0^\tau 2N\alpha((s))ds = -\frac{1}{\tau}\int_0^\tau\beta((s))\dot{Q}(s)ds
\tag{14.7}
$$

$\bar{\Omega}$ is thus the average of $-\beta\dot{Q}$. Since $\dot{Q}$ is the rate at which heat is absorbed, averaging $-\beta\dot{Q}$ over the time interval τ yields the average entropy production rate.

Finally, we plug in the equation for α (Eq. 14.3) to obtain

$$\bar{\Omega}_\tau = \frac{-\gamma V}{\tau} \int_0^\tau \beta(\Gamma(s))P_{xy}(\Gamma(s))ds \tag{14.8}$$

In other words, heat fluctuations, and thus entropy production fluctuations, are closely related to work fluctuations here. Applied to our system, the FT then follows as

$$\frac{p(\Omega_\tau)}{p(-\Omega_\tau)} = \exp\left(2N\langle\alpha\rangle_\tau \tau\right) \tag{14.9}$$

14.1.2 NEGATIVE ENTROPY PRODUCTION TRAJECTORIES

We discuss in this section how to obtain the probability distribution for an average dissipation Ω_τ over trajectory segments of length τ. To simulate a system under shear, we integrate the nonequilibrium equations of motion [106] given in Eq. 14.1 using a fourth-order Gear predictor-corrector algorithm and the Lees-Edwards periodic boundary conditions. Since the analysis is carried out in the steady state, we generate a long nonequilibrium trajectory which, when divided into trajectory segments of length τ, will yield many such segments (of the order of 10^6).

We first examine the probability distribution $p(P_{xy\tau})$. Here $P_{xy\tau}$ denotes the average shear stress over a trajectory segment of length τ. We show in Figure 14.1 examples of $p(P_{xy\tau})$ for different τ values. As expected, the distribution is centered around a negative value for the shear stress, characteristic of the response of the fluid to the applied shear. However, the plot shows that the

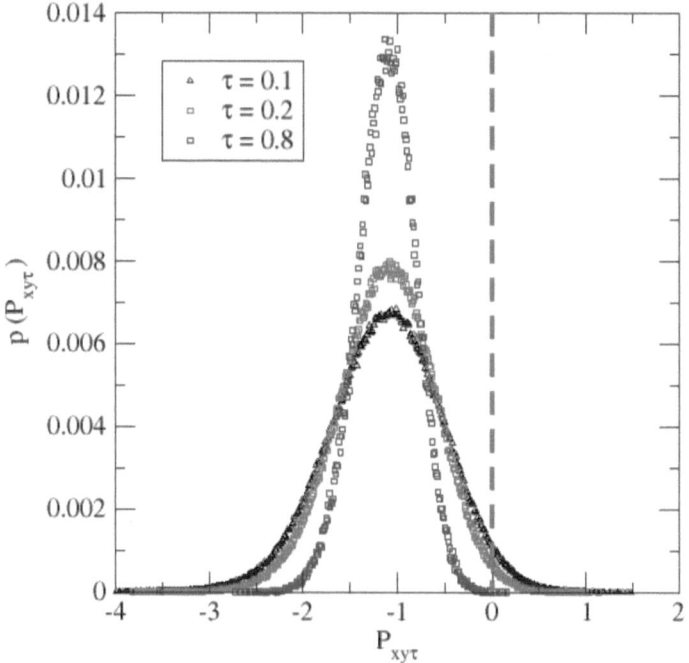

Figure 14.1 Probability distribution $p(P_{xy\tau})$ for segments of duration τ in a system of $N = 56$ disks under shear. The distribution is shown for τ ranging from $\tau = 0.1$ to $\tau = 0.8$. The dashed line indicates the sign change for $P_{xy\tau}$.

distribution does not vanish for $P_{xy\tau} > 0$. This means that $P_{xy\tau}$ can take positive values and reverse sign over length segments τ. Such segments represent a fraction of 4.5% of all segments for $\tau = 0.1$. It is, of course, much more probable to observe a negative value for $P_{xy\tau}$ with, for instance, a ratio of $1 : 4$ when comparing $p(P_{xy\tau}) = 0.2$ to $p(P_{xy\tau}) = -0.2$ and a ratio of about $1 : 8$ when comparing $p(P_{xy\tau}) = 0.3$ to $p(P_{xy\tau}) = -0.3$. The probability of observing segments with positive $P_{xy\tau}$ also decreases with τ as its fraction of the overall number of segments decreases from 4.5% to 1.2% when τ increases from 0.1 to 0.2.

The second quantity we examine is the probability distribution $p(\bar{\Omega}_\tau)$ for the average dissipation $\bar{\Omega}_\tau$ over a segment of length τ, or, equivalently, the average rate of entropy production over this segment. The distribution $p(\bar{\Omega}_\tau)$ shown in Figure 14.2 confirms the existence of trajectory segments associated with a negative rate of entropy production. As observed for $p(P_{xy\tau})$, the fraction of such segments decreases as their length τ increases. Indeed, the fraction of trajectory segments with a negative entropy production decreases from 2.76% at $\tau = 0.1$, to 1.18% at $\tau = 0.2$, and to 0.2% at $\tau = 0.4$. To test the expression of the fluctuation theorem obtained in Eq. 14.9, we compute the function $\Pi(\bar{\Omega}_\tau)$, which denotes the ratio of the probability of observing a given positive value for $\bar{\Omega}_\tau$ over a segment of length to the probability of observing its opposite value $-\bar{\Omega}_\tau$. This function can be written as

$$\Pi(\bar{\Omega}_\tau) = \frac{1}{2N\tau} \ln \left[\frac{p(\bar{\Omega}_\tau)}{p(-\bar{\Omega}_\tau)} \right] \tag{14.10}$$

The plot in Figure 14.3 provides a graphical verification of the fluctuation theorem of Eq. 14.9, as we observe an excellent agreement between the probability ratio (left-hand side of Eq. 14.9) and the average value of $- <\alpha>_\tau$ (right-hand side of Eq. 14.9).

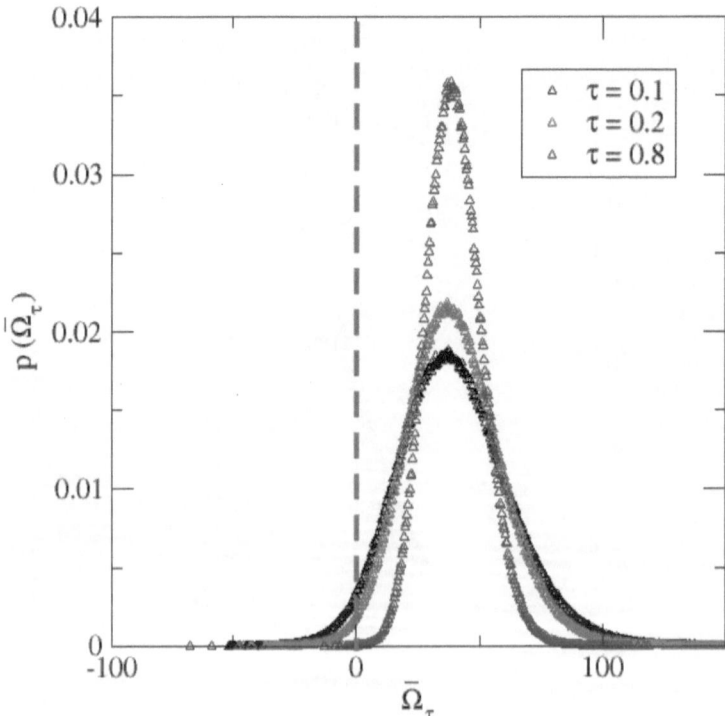

Figure 14.2 Probability distribution $\bar{\Omega}_\tau$ for segments of duration τ, with τ ranging from $\tau = 0.1$ to $\tau = 0.8$. The dashed line corresponds here to an entropy production of zero.

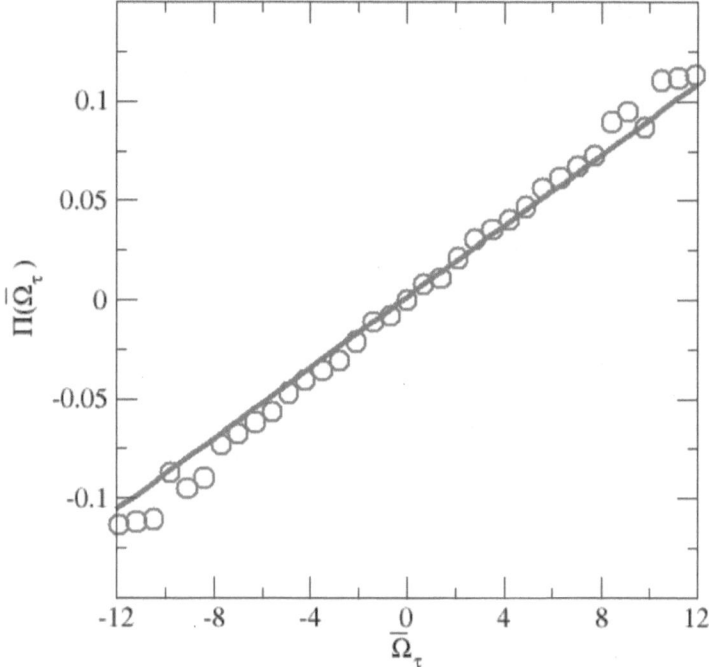

Figure 14.3 $\Pi(\bar{\Omega}_\tau)$ against $\bar{\Omega}_\tau$ (circles), together $\langle \alpha \rangle_\tau$ (line) for $\tau = 0.4$.

We finally assess the impact of the system size on the probability of observing trajectory segments associated with a negative entropy production. As shown in Eq. 14.8, $\bar{\Omega}_\tau$ depends on N and V and thus increases with system size. Figure 14.4 shows that the distribution for $N = 56$ exhibits a negative tail with $\bar{\Omega}_\tau < 0$ and that no such tail for $N = 448$. Integrating the distributions for $\bar{\Omega}_\tau$ leads to a fraction of trajectory segments with a negative entropy production of 0.2% for $N = 56$.

We have seen so far that the fluctuation theorem predicts the existence of trajectory segments associated with a negative entropy production that does not conform to the second law of thermodynamics. The simulation results on simple molecular systems outlined in this section confirm the FT predictions and point to an enhanced probability of observing such negative entropy production trajectories on short length- and timescales, which is especially relevant to nanosized systems and molecular machines as we will discuss in the following sections.

14.1.3 FREE ENERGY DIFFERENCES

The fluctuation theorem we discussed in the previous section belongs to a series of nonequilibrium relations that apply to a wide range of non-equilibrium systems. Among others, fluctuation theorems have been derived for deterministic systems either in the steady state [148] or in the transient regime [324], and for stochastic systems [213,219,236]. Another related expression is known as the Jarzynski equality [201]. This equality allows us to determine the difference in free energy between two equilibrium ensembles through a series of nonequilibrium trajectories. Indeed, if we denote by ΔA the free energy difference between two equilibrium ensembles and by W the amount of work necessary to switch between these two ensembles over a finite time interval, the Jarzynski equality predicts that

$$\left\langle e^{-\beta W} \right\rangle = e^{-\beta \Delta A} \tag{14.11}$$

where $\langle .. \rangle$ indicates an average performed over the series of nonequilibrium switching trajectories.

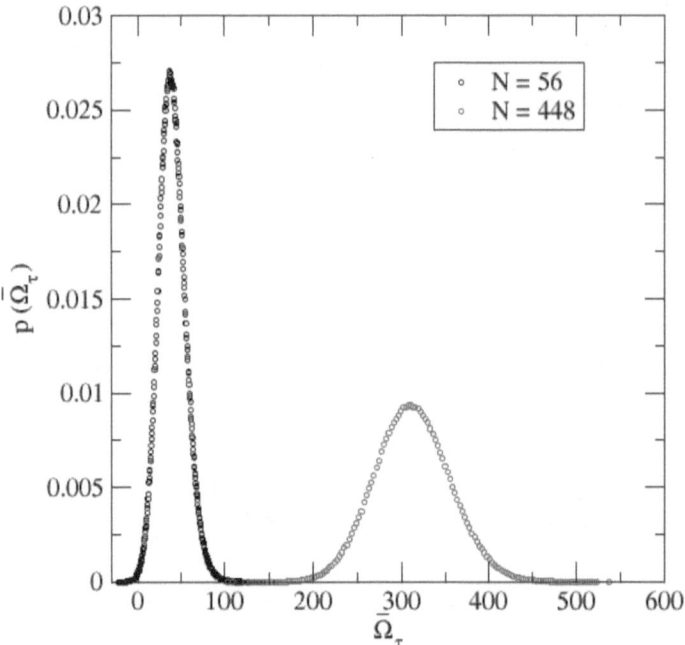

Figure 14.4 Probability distributions of $\bar{\Omega}_\tau$ for $\tau = 0.4$ and system sizes of $N = 56$ and $N = 448$.

Crooks [66] proposed a generalization of the fluctuation theorem, as summarized by the following identity

$$\frac{p_f(+\omega)}{p_r(-\omega)} = e^\omega \tag{14.12}$$

where ω denotes the entropy production of a non-equilibrium system over a given time interval, $p_f(\omega)$ is the probability for the entropy production, and $p_r(\omega)$ is the probability for the entropy production when the nonequilibrium field is applied in a time-reversed manner. Crooks also provides an example, based on isothermal compressions and expansions of gas, to shed light on how the two probabilities p_f and p_r can be evaluated. In this example, the gas is enclosed in a chamber, whose walls allow for heat exchanges with the surroundings and maintain the temperature of the gas constant throughout compression or expansion processes. Initially, the gas is in equilibrium, occupying a volume V and having a temperature T. If we push on the piston, we carry out an isothermal compression of the gas to a smaller volume V'. The compression corresponds to applying the perturbation in the forward sense, and the amount of entropy production can be collected over many repetitions of the compression process. Similarly, starting from the gas in equilibrium at the new volume V', pulling on the piston will result in an isothermal expansion of the gas, *i.e.*, the application of the perturbation in the reverse sense. As for the compression process, the entropy produced can also be evaluated over many repetitions of the expansion process, yielding the two probability distributions p_f and p_r. Interestingly, this example also connects with the steps outlined by the Jarzynski equality since it describes a switching process between two equilibrium states at V and V', respectively. Just as we describe the collection of entropy production over many repetitions of the switching process, we can envision collecting the work during many repetitions. While the (average) macroscopic work involved when switching isothermally from V to V' is constant and equal, *e.g.*, for an ideal gas, to $W = -nRT \ln \left[\frac{V'}{V}\right]$, the microscopic trajectories and thus, the work, heat exchanged, and entropy production take different values for each repetition of the switching process. This is because the initial conditions of the system, that is, the positions and velocities for each of the atoms belonging to the system, all vary with time and thus result in different nonequilibrium switching trajectories. Crooks showed that the generalized fluctuation theorem and the Jarzynski equality were related. Starting from Gallavotti's result [145], we can write that

$$\langle e^{-\omega} \rangle = \int_{-\infty}^{+\infty} P_F(+\omega)e^{-\omega}d\omega = \int_{-\infty}^{+\infty} P_R(-\omega)e^{-\omega}d\omega = 1 \tag{14.13}$$

If the system starts from an equilibrium state, we can apply the generalized fluctuation, and the entropy production is given by $\omega = -\beta\Delta A + \beta W$ for a system initially at equilibrium. Plugging the equation for ω into Eq. 14.13 yields

$$\left\langle e^{\beta\Delta A - \beta W} \right\rangle = 1 \tag{14.14}$$

Since ΔA is state-dependent-only, it can be moved outside the average to yield

$$e^{\beta\Delta A} \cdot \left\langle e^{-\beta W} \right\rangle = 1 \tag{14.15}$$

and we recover the Jarzynski equality as

$$\left\langle e^{-\beta W} \right\rangle = e^{-\beta\Delta A} \tag{14.16}$$

14.2 TOWARDS A NEW PHYSICS OF LIVING SYSTEMS

14.2.1 WORK RELATIONS AND RNA FOLDING

The fluctuation relations rapidly emerged as a promising tool to study biological objects and further our understanding of living systems. In 2002, Liphardt et al. [226] performed the first tests of the Jarzynski equality and the Crooks generalized fluctuation theorem. Their work involved mechanically stretching a single molecule of RNA between two conformations. Specifically, they focused on the mechanical unfolding of single RNA molecules derived from the P5abc domain of the *Tetrahymena thermophila* group 1 intron. This domain is of particular interest for testing the fluctuation relations since it is possible for P5abc to unfold reversibly when stretched slowly and irreversibly when stretched rapidly. It is possible to model this "switching" process with a single one-dimensional variable. Indeed, an appropriate choice for the reaction coordinate spanning the switching pathway is the molecular end-to-end extension. From a practical standpoint, to implement the switching process, Liphardt et al. attached a RNA molecule to polystyrene beads via RNA-DNA hybrid handles. The two beads to which the RNA molecule is attached play different roles. Using a force-measuring optical trap [45,227], the left bead is held in a fixed position while the right bead, connected to a piezoelectric actuator through a micropipette tip, is moved to stretch the RNA molecule. Applying Jarzynski's equality to the stretching of the RNA molecule, we can relate the free energy difference associated with the two states 0 and z along a reaction coordinate, $\Delta G(z)$, to the work done to switch the system between two states irreversibly,

$$\exp[-\beta\Delta G(z)] = \lim_{N \to \infty} \langle \exp[-\beta w_i(z,s)] \rangle \tag{14.17}$$

where the free energy difference is here measured in terms of the Gibbs free energy since the experiment is carried out under constant temperature and pressure, $\langle .. \rangle$ indicates that the average is taken over N switching trajectories, $w_i(z,s)$ represents the amount of work for trajectory i carried out at the switching rate s. $w_i(z,s)$ required to switch is given by

$$w_i(z,s) \approx \int_0^z f_i(z',s)dz' \tag{14.18}$$

where f_i is the external force applied to the system and is measured through the deflection of the trapping laser beams with position-sensitive photodetectors [226]. During the experiments, the P5abc domain was unfolded either slowly, with stretching rates ranging from 2 to 5 pN/s, or rapidly, with stretching rates between 34 and 52 pN/s rates, with data collected with different RNA molecules and several tens of unfolding-refolding cycles per switching rate. The slow stretching rate provided an estimate of the change in Gibbs free energy during a reversible switching of $60.2 \pm 1.6\,k_BT$. On the other hand, carrying out the unfolding process at a fast stretching rate allowed for the test of the Jarzynski equality under far-from-equilibrium conditions and provided a value for ΔG via Eq. 14.17 of $59.6 \pm 0.2\,k_BT$. The excellent agreement between estimates allowed for the first experimental validation of the Jarzynski equality and opened the door to studying folding free energy landscapes via applying fluctuation relations to stretching experiments.

14.2.2 MUTATING, STRETCHING, BINDING, AND UNBINDING

The fluctuation relations have then been used to study the impact of several biological processes [192], including mutations. In their 2005 study, Collin et al. determined experimentally the difference in free energy between an RNA molecule and a mutant that differs by a single base pair [62]. The idea of the study is to leverage the Crooks fluctuation relation and, more particularly, the symmetry between the work fluctuations associated with the forward and reverse changes a system undergoes during switching from one equilibrium state to another. As with the unfolding of RNA discussed in the previous section, the experiments use optical tweezers to apply a mechanical force to a biomolecule and carry out the unfolding or refolding processes. Collin et al. apply their approach to evaluating the work associated with unfolding and refolding an RNA hairpin and an RNA three-helix junction [62]. The unfolding and refolding experiments are carried out by moving the optical trap. Suppose the optical trap moves in the direction that stretches the biomolecule. In that case, the biomolecule unfolds during the experiment, and the work done on the system can be collected according to the protocol discussed in the previous section. On the other hand, if the optical trap moves in the opposite direction, the biomolecule refolds, and similarly, the corresponding refolding work can be collected during the experiment. Furthermore, the optical trap is moved at the same speed in both directions to ensure the time-reversal symmetry of the unfolding-refolding trajectories. Repeating the unfolding-refolding experiments often allows us to gather the two probability distributions $p_u(W)$ and $p_r(W)$ corresponding to the work during unfolding and refolding, respectively. Using the Crooks generalized fluctuation theorem, Collin et al. obtain the following relation between the two probability distributions

$$\frac{p_u(W)}{p_r(-W)} = \exp\left(\frac{W - \Delta G}{k_BT}\right) \qquad (14.19)$$

where ΔG is the change in free energy between the two equilibrium states connected by the switching pathway. When the two distributions intersect, we have $p_u(W) = p_r(-W)$ and thus $W = \Delta G$ (see Figure 14.5). This approach was used to analyze the work distributions of unfolding-refolding for the siRNA hairpin that targets the mRNA of the CD4 receptor of the Human Immunodeficiency Virus [252] and found that the work for which the two work distributions intersect is independent of the pulling speed [62]. In addition, this approach was also used to determine the difference in folding free energy between mutants of an RNA three-helix junction, *i.e.*, a wild type and a $C \cdot G$ to $G \cdot C$ mutant ($C754G - G587C$). The results showed that the Crooks generalized fluctuation theorem accurately estimated the difference in folding free energy for two RNA molecules that only differed by a single base pair in 34 base pairs.

Finally, fluctuation relations can also provide insight into the stabilization associated with binding ions to biomolecules. For instance, determining the folding free energy for a biomolecule with (and without) a specific ion gives access to the free energy of stabilization by the ion.

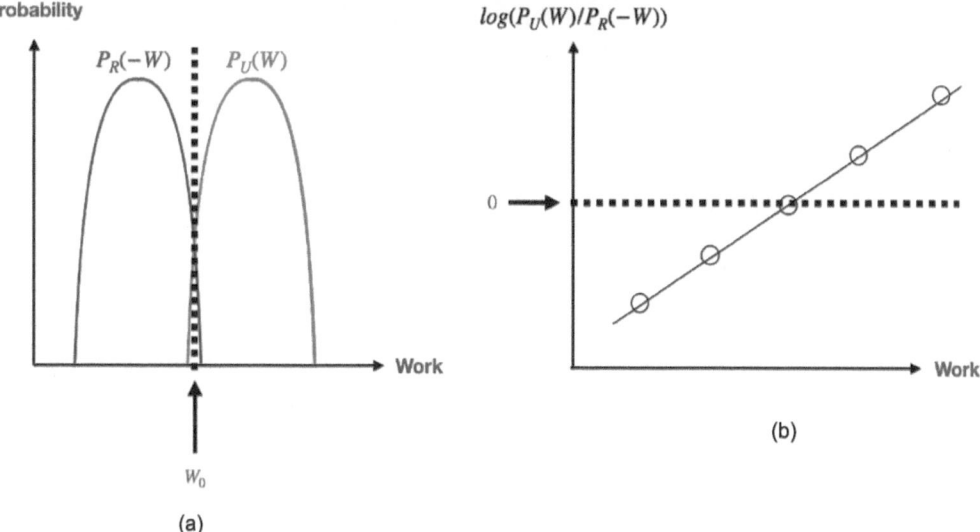

Figure 14.5 Estimating the folding free energy from work probability distributions. (a) The crossing of the two distributions (refolding $P_R(-W)$ and unfolding $P_U(W)$) at $W = W_0$ yields the Gibbs free energy ΔG. (b) The linear variation of the experimental $\log[p_u(W)/p_r(-W)]$ is used to confirm the fluctuation theorem.

14.2.3 FREE ENERGY CALCULATIONS VIA STEERED MD

As discussed by Park et al. [283], molecular trajectories may be used to compute the free energy profile of the molecular system as a function of a given reaction coordinate. Such a free energy profile is termed a potential of mean force. The potential of mean force $\mathscr{F}$ can be written as

$$\exp[-\beta\mathscr{F}(\xi_0)] = \int d\mathbf{r}d\mathbf{p}\delta[\xi(\mathbf{r}) - \xi_0]\exp[-\beta\mathscr{H}] \tag{14.20}$$

where ξ denotes the reaction coordinate, ξ_0 is a specific value of the reaction coordinate for which we wish to evaluate the system's free energy, and $\mathscr{H}$ is the Hamiltonian. In the rest of the section, the dependence of ξ upon the positions $(\mathbf{r})$ is implied, and we denote $\xi(\mathbf{r})$ as simply ξ. The computation of free energy via determining the probability distribution of work can also be implemented in nonequilibrium trajectories. Examples of numerical analogs to the pulling experiments include the simulation method known as steered molecular dynamics [283]. To this end, steered molecular dynamics incorporates an additional potential energy h into the Hamiltonian for the system. The potential energy is usually chosen as a harmonic function of the reaction coordinate ξ spanning the switching pathway

$$h_{\xi_0} = \frac{k}{2}[\xi - \xi_0]^2 \tag{14.21}$$

where k is the spring constant and the potential h drives the system towards a target value ξ_0 for the reaction coordinate ξ. With this new extra potential energy, the total Hamiltonian of the system becomes $\tilde{\mathscr{H}} = \mathscr{H} + h(\xi_0)$, where $\mathscr{H}$ stands for the Hamiltonian of our original system. To mimic a "pulling" experiment and the motion of the optical trap, the target value ξ_0 needs to be time-dependent and, for instance, can be chosen to vary linearly with time

$$\xi_0(t) = \xi_0(0) + vt \tag{14.22}$$

Applying the Jarzynski equality yields the following relation [283] between the free energy change $F(\xi_0(t)) - F(\xi_0(0))$ and the work $W(0 \to t)$ along the switching pathway

$$\exp\{-\beta[F(\xi_0(t)) - F(\xi_0(0))]\} = \langle \exp(-\beta W(0 \to t)) \rangle \qquad (14.23)$$

where the average $\langle .. \rangle$ is taken over a set of trajectories with initial conditions drawn from the canonical ensemble, and the work $W_{0 \to t}$ is defined as

$$W(0 \to t) = \int_0^t dt' \frac{\partial \xi_0 t'}{\partial t'} \left[\frac{\partial \tilde{H}(\xi_0)}{\partial \xi_0} \right]_{t'} \qquad (14.24)$$

We finally need to relate the free energy F obtained over the steered MD trajectories to the system's potential of mean force $\mathscr{F}$. We start from the definition of F given by

$$\exp[-\beta F(\xi_0)] = \int d\mathbf{r} d\mathbf{p} \exp[-\beta \tilde{H}] \qquad (14.25)$$

which becomes, when recast as a function of $\mathscr{F}$,

$$\begin{aligned}
\exp[-\beta F(\xi_0)] &= \int d\mathbf{r} d\mathbf{p} \exp\left\{ -\beta H - \frac{\beta k}{2}[\xi - \xi_0)^2] \right\} \\
&= \int d\xi \exp\left[-\beta \Phi(\xi) - \frac{\beta k}{2}(\xi - \xi_0)^2 \right]
\end{aligned} \qquad (14.26)$$

If k is sufficiently large (stiff spring approximation), ξ is maintained very close to ξ_0 thus leading to $F(\xi_0) \sim \mathscr{F}(\xi_0)$. Plugging this result into Eq. 14.23 yields the potential of mean force as

$$\mathscr{F}(\xi_0(t)) = \mathscr{F}(\lambda_0) - \frac{1}{\beta} \langle \exp(-\beta W_{0 \to t}) \rangle \qquad (14.27)$$

This expression, as well as its expansion in terms of the cumulants for W, has been leveraged to determine free energy changes relevant to biological processes including, among others, helix-coil transitions and other conformational changes in model proteins [183,283] and drug discovery [70].

14.3 EMERGENCE IN ACTIVE MATTER

14.3.1 DRY ACTIVE MATTER

Active matter [122,242,301] emerged in recent years as an essentially new field in nonequilibrium statistical mechanics. In active matter, each particle runs its engine to achieve self-propulsion and is thus out-of-equilibrium itself [238]. The self-propulsion process involves transducing chemical energy, induced by a chemical reaction into mechanical energy associated with the motion of active particles. Energy transduction is a hallmark of living systems and accounts for the motion of, among others, *E. coli* bacteria [21], arguably the most famous example of active particles. Bacteria often form active clusters and assemblies, giving rise to the onset of collective behavior and emergence in living matter. Before focusing on the properties and behavior of biological "swimmers", or wet active matter, like bacteria, we discuss a set of minimal nonequilibrium equations of motion for dry active matter, known as the Vicsek model [382]. This model sheds light on the transition from isolated active particles to collective motion or flocking [51,52,160,336]. Flocking takes place on a wide range of length and time scales, with the formation, for instance, of bacterial colonies at the microscopic level [400] to the formation of a herd of mammals on the macroscopic level [161]. The Vicsek equations of motion model the dynamics for a system of N self-propelled particles that

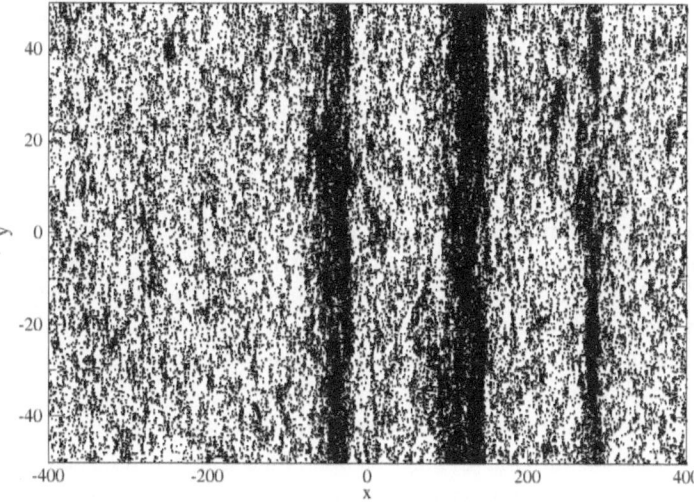

Figure 14.6 Flocking transition in the Vicsek model, with the onset of dark traveling bands in the system.

interact through a local alignment rule. The equations of motion for the two-dimensional active particle j are given by

$$\mathbf{r}_j(t+\Delta t) = \mathbf{r}_j(t) + \Delta t v_0 \mathbf{s}_j(t+\Delta t)$$

$$\theta_j(t+\Delta t) = \text{Arg}\left[\sum_{k\in D_j} e^{i\theta_k(t)}\right] + N_{D_j}\eta\xi_j \qquad (14.28)$$

$$\mathbf{v}_j(t+\Delta t) = v_0 e^{i\theta_j(t+\Delta t)}$$

where $\mathbf{r}_j$ denotes the position of atom j, $\mathbf{v}_j = v_0\mathbf{s_j}$ its velocity, $\mathbf{s}_j = (\cos\theta_j, \sin\theta_j)$ its heading, ξ_j a random angle between $-\pi$ and π, and η the noise intensity. In the equation for θ_j, the average heading at time t is evaluated over all neighboring particles k within a unit disk D_j around particle j. The random angle ξ_j is a zero-average, delta-correlated scalar white noise [160]. Recent work has identified the key features of the Vicsek model that trigger the onset of collective motion, which includes (i) the self-propelling velocity v_0, (ii) the complex changes in their relative positions as a result of velocity fluctuations, and (iii) the absence of momentum conservation [160]. The latter is specific to the Vicsek case since momentum transfers from the microswimmers to the surrounding fluid in wet active matter.

This simple set of nonequilibrium equations illustrates the impact of noise on the transition to collective motion (see Figure 14.6). A system modeled with the Vicsek equations exhibits a disordered gas-like phase for a high noise intensity η, a microphase separation with propagating bands of high density for intermediate noise levels, and a polar liquid for a low noise intensity [51,52,160]. Similarities between the phase transitions in the Vicsek model with the liquid-gas transition have recently been highlighted [336,370], in agreement with results obtained with hydrodynamic equations for flocking models [47,195,196].

14.3.2 ACTIVE BROWNIAN MATTER & MIPS

As discussed in the previous section, the Vicsek equations of motion are not well suited to simulate nonequilibrium trajectories for biological swimmers and, more generally, for wet active matter. This led to the development of the equations of motion for the Active Brownian Particle (ABP) model [46,115,150,311,338,340,341,385]. According to the ABP model [136,137], particles are modeled as soft repulsive disks, characterized by their positions $\mathbf{r}$ and their axis $\hat{v} = (\cos\theta, \sin\theta)$. The equations of motion for a particle i are as follows (Figure 14.7)

$$\partial_t \mathbf{r}_i = v_0 \hat{v}_i + \mu \sum_{j \neq i} \mathbf{F}_{ij} + \eta_i^T(t)$$

$$\partial_t \theta_i = \eta_i^R(t)$$

(14.29)

in which v_0 denotes the self-propulsion velocity, μ the mobility, $\mathbf{F}_{ij}$ the force between two particles i and j, $\eta_i^T(t)$ and $\eta_i^R(t)$ are the translational and rotational white noise terms, respectively. There is no specific functional form for the interparticle potential and forces. In most cases, the interparticle potential is short-range and repulsive. A possible choice for the force $\mathbf{F}_{ij}$ is the following: $\mathbf{F}_{ij} = -k(\frac{2a}{r_{ij}} - 1)\mathbf{r}_{ij}$ when $r_{ij} < 2a$ and 0 otherwise. In this equation, k denotes a force constant, and a is the radius of the self-propelled particle.

The phase diagram for ABP can then be determined by running simulations for various sets of (ϕ, v_0), where $\phi = \frac{N\pi a^2}{L^2}$ denotes the packing fraction of the system and L is the side of the (square) simulation cell. Unlike the Vicsek model, the ABP equations of motion do not invoke any "alignment" rule in determining the heading of the particle. Nonetheless, the ABP model forms clusters for specific (ϕ, v_0) values, as shown in Figure 14.7. This phenomenon resembles a liquid-gas transition called motility-induced phase separation [349,360]. The term "motility-induced" indicates here that the activity of the particles triggers the phase transition since purely repulsive and passive particles do not undergo a liquid-gas transition. This result is also reminiscent of the onset of flocking with the Vicsek model. For the Vicsek model, flocking occurs for a low enough noise intensity compared to the (fixed) self-propelling velocity. Similarly, for the ABP model, phase separation occurs

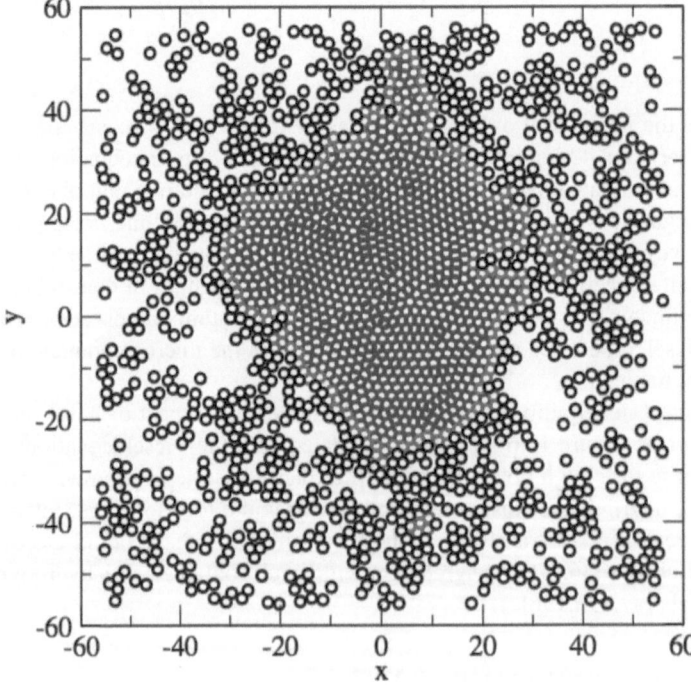

Figure 14.7 Example of the cluster formed during motility-induced phase separation. The gas-like particles are shown as isolated circles, particles inside the cluster are shown as densely-packed circles at the center of the plot, and the remaining particles are located at the cluster-gas interface.

for a large enough self-propelling velocity and density when compared to the (fixed) intensity of the noise (see Figure 14.8). Since active matter is a far-from-equilibrium system, there is no straightforward criterion to predict the conditions for phase separation. Examples of methods used to detect the formation of inhomogeneities and the onset of the motility-induced phase separation include determining the onset of giant fluctuations in the number of particles [137].

Recent work has focused on extending the ABP model to include different interparticle interactions and particle geometries. For instance, ABP models with more steeply repulsive interactions [307] or attractive interactions [278] exhibit similar clustering and phase separation. In addition, the ABP equations of motion have been modified to account for active particles of different shapes [238], such as, for instance, dumbells [18,58,110,330] for the simulations of active nematics [81,111,206,292,299].

14.3.3   ENTROPY PRODUCTION: FROM ACTIVE MATTER TO MOLECULAR MACHINES

Given the intrinsic non-equilibrium nature of active particles, determining which equilibrium thermodynamic concepts and properties apply to active matter remains an outstanding challenge [350]. For instance, while pressure may be used to locate coexistence between phases in active matter, it yields an incorrect binodal through a Maxwell construction [340]. This has led to the proposal of alternative concepts, such as the property known as swim pressure [274,352] to account for the impact of the swimmer self-propulsion and reorientation [351], and to the development of a generalized thermodynamics for active systems [162,309,337].

As previously discussed, entropy and entropy production are critical for characterizing non-equilibrium systems. We discuss in this section an approach drawn from data science that prefigures the fully fledged machine learning exploration of dynamical systems of Chapter 15. We start by determining a data science-based entropy applicable to systems in a wide range of settings, from nonequilibrium molecular trajectories to images collected experimentally. Entropy is generally

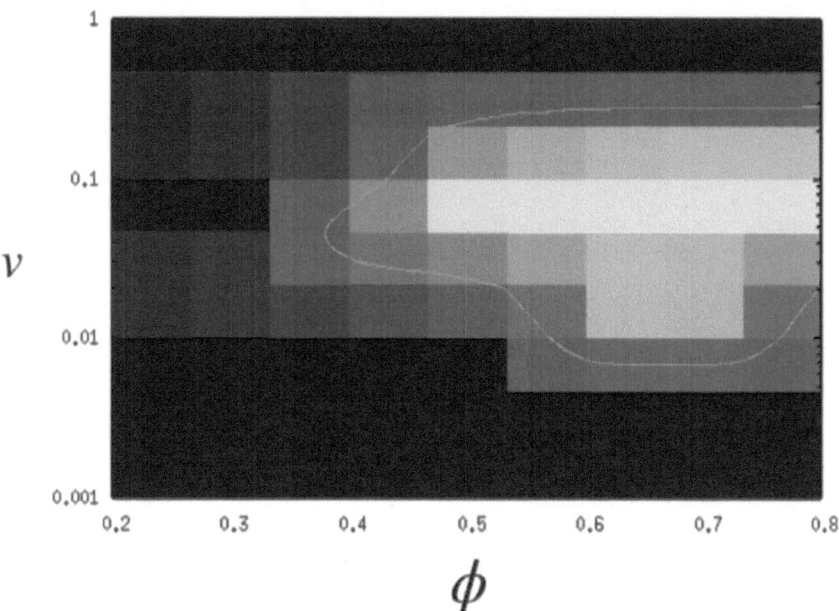

Figure 14.8   Schematic phase diagram for an ABP system, where v and ϕ indicate the self-propelling velocity and ϕ the packing fraction. The bright regions inside the green line indicate the phase-separated region.

associated with a measure of the amount of information encoded in a system. The idea underlying the data science-based approach to entropy consists in using a lossless data compression algorithm to encode the information contained in the configurations of the system in a shorter binary representation. To this end, the Lempel-Ziv 77 coding algorithm (LZ77) [64] was used to determine the computable information density (CID) [243]

$$CID = \frac{\mathscr{L}(x)}{L} \tag{14.30}$$

where $\mathscr{L}(x)$ is the total binary code length of the compressed sequence, and L is the length of the original sequence x. We show in Figure 14.9 an example of how LZ77 operates to generate a compressed sequence [243].

To compute the CID for a given configuration, coordinates of ABPs are discretized or quantized according to the RandOrg protocol [291], with a grid size such that there is at most one particle center in each bin. The variation of CID is computed for each configuration and then averaged over many configurations. The variation of the CID is shown against the packing fraction ϕ of the system in Figure 14.9. The system is in a homogeneous gas-like state for a low packing fraction ϕ. In this regime, the CID increases steadily with ϕ. Then, when the packing fraction reaches ϕ_c, the system undergoes a motility-induced phase separation. The onset of the motility-induced phase separation coincides with a discontinuity in the CID. A further increase in ϕ beyond ϕ_c results in a steady and continuous increase of the CID after that. The behavior of the CID is thus consistent with the expected behavior for entropy during a first order. Since existing theoretical results point to MIPS being akin to a first-order phase transition, this result further confirms that the CID is a reliable estimate for a nonequilibrium entropy in active matter systems [48].

Coding a sequence with LZ77

Consider the sequence x of length L=12 and an alphabet with 3 characters (a, b, c) of size $\alpha = 3$

x=abcbabababab

The LZ77 algorithm works by finding repetitions of the longest previous factor (LPF)

In the sequence x, the first character (a) appears once and is not repeated (0 repeats). This is coded into a tuple (a, 0)

The second and third characters are not repeated either and lead to tuples, or LPFs, (b, 0) and (c,0)

The next character in x (b) is part of a LPF starting at position #2 copied for 1 character, leading to LPF (2, 1)

The next group (ab) is part of a LPF starting at position #1 copied for 2 characters, leading to LPF (1, 2)

We finally take LPF#5 and copy if for the next 6 characters, leading to tuple (5, 6)

$$LZ77(x) = [(a,0); (b,0); (c,0); (2,1); (1,2); (5,6)]$$

The code thus contains C=6 tuples, and the CID is

$$CID = \frac{ClogC + 2Clog(C/L)}{L}$$

(a)

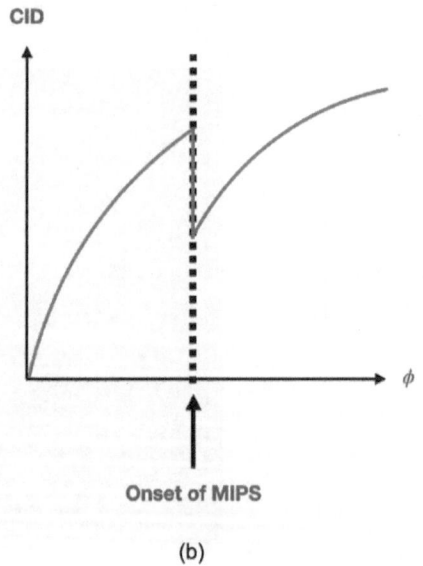

CID

Onset of MIPS

(b)

Figure 14.9 Data-based entropy from lossless compression algorithms: (a) Comparison between the original sequence and the corresponding LZ77 compressed sequence [243], and (b) Schematic variation of the CID as a function of packing fraction in an ABP system.

We conclude this section by emphasizing that nonequilibrium living systems are inherently associated with irreversibility and time-reversal symmetry breaking. In the case of active matter, the continuous injection of energy into the system gives rise to steady-state currents and entropy production. Methods to determine the rate of entropy production rate in living systems and molecular machines are currently under development using either thermodynamics [239,267,306] or data science-based approaches [308] and promise to shed light shortly on these complex far-from-equilibrium systems [308]

15 Learning Evolution and Transport

"The theory of automata (...) is an interdisciplinary science bordered mathematically by symbolic logic and Turing machine theory, bordered engineering-wise by the theory and the use (...) of large-scale computing machines, and bordered biologically by neurophysiology."

Claude E. Shannon,
Von Neumann's contributions to automata theory [328]

Thermodynamics and statistical mechanics provide accurate predictions for the features of equilibrium systems. Equilibrium states are associated with an extremum of a thermodynamic potential such as, for instance, a maximum in entropy under (N, V, E) (microcanonical) conditions or a minimum in Gibbs free energy under (N, P, T) (isothermal-isobaric) conditions. Similarly, the properties of equilibrium states are encompassed in partition functions, which provide a direct relation between the microscopic information (or probabilities) and thermodynamic state functions with, e.g., $A(N, V, T) = -k_B T \ln Q(N, V, T)$ for the Helmholtz free energy. For equilibrium systems, data science, and machine learning may be used to interpolate or extrapolate data to extend the range of predictions to various thermodynamic conditions. Moreover, machine learning can accelerate the exploration of the pathways connecting two states and shed light on the mechanisms underlying many phenomena, including chemical reactions, phase transitions, conformational changes in biomolecules, self-organization processes, and self-assembly. Compared to equilibrium thermodynamics and statistical mechanics, theoretical frameworks for non-equilibrium systems are relatively recent, with, for instance, the introduction of fluctuation theorems and nonequilibrium free energy relations in the 1990s and their experimental verification in the 2000s. This means that nonequilibrium data, obtained by integrating the nonequilibrium trajectories during molecular dynamics simulations or collected during time-resolved experiments, is uniquely poised to benefit from a data-driven treatment. This chapter discusses how machine learning can help analyze time-series data, discover governing equations for nonequilibrium systems, or elucidate the system's time-dependent response to environmental cues. As we will see shortly, this will allow us to identify unusual transport mechanisms through the use of nerve nets mentioned by Shannon in the opening citation, lay the groundwork for a new far-from-equilibrium physics from the data, but also consider particles as intelligent machines that sense external fields, act on the environment, and adapt by finding the most efficient navigation strategies.

15.1 LEARNING TRANSPORT

15.1.1 RATIONALE FOR DIFFUSION LEARNING

The first application focuses on analyzing the diffusion behavior from time-series data. We previously discussed the case of Brownian motion or standard diffusion. In this case, we recall that the mean-square displacement of a labeled, or tracer, particle in a liquid varies quadratically with time at short times. The long-time behavior for the mean-square displacement can be defined as

$$\lim_{t \to \infty} < \Delta x^2(t) >= 2Dt \tag{15.1}$$

where $< \Delta x^2(t) >=< (x(t) - x(0))^2 >$ denotes the 1D mean-square displacement from the initial position $x(0)$ measured over the time interval t, and D is the diffusion coefficient. Many systems,

DOI: 10.1201/9781003006411-18

however, do not exhibit a linear dependence on time [25]. In such cases, the mean-square displacement can be given by the generalized equation

$$< \Delta x^2(t) > = Kt^{\alpha} \qquad (15.2)$$

where α is an exponent characteristic of the type of diffusion behavior, and K is a generalized diffusion coefficient. When $\alpha = 1$, we recover the usual Brownian diffusion behavior. The diffusion process is termed anomalous diffusion for any exponent such as $\alpha \neq 1$. Specifically, $\alpha < 1$ corresponds to a subdiffusive process and $\alpha > 1$ to a superdiffusive process. We summarize in Figure 15.1 a comparison between the various broad classes of diffusion behavior.

What is the origin of anomalous behavior? The environment plays a significant role in the onset of anomalous diffusion behavior. In a uniform liquid, a tracer particle will undergo a standard diffusion. On the other hand, subdiffusion takes place when the tracer particle is immersed in a solution crowded with macromolecules. The value for the exponent α is also found to decrease with the concentration in crowding macromolecules [11]. Furthermore, depending on the nature of the environment, subdiffusive behavior often results from a complex combination of factors. For instance, the exponent α is lower in the cytoplasm than in similarly crowded fluids. This is due to transient associations with the cytoskeleton-shaken endoplasmic reticulum network. Together with molecular crowding, these binding interactions contribute to subdiffusion in the cytoplasm. Superdiffusive behavior can also be observed in biological systems such as, for instance, T cells in the brain [177]. Several physical models have been developed to account for anomalous diffusion behaviors. These include, among others, continuous time random walks, fractional Brownian motion, and Levy walks. In continuous time random walks, the tracer particle waits for a given (random) time at a given position before traveling over a segment of random length. Fractional Brownian motion is a generalization of Brownian motion that includes antipersistent memory effects due to the environment. Levy walks are similar to continuous random walks but with a segment length determined by the waiting time.

The identification of a specific diffusion behavior is often complex. While it seems *a priori* straightforward to determine the diffusion behavior from a large dataset for the mean-square displacements, data analysis is often difficult to obtain. For instance, the diffusion of tracer particles in the cytoplasm often exhibits a notable heterogeneity and strongly depends on the tracer size [320]. Moreover, the exponent may undergo sudden changes and switch from one type of diffusive behavior to another. This has prompted machine learning applications to characterize anomalous diffusion

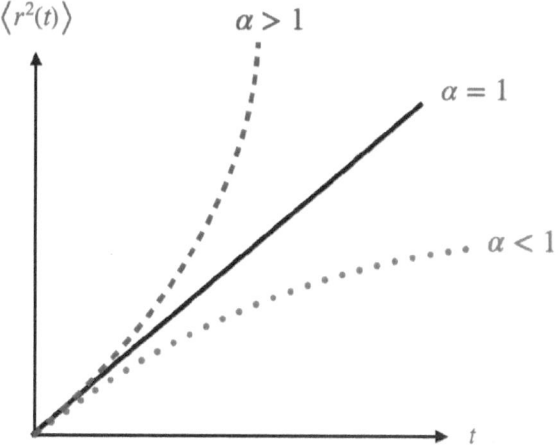

Figure 15.1 Mean-square displacements for diffusive ($\alpha = 1$), subdiffusive ($\alpha < 1$), and superdiffusive behavior ($\alpha > 1$).

features and identify the mechanisms accounting for a specific type of diffusion behavior. In the following sections, we discuss a machine learning-based approach to analyze time-series data and extract the nature of diffusion for a given system.

15.1.2   RNNs AND LSTMs IN ACTION

Neural networks can be trained to learn how a system evolves with time and any time-dependent information from time-series data. As discussed in Chapter 5, such studies require using a specific type of network known as recurrent neural networks (RNNs). The idea behind RNNs is to introduce cycles in the computation graph and use as input for a unit of the network a value computed from its output at an earlier step in the computation. By doing so, we build a "memory" effect into RNNs and enable the learning of time-dependent information. In practical applications, a slightly modified version of RNNs, known as long short-term memory RNNs (LSTMs), is preferred for numerical stability. Unlike RNNs, LSTMs contain memory cells that pass information from one time step to the next, and network units function as gates that control whether the contents from a memory cell are copied to the next time step (remembered) or reset to zero (forgotten). LSTMs have a distinct advantage over other types of neural networks, such as CNNs, in handling trajectories of different lengths. This means that there is no need to standardize trajectories to a specific, fixed, length via the use of padding or filling the dataset with, *e.g.*, an arbitrary value for the position every time data for the trajectory is missing, or via the use of cutting and slicing, *i.e.*, by discarding trajectory data for times that exceed the pre-set trajectory length.

Like all neural networks, RNNs can be used to perform different tasks. Here, we discuss the application of RNNs to achieve two other objectives. First, RNNs will allow for the classification of different types of diffusive behavior. Second, RNNs can be used for inference tasks and lead to the prediction of the value for the diffusion exponent α in Eq. 15.2. In recent years, Volpe and co-workers [4,25,265] proposed developing LSTM architectures to perform classification and inference tasks. For classification tasks, the overall network employs two layers of LSTM networks, with dimensions 250 and 50, respectively, followed by a dense layer (20 nodes) and an output layer of 5 nodes which correspond to the five classes or diffusion models considered in Ref. [4]. The five diffusion models are: (i) continuous time random walks (CTRW), (ii) annealed transient time motion (ATTM), (iii) fractional Brownian motion (FBM), (iv) Lévy walks (LW), and (v) scaled Brownian motion (SBM). For inference tasks, the overall network employs two layers of LSTM networks, dimensions 250 and 50, respectively, followed by the output layer with a single node corresponding to the diffusion exponent α value. We summarize in Figure 15.2 the architectures for classification networks.

Argun et al. generated a balanced dataset [4], with the same number of trajectories for each of the five diffusion models considered in the classification task and with a uniform sampling of the anomalous diffusion exponents for the inference task. We summarize below some of the critical ingredients of this machine learning approach, with the full details given in ref. [4]. The first step involves collecting a relatively large number (3×10^6) of trajectories with different amounts of noise. These trajectories are generated through simulations and gathered into a dataset. The dataset is split into 30 subsets of 10^5 trajectories (with 90% for training and 10%, *i.e.*, 10^4 trajectories, of the data subset set aside for validation). The training subsets are progressively fed to the overall network to avoid overfitting. For instance, the first subset is used to train the network over five epochs before a second subset is used to train the network over four epochs, and so on. A dropout rate of 20% was used in both recurrent network layers. For the inference task, the networks are trained by minimizing the mean squared error as the loss function. In contrast, the loss function is chosen as the categorical cross-entropy for classification. We examine below the performance of this type of approach on the example of the classification between diffusion behaviors.

15.1.3 CLASSIFYING DIFFUSION BEHAVIORS

The LSTM-based approach [4] allows to classify accurately data for a single trajectory (hereafter termed as test trajectory) into the five types of diffusion behavior considered in the study (CTRW, ATTM, FBM, LW, and SBM). This performance is evaluated by calculating the fraction of trajectories for which the diffusion behavior has been correctly assigned. However, several factors impact the accuracy of the neural networks for the classification task. The first factor is the length of the trajectories used in training the network. The accuracy of the network is, of course, very good when it is trained on a trajectory length similar to the test trajectory. If such a network is unavailable, adopting an approach akin to ensemble learning also yields excellent results. In this case, the prediction is obtained by combining the predictions from networks trained on trajectory lengths close to the test trajectory. Finally, as with any other method for the determination of the features of the diffusion process, training on short trajectory lengths results in a less accurate model. This is because training on short trajectories essentially means using less data to carry out this process and, thus a decreased ability to filter out the effects of noise. The second factor impacting the classification accuracy is related to the types of diffusion behavior, or classes, included in the classification task. As reported by Argun et al., the CTRW and LW diffusion processes are generally correctly classified, unlike FBM, SBM, and ATTM. The third factor is the noise level, with the noisier trajectories being the hardest to classify despite including trajectories with varying noise levels in the training dataset.

We finally conclude this section by commenting on the ability of the LSTM-based model to carry out inference and predict the value for the exponent α. Building five different machine learning models to predict α (one for each diffusion behavior) is a definite possibility since it increases accuracy for α when the diffusion model is correctly identified. However, when applied to an unseen test trajectory, this approach adds two uncertainties, the error in classifying adding up to the error in predicting α. From a practical standpoint, such an approach does not improve the overall accuracy. This led to training a single network for the prediction of α using data from all diffusion models [4],

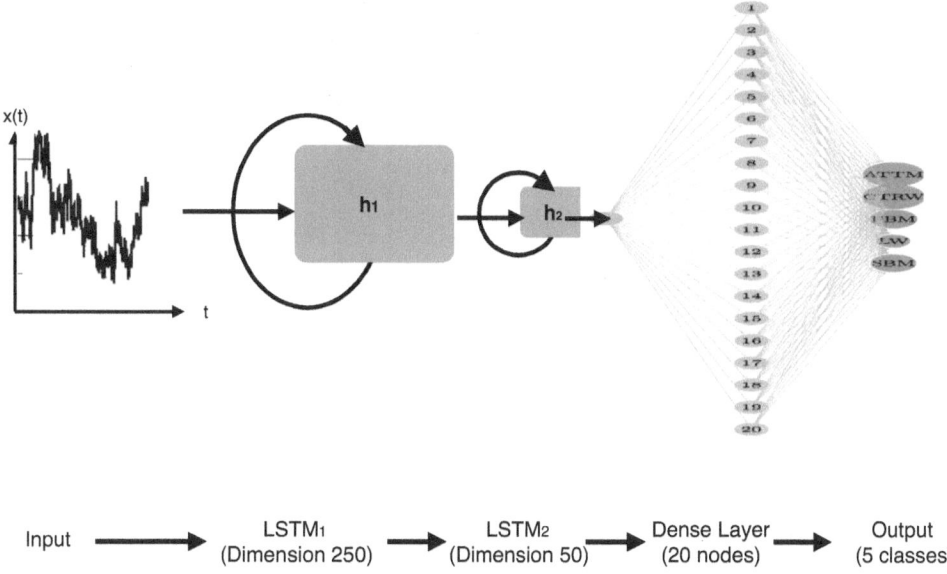

Figure 15.2 RNNs and LSTMs architecture for the classification of diffusion behaviors.

resulting in prediction with mean absolute errors as low as 0.1 for single long trajectories. The factors impacting the accuracy of the predictions were the same as for the classification task, with lower accuracy obtained when the test trajectory is short, when the noise level is high, or for certain types of diffusion models.

15.2 LEARNING DYNAMICS

15.2.1 LEARNING EQUATIONS OF MOTION FOR MESOSCOPIC AND STRUCTURED SYSTEMS

In the preceding chapters, we introduced sets of microscopic equilibrium and nonequilibrium equations of motion to determine the time-dependent properties of molecular systems. At the other end of the spectrum, equations of motion, such as, for instance, the Navier–Stokes equations, predict the properties of macroscopic flowing systems. However, do macroscopic equations apply to mesoscopic domains with a size between micro- and macro-sized systems? Similarly, the Navier–Stokes equations apply to fluids composed of spherical particles. Are these equations well suited to capture the intricacies of structured particle motion? How does molecular structure impact the properties of flowing molecular systems?

We consider a simple type of flow known as Poiseuille flow to address these questions. A Poiseuille flow is typically driven by a constant field such as gravity. This flow type is also often used to model flows triggered by a difference in pressure between two regions or pressure gradients. Mesoscopic Poiseuille flows occur in nanotubes, mesoscopic cylindrical pores, or confined systems within a slit-pore geometry, i.e., between two confining planes. Considering at this stage an atomic system for simplicity, the microscopic equations of motion for a fluid atom i are given by

$$\dot{\mathbf{r}}_{\mathbf{i}} = \mathbf{p}_{\mathbf{i}}/m_i$$
$$\dot{\mathbf{p}}_{\mathbf{i}} = -\nabla_{\mathbf{r}_{\mathbf{i}}} U + \mathbf{F}_e \tag{15.3}$$

where $\mathbf{r}_i$, $\mathbf{p}_i$, and m_i denote the position, momentum, and mass for atom i, $-\nabla_{\mathbf{r}_i} U$ is the force arising from the interatomic interactions, and $\mathbf{F}_e$ is the external field that drives the Poiseuille flow. The fluid reaches a steady state as heat is dissipated to the boundaries. In other words, the confining walls also act as a thermostat.

The corresponding macroscopic equations of motion for the fluid are the Navier–Stokes equations, given by

$$\rho \frac{d\mathbf{u}}{dt} = -\nabla \cdot \mathbf{P} + n\mathbf{F}_e \tag{15.4}$$

where ρ is the mass density, $\mathbf{u}$ the flow velocity, $\mathbf{P}$ the pressure tensor, n the number density, and $\mathbf{F}_e$ the external field used in the microscopic equations. the pressure tensor is generally written as a sum of the stress tensor Π and the equilibrium pressure tensor $P\mathbf{1}$, where P is a scalar. In the steady state, the Navier-Stokes equations provide the following equation for the shear stress Π_{xy}

$$\frac{d\Pi_{xy}}{dy} = n\mathbf{F}_e \tag{15.5}$$

where the x axis is parallel to the external force and the y axis is normal to the two confining planes. This implies that the shear stress profile along a direction normal to the confining walls is linear, provided that the number density of the fluid is constant [32,369]. This assumption, however, breaks down in mesoscopic systems since the fluid density profile is far from constant close to the confining walls. This, in turn, leads to a nonlinear profile for the shear stress close to the walls (see Figure 15.3)

The properties of mesoscopic systems thus cannot be entirely captured by the macroscopic Navier-Stokes. The other aspect we mentioned at the beginning of this section is the existence of molecular structure, which is neglected in the Navier–Stokes equation. For instance,

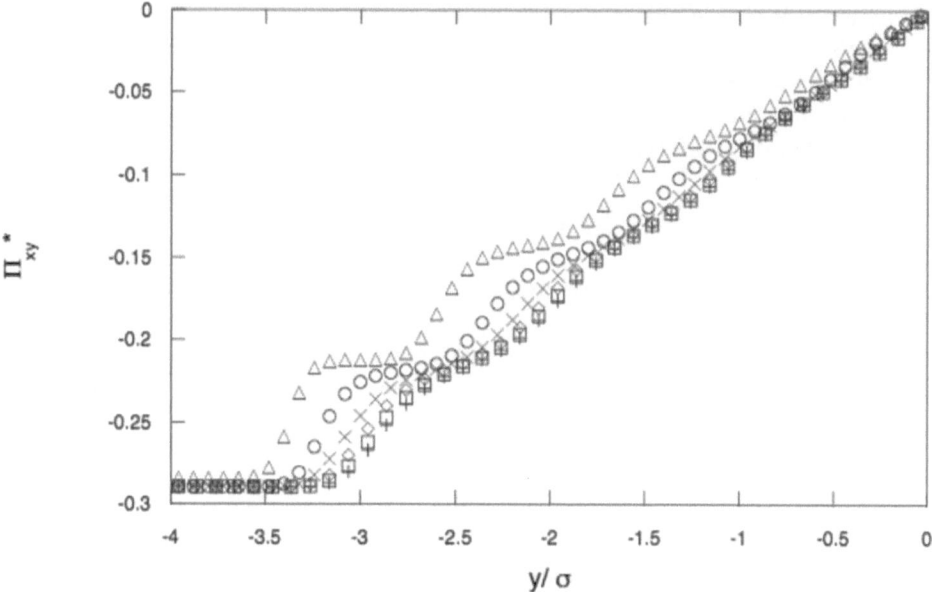

Figure 15.3 Effect of mesoscopic confinement on a fluid undergoing Poiseuille flow. The plot shows how changes in local density close to the confining walls lead to a nonlinear shear stress profile.

a fluid composed of, for example, diatomic molecules is termed a micropolar fluid and requires additional equations to model the rotation of the molecules that takes place as the system flows [71,77,176,368]. In other words, every additional symmetry element in the molecular structure will require determining a new set of macroscopic equations of motion. How can we learn new equations of motion for mesoscopic and structured systems? Are there data-driven approaches that can complement theory to obtain equations of motion for complex systems? This is the challenge we address in the following sections.

15.2.2   LEARNING DIFFERENTIAL EQUATIONS

Nonlinear dynamical systems theory has had great success in modeling a wide range of phenomena. The approach traditionally followed consists of proposing a simplified model or differential equation that accounts for the system's behavior. However, developing a model from time-series data alone is now possible. Several questions arise at the outset of this process, most notably in the choice of the model. Which terms should be included in the differential equations? What should be the order for each term? And finally, how do we assign coefficients to each term? To answer these questions, we need to define criteria that ensure that the model is complex enough to capture the features of the phenomena but simple enough to be interpretable and tractable. A candidate model with zero terms leads to a maximum error. On the other hand, increasing the number of terms in a candidate model will result in a very small error but run the risk of overfitting the data. The selection process thus bears many similarities with a topic we discussed in Part I of this book, with the choice of a specific force field or interaction model. We previously invoked Ockham's razor to find the best match between the level of detail included in a potential model and the accuracy required to account for a given property. Similarly, obtaining a parsimonious differential equation to model a phenomenon with as few terms as necessary should be an essential guiding principle in the learning process. Recent studies have focused on using sparse regression to achieve the selection from large sets of candidate models using a method known as sparse identification of nonlinear dynamics or SINDy [39,315], summarized in Figure 15.4.

This approach has also been connected to the Akaike information criteria (AIC) to identify candidate models exhibiting the highest performance. AIC builds on prior work on the measure of information loss such as *e.g.*, the Kullback–Leibler divergence, quantifying the difference between the data distribution and the model-generated distribution [376]. For a candidate model, the AIC takes the following value [1,241]

$$\text{AIC} = m \ln \left[\frac{\sum_{i=1}^{m} (y_i - g(x_i; \mu))^2}{m} \right] + 2k \tag{15.6}$$

where y_i denote one of the m observed outcomes, x_i the observed independent variables, g the candidate model, $\hat{\mu}$ the best-fit parameter values for the data, and k the number of free parameters to be estimated. The SINDy approach starts from the fact that dynamical systems of a state $x \in \mathbb{R}^n$ have few active terms in the dynamics and are thus sparse. The generic form of the dynamical system is

$$\frac{d}{dt} \mathbf{x}(t) = f(\mathbf{x}(t)) \tag{15.7}$$

To implement SINDy, one first needs to build a library of candidate symbolic functions $\Theta(\mathbf{x}) = [\theta_1(\mathbf{x}) \ldots \theta_p(\mathbf{x})]$, where each column [39] represents a candidate function for the right-hand side of Eq. 15.7.

Denoting by $\mathbf{X} \in \mathbb{R}^{m \times n}$ the time-series data, where each row of $\mathbf{X}$ is a time-measurement of the state $\mathbf{x}^T(t_k)$, and its time derivative $\dot{\mathbf{X}}$, the following relation is obtained

$$\dot{\mathbf{X}} = \Theta(\mathbf{X})\Xi \tag{15.8}$$

where $\Xi = [\xi_1 \ \xi_2 \ \cdots \ \xi_n]$ is the matrix containing the sparse vectors coefficients. Sparse regression allows us to identify the active terms in the dynamics, *i.e.*, the non-zero entries in the columns of Ξ. An optimization is performed for each column of Eq. 15.8 to find the sparse vector of coefficients ξ_k for the kth row equation and identify the differential equation that models the best the time-series data.

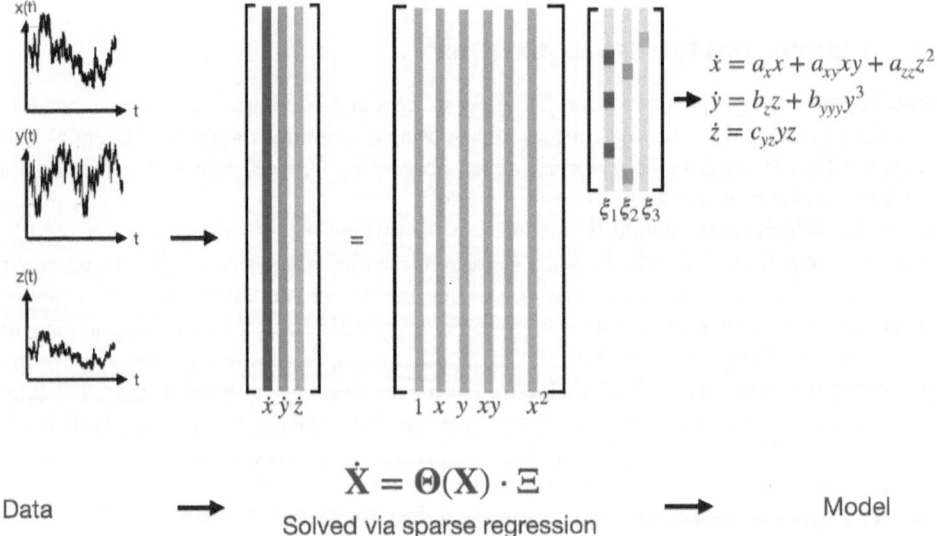

Figure 15.4 Summary of the SINDy approach [39,315], where data (left) is fed to SINDy (middle), which identifies the active coefficients (shown as bright rectangles in matrix Ξ) via sparse regression in the partial differential equations defining the model (right).

15.2.3   DATA-DRIVEN IDENTIFICATION OF GOVERNING EQUATIONS

We discuss here a few examples of applications of the approach to determine from time-series data a parametrized and nonlinear partial differential equation of the form

$$u_t = f(u, u_x, u_{xx}, \ldots, x, \mu) \tag{15.9}$$

where the subscripts denote partial differentiation in either time or space, and $f(\cdot\cdot)$ is an unknown nonlinear function of $u(x,t)$, its derivatives, and parameters in μ. For instance [50,315], the convection-diffusion equation, known as Burgers' equation, takes the general form

$$\frac{\partial u}{\partial t} + u\frac{\partial u}{\partial x} = \mu\frac{\partial^2 u}{\partial x^2} \tag{15.10}$$

This leads to the following right-hand-side function $f = -uu_x + \mu u_{xx}$. As discussed in the previous section, the function $f(.)$ includes very few terms and can thus be determined via sparse regression.

We discuss two examples of partial differential equations determined via SINDy. The first case we examine [241] is a set of equations introduced by Lorenz [229] as

$$\begin{aligned}
\dot{x} &= 10(y - x) \\
\dot{y} &= x(28 - z) - y \\
\dot{z} &= xy - \frac{8}{3}z
\end{aligned} \tag{15.11}$$

The library for the model candidates is built using functions that are polynomials up to second order ($d = 3$) for this three-state variable system ($n = 3$). This leads to a number of candidate models $N_p = \sum_{i=1}^{N_m} \binom{N_m}{i} = 1023$, where N_m is the number of possible monomials $N_m = \binom{n+d}{n}$. The SINDy approach ranks the correct model first, *i.e.*, with the lowest AIC. The second-best model, with the second lowest AIC, includes an additional constant in the $\dot{x}$ equation as $\dot{x} = 8.5 \times 10^{-6} + 10(y - x)$.

The second example we examine is the extraction of the Navier–Stokes equations from time-series data. To generate the time-series data, fluid flow is simulated around a cylinder at a given Reynolds number, and measurements of the vorticity and velocity are carried out to provide training and validation datasets for the SINDy approach. In this case, the Navier–Stokes equation is given by

$$\omega_t + (\mathbf{u} \cdot \nabla)\omega = \frac{1}{R_e}\nabla^2\omega \tag{15.12}$$

where $\mathbf{u}$ denotes the velocity, ω the vorticity, and $R_e = 100$ the Reynolds number. The candidate model with the lowest AIC is identified as [315]

$$\omega_t + 0.9931u\omega_x + 0.9910u\omega_y = 0.0099\omega_{xx} + 0.0099\omega_{yy} \tag{15.13}$$

This confirms the approach's ability to learn from time-series data and extract a high-accuracy model for partial differential equations.

15.3   LEARNING NAVIGATION

15.3.1   ADAPTING TO THE ENVIRONMENT

Finally, we apply reinforcement learning, especially deep reinforcement learning, to model the behavior of a far-from-equilibrium system like active matter and understand navigation strategies in such systems. Rather than mapping input and output pairs, the idea here is to consider a flow of particles, track how many particles successfully move from the starting point to the endpoint within a given time interval, and use the outcome of the trajectory (positive if the particle has reached the target promptly) to define a system of rewards, or "reinforcements" in psychology, to guide the particle towards the target.

Let us consider the example of self-propelled particles. Depending on their size, such particles are researched as potential microrobots and nanorobots operating in complex and crowded environments in applications such as drug delivery or environmental remediation [15]. For instance, in the case of drug delivery, active particles would need to travel through different types of environments, such as tissues or blood vessels, avoid obstacles and, more generally, adapt to spatially and temporally varying conditions. This means that finding efficient navigation strategies is a challenging task.

An intriguing idea was developed in a series of recent studies that leverage deep reinforcement learning to learn efficient navigation strategies in a complex environment filled with obstacles [397–399,401]. The proposed network mimics animal navigation strategies in two respects, with (i) the processing of high-dimensional raw sensory information and (ii) the dynamic splitting of the long-range goal, *i.e.*, reaching the target, into a series of short-range goals, *i.e.*, taking the immediate next steps, to match animal navigation behavior. In this approach, objective (i) is ensured by a deep convolutional network learning images of the environment, while objective (ii) is achieved via a reinforcement strategy. In the rest of the section, we discuss the model employed for the active particles and the environment before focusing on the details of the learning algorithm in the next section.

The model for the active particles is similar to the 2D Brownian active particles discussed in the previous chapter. The equations of motion are as follows:

$$\partial_t x = \xi_x(t) + v\cos\theta \tag{15.14}$$

$$\partial_t y = \xi_y(t) + v\sin\theta \tag{15.15}$$

$$\partial_t \theta = \xi_\theta(t) \tag{15.16}$$

where x, y, and θ denote the position and orientation of the Brownian self-propelled particle, and v is the self-propulsion velocity which is either equal to 0 and v_{max}. The self-propelled particle can thus switch on or off its activity in response to the presence of obstacles in the environment. The other functions (ξ_x, ξ_y, and ξ_τ) are zero-mean Gaussian noise processes.

To determine efficient navigation strategies [398], the swimmer needs to learn images of the environment and of the obstacles (see Figure 15.5). The local environment around a swimmer is mapped to a binary image of size W (also called vision field size) and resolution U. By binary image, it is implied that the surroundings of a swimmer are associated with a $W \times W$ matrix, with each of the matrix elements taking a value of either 0 or 1. In other words, each matrix element is a region of size $U \times U$, with a value of 1 if this region overlaps with an obstacle and 0 otherwise. W corresponds to the spatial range perceived and needs to be large enough, typically twice the dimension of the largest obstacle, to allow for the design of a reliable navigation strategy. Similarly, the resolution must be high enough to allow for the detection of the smallest obstacle. The swimmer's position is advanced using Eq. 15.16 as long as the predicted position does not overlap with an obstacle. Otherwise, the swimmer bounces back on the obstacle as a collision occurs, and a set of boundary conditions needs to be applied [398].

15.3.2   IDENTIFYING NAVIGATION STRATEGIES

We discuss how a model-free reinforcement learning algorithm can help elucidate an appropriate navigation strategy. There are two essential functions. The first important function is the quality function, or Q-function $Q(s,a)$, which denotes the sum of rewards from state s onward if action a is taken. The second quantity of significance is the policy Π that captures the mapping from states to actions.

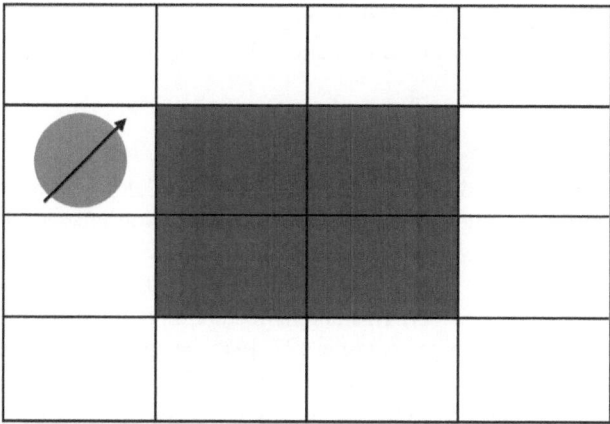

Figure 15.5 Navigating in a grid world. The 2D space is pixelated into squares, and the particle "sees" its local environment. The state is represented by the particle's location in a square and orientation, and dark squares denote regions occupied by an obstacle).

The state of a particle is defined by its position coordinate (x_i, y_i) and direction (θ_i) at a given time $t_i = it$ (t_i is thus the time after i time steps). In other words, we have a state s_i at time t_i given by the triplet (x_i, y_i, θ_i). The policy π maps the state to an action a, corresponding to switching on or off the particle's "engine" in the case of a self-propelled particle. In other words, we have for the action $a = 0$ ("engine" switched off) or $a = v_{\max}$ ("engine" switched on). The idea behind model-free reinforcement learning is to maximize the expected total reward over the entire navigation calculated as $\mathscr{E}\left[\sum_{i=1}^{\infty} \gamma^i [r(s_i)]\right]$. In this equation, r is a one-step reward function. A possible choice for r is to assign a reward of 1 for states close to the target and 0 for states far from the target. With such a choice, maximizing the rewards will promote the system to reach states close to the target in as few time steps as possible, thereby minimizing the time necessary to reach the target. The other parameter in the equation defining the expected total reward is γ. γ is a weight factor that balances the contributions from short-term ($\gamma \to 0$) and long-term rewards ($\gamma \to 1$). The neural network is trained to obtain the optimal quality function $Q^*(s_n, a)$ given by

$$Q^*(s_n, a) = \mathscr{E}\left[r(s_1) + \gamma^1 r(s_2) + \gamma^2 r(s_3) + \cdots + \gamma^{(n-1)} r(s_n) | s_0, \Pi^*\right] \qquad (15.17)$$

where π^* denotes the optimal propulsion policy. The Q-learning process determines iteratively the Q^* function, leading to the optimal propulsion decision ($a = 0$ or $a = v_{max}$) given by $\Pi^* = \arg\max_a Q^*(s, a)$. In practice, the neural network is trained through multiple navigation episodes, with an episode denoting a trajectory that starts at a random position and ends when the particle arrives at the target.

15.3.3 LEARNING COLLECTIVE MOTION

In the previous section, we examined how deep reinforcement learning allows us to identify navigation strategies for an isolated self-propelled particle in an environment containing one or more obstacles. The next challenge consists in controlling the collective behavior of interacting active particles. To this end, the fundamental concepts of artificial intelligence, *i.e.*, sensing, and actuating, are leveraged to guide assembly. One of the first applications of a feedback mechanism to control the group behavior for such particles was carried out in 2019 [218]. Instead of focusing on visualizing obstacles, the idea here is to use the concept of vision for the perception of neighboring particles. When other self-propelled particles are located in the local environment, a feedback loop changes the motility or self-propelled velocity, which, in turn, triggers the formation of clusters of particles. This hinges on an intrinsic property of active matter, which the following equation can capture

$$\frac{\rho_2}{\rho_1} = \frac{v_1}{v_2} \tag{15.18}$$

where ρ_1 and ρ_2 are the number densities of self-propelled particles with a self-propelled velocity of v_1 and v_2, respectively. In other words, when there is a region labeled "Region 1" where the self-propelled velocity is lower (*i.e.*, v_1) than everywhere else (*i.e.*, the self-propelled velocity is v_2 in the rest of the system, labeled "Region 2"), then the density ρ_1 in "Region 1" is greater than the density ρ_2 in "Region 2" [5,6,108,342]. This unique feature of active matter can be used to trigger the assembly of synthetic self-propelled particles [277] via the application of a specific light pattern to control the assembly of active rectification devices [342] and, more generally, to program the formation of patterned materials via templated assembly and/or inverse assembly [108]. A similar strategy has also been employed for biological systems [5,6,140] since, under anaerobic conditions, *E. coli* bacteria can express proteorhodopsin, a green-photon-driven proton pump [388] and undergo a light-activated assembly. In line with applications of computer vision to obstacle recognition, feedback-assisted assembly is a complex process resulting from a subtle interplay between the range of the field of vision and how motility is modulated. The controlled assembly into a reconfigurable swarm of self-propelled particles is the capture and delivery of cargo can also lead to a new way to perform nano- and microscale applications in, for instance, drug delivery [397].

Assembly is generally controlled by thermodynamics by leveraging the difference in stability between structures and phases. This, in turn, can be controlled by carefully designing particles and identifying the right balance between attractive, repulsive, and orientation-dependent interactions. The nonequilibrium assembly processes we discussed in this section markedly differ from conventional mechanisms for the formation of structures and phases. Here assembly and inverse assembly are induced by leveraging external stimuli, such as light and external fields. Together with the fast-paced development of nonequilibrium statistical mechanics for small systems, as illustrated by the fluctuation theorems presented in Chapter 14, it opens the door to designing novel objects and new applications involving physics-informed and physics-guided soft, autonomous, nano-, and micro-robotics.

Bibliography

1. Hirotugu Akaike. A new look at the statistical model identification. *IEEE Trans. Autom.*, 19(6):716–723, 1974.
2. Berni J. Alder and Thomas Everett Wainwright. Studies in molecular dynamics. I. General method. *J. Chem. Phys.*, 31(2):459–466, 1959.
3. Michael P. Allen and Dominic J. Tildesley. *Computer Simulation of Liquids*. Clarendon Press, Oxford, 1987.
4. Aykut Argun, Giovanni Volpe, and Stefano Bo. Classification, inference and segmentation of anomalous diffusion with recurrent neural networks. *J. Phys. A*, 54(29):294003, 2021.
5. Jochen Arlt, Vincent A. Martinez, Angela Dawson, Teuta Pilizota, and Wilson C. K. Poon. Painting with light-powered bacteria. *Nat. Commun.*, 9(1):1–7, 2018.
6. Jochen Arlt, Vincent A. Martinez, Angela Dawson, Teuta Pilizota, and Wilson C. K. Poon. Dynamics-dependent density distribution in active suspensions. *Nat. Commun.*, 10(1):1–7, 2019.
7. Stefan Auer and Daan Frenkel. Prediction of absolute crystal-nucleation rate in hard-sphere colloids. *Nature*, 409(6823):1020–1023, 2001.
8. Lee Badger. Lazzarini's lucky approximation of π. *Math. Mag.*, 67(2):83–91, 1994.
9. Peng Bai and Jörn Ilja Siepmann. Assessment and optimization of configurational-bias Monte Carlo particle swap strategies for simulations of water in the Gibbs ensemble. *J. Chem. Theory Comput.*, 13(2):431–440, 2017.
10. J. Balatoni and Alfred Renyi. On the notion of entropy. *Publ. Math. Inst. Hung. Acad. Sci*, 1(9):9–40, 1956.
11. Daniel S. Banks and Cécile Fradin. Anomalous diffusion of proteins due to molecular crowding. *Biophys. J.*, 89(5):2960–2971, 2005.
12. Samir Barman, Anupam Khutia, Ralph Koitz, Olivier Blacque, Hiroyasu Furukawa, Marcella Iannuzzi, Omar M. Yaghi, Christoph Janiak, Jürg Hutter, and Heinz Berke. Synthesis and hydrogen adsorption properties of internally polarized 2, 6-azulenedicarboxylate based metal–organic frameworks. *J. Mater. Chem. A*, 2(44):18823–18830, 2014.
13. Christian Bartels and Martin Karplus. Multidimensional adaptive umbrella sampling: Applications to main chain and side chain peptide conformations. *J. Comput. Chem.*, 18(12):1450–1462, 1997.
14. Yasemin Basdogan, Mitchell C. Groenenboom, Ethan Henderson, Sandip De, Susan B. Rempe, and John A. Keith. Machine learning-guided approach for studying solvation environments. *J. Chem. Theory Comput.*, 16(1):633–642, 2019.
15. Clemens Bechinger, Roberto Di Leonardo, Hartmut Löwen, Charles Reichhardt, Giorgio Volpe, and Giovanni Volpe. Active particles in complex and crowded environments. *Rev. Mod. Phys.*, 88(4):045006, 2016.
16. Jörg Behler. Neural network potential-energy surfaces in chemistry: A tool for large-scale simulations. *Phys. Chem. Chem. Phys.*, 13(40):17930–17955, 2011.
17. Jörg Behler. Constructing high-dimensional neural network potentials: A tutorial review. *Int. J. Quantum Chem.*, 115(16):1032–1050, 2015.
18. Sergey Belan and Mehran Kardar. Active motion of passive asymmetric dumbbells in a non-equilibrium bath. *J. Chem. Phys.*, 154(2):024109, 2021.
19. Giancarlo Benettin, Luigi Galgani, Antonio Giorgilli, and Jean-Marie Strelcyn. Lyapunov characteristic exponents for smooth dynamical systems and for hamiltonian systems; a method for computing all of them. Part 1: Theory. *Meccanica*, 15:9–20, 1980.
20. Herman J. C. Berendsen, J. R. Grigera, and T. P. Straatsma. The missing term in effective pair potentials. *J. Phys. Chem.*, 91(24):6269–6271, 1987.
21. Howard C. Berg. *E. coli in Motion*. Springer, Berlin, Heidelberg, 2004.
22. Bruce J. Berne. *Statistical Mechanics, Part B: Time-Dependent Processes*, vol. 233. Springer, Berlin, Heidelberg, 1977.
23. Ajeevsing Bholoa, Steven D. Kenny, and Roger Smith. A new approach to potential fitting using neural networks. *Nucl. Instrum. Methods Phys. Res. B*, 255(1):1–7, 2007.

24. Thomas B. Blank, Steven D. Brown, August W. Calhoun, and Douglas J. Doren. Neural network models of potential energy surfaces. *J. Chem. Phys.*, 103(10):4129–4137, 1995.

25. Stefano Bo, Falko Schmidt, Ralf Eichhorn, and Giovanni Volpe. Measurement of anomalous diffusion using recurrent neural networks. *Phys. Rev. E*, 100(1):010102, 2019.

26. Peter G. Bolhuis and David A. Kofke. Monte Carlo study of freezing of polydisperse hard spheres. *Phys. Rev. E*, 54(1):634, 1996.

27. Ludwig Boltzmann. Ueber die Eigenschaften monocyclischer und anderer damit verwandter systeme. *J. Reine Angew. Math.*, 1885(98):68–94, 1885.

28. Luigi Bonati, Valerio Rizzi, and Michele Parrinello. Data-driven collective variables for enhanced sampling. *J. Phys. Chem. Lett.*, 11(8):2998–3004, 2020.

29. Luigi Bonati, Yue-Yu Zhang, and Michele Parrinello. Neural networks-based variationally enhanced sampling. *Proc. Natl. Acad. Sci. U. S. A.*, 116(36):17641–17647, 2019.

30. Ernesto E. Borrero, Marcus Weinwurm, and Christoph Dellago. Optimizing transition interface sampling simulations. *J. Chem. Phys.*, 134(24):244118, 2011.

31. Istvan Borzsak, Peter T. Cummings, and Denis J. Evans. Shear viscosity of a simple fluid over a wide range of strain rates. *Mol. Phys.*, 100(16):2735–2738, 2002.

32. Alexandru Botan, Benjamin Rotenberg, Virginie Marry, Pierre Turq, and Benoît Noetinger. Hydrodynamics in clay nanopores. *J. Phys. Chem. C*, 115(32):16109–16115, 2011.

33. Carlos Braga and Karl P. Travis. A configurational temperature Nosé-Hoover thermostat. *J. Chem. Phys.*, 123(13):134101, 2005.

34. Leo Breiman. Bagging predictors. *Mach. Learn.*, 24(2):123–140, 1996.

35. Donald W. Brenner. Empirical potential for hydrocarbons for use in simulating the chemical vapor deposition of diamond films. *Phys. Rev. B*, 42(15):9458, 1990.

36. Bernard R. Brooks, Charles L. Brooks III, Alexander D. Mackerell Jr, Lennart Nilsson, Robert J. Petrella, Benoit Roux, Youngdo Won, Georgios Archontis, Christian Bartels, Stefan Boresch, et al. Charmm: The biomolecular simulation program. *J. Comput. Chem.*, 30(10):1545–1614, 2009.

37. Robert Brown. XXVII. A brief account of microscopical observations made in the months of June, July and August 1827, on the particles contained in the pollen of plants; and on the general existence of active molecules in organic and inorganic bodies. *Phil. Mag.*, 4(21):161–173, 1828.

38. W. Byers Brown. Constant pressure ensembles in statistical mechanics. *Mol. Phys.*, 1(1):68–82, 1958.

39. Steven L. Brunton, Joshua L. Proctor, and J. Nathan Kutz. Discovering governing equations from data by sparse identification of nonlinear dynamical systems. *Proc. Natl. Acad. Sci. U. S. A.*, 113(15):3932–3937, 2016.

40. Giovanni Bussi, Alessandro Laio, and Michele Parrinello. Equilibrium free energies from nonequilibrium metadynamics. *Phys. Rev. Lett.*, 96(9):090601, 2006.

41. Carlos Bustamante, Jan Liphardt, and Felix Ritort. The nonequilibrium thermodynamics of small systems. *Phys. Today*, 58(7):43, 2005.

42. B. D. Butler, Gary Ayton, Owen G. Jepps, and Denis J. Evans. Configurational temperature: Verification of Monte Carlo simulations. *J. Chem. Phys.*, 109(16):6519–6522, 1998.

43. Angelo Cacciuto, Stefan Auer, and Daan Frenkel. Onset of heterogeneous crystal nucleation in colloidal suspensions. *Nature*, 428(6981):404–406, 2004.

44. Dapeng Cao, Jianhui Lan, Wenchuan Wang, and Berend Smit. Lithium-doped 3D covalent organic frameworks: High-capacity hydrogen storage materials. *Angew. Chem. Intl. Ed.*, 48(26):4730–4733, 2009.

45. Jamie H. Cate, Anne R. Gooding, Elaine Podell, Kaihong Zhou, Barbara L. Golden, Craig E. Kundrot, Thomas R. Cech, and Jennifer A. Doudna. Crystal structure of a group I ribozyme domain: Principles of RNA packing. *Science*, 273(5282):1678–1685, 1996.

46. Michael E. Cates and Julien Tailleur. When are active Brownian particles and run-and-tumble particles equivalent? Consequences for motility-induced phase separation. *Europhys. Lett.*, 101(2):20010, 2013.

47. Jean-Baptiste Caussin, Alexandre Solon, Anton Peshkov, Hugues Chaté, Thierry Dauxois, Julien Tailleur, Vincenzo Vitelli, and Denis Bartolo. Emergent spatial structures in flocking models: A dynamical system insight. *Phys. Rev. Lett.*, 112(14):148102, 2014.

48. Andrea Cavagna, Paul M. Chaikin, Dov Levine, Stefano Martiniani, Andrea Puglisi, and Massimiliano Viale. Vicsek model by time-interlaced compression: A dynamical computable information density. *Phys. Rev. E*, 103(6):062141, 2021.

49. D. J. Chadi and Marvin L. Cohen. Tight-binding calculations of the valence bands of diamond and zincblende crystals. *Phys. Status Solidi B*, 68(1):405–419, 1975.

50. Kathleen Champion, Bethany Lusch, J. Nathan Kutz, and Steven L. Brunton. Data-driven discovery of coordinates and governing equations. *Proc. Natl. Acad. Sci. U. S. A.*, 116(45):22445–22451, 2019.

51. Hugues Chaté, Francesco Ginelli, Guillaume Grégoire, Fernando Peruani, and Franck Raynaud. Modeling collective motion: Variations on the vicsek model. *Eur. Phys. J. B*, 64(3–4):451–456, 2008.

52. Hugues Chaté, Francesco Ginelli, Guillaume Grégoire, and Franck Raynaud. Collective motion of self-propelled particles interacting without cohesion. *Phys. Rev. E*, 77(4):046113, 2008.

53. Bin Chen, Jörn Ilja Siepmann, and Michael L. Klein. Direct Gibbs ensemble Monte Carlo simulations for solid–vapor phase equilibria: Applications to Lennard-Jonesium and carbon dioxide. *J. Phys. Chem. B*, 105(40):9840–9848, 2001.

54. Wei Chen and Andrew L. Ferguson. Molecular enhanced sampling with autoencoders: On-the-fly collective variable discovery and accelerated free energy landscape exploration. *J. Comput. Chem.*, 39(25):2079–2102, 2018.

55. T. Chou, Kirone Mallick, and R. K. P. Zia. Non-equilibrium statistical mechanics: From a paradigmatic model to biological transport. *Rep. Prog. Phys.*, 74(11):116601, 2011.

56. Kamal Choudhary, Taner Yildirim, Daniel W. Siderius, A. Gilad Kusne, Austin McDannald, and Diana L. Ortiz-Montalvo. Graph neural network predictions of metal organic framework CO_2 adsorption properties. *Comput. Mater. Sci.*, 210:111388, 2022.

57. S. Ciliberto and C. Laroche. An experimental test of the Gallavotti-Cohen fluctuation theorem. *J. Phys. IV*, 8(PR6):Pr6–215, 1998.

58. Judit Clopés, Gerhard Gompper, and Roland G. Winkler. Hydrodynamic interactions in squirmer dumbbells: Active stress-induced alignment and locomotion. *Soft Matter*, 16(47):10676–10687, 2020.

59. Ezechiel Godert David Cohen. Kinetic theory of non-equilibrium fluids. *Physica A*, 118(1–3):17–42, 1983.

60. Ronald E. Cohen, Michael J. Mehl, and Dimitrios A. Papaconstantopoulos. Tight-binding total-energy method for transition and noble metals. *Phys. Rev. B*, 50(19):14694, 1994.

61. Simona Colabrese, Kristian Gustavsson, Antonio Celani, and Luca Biferale. Flow navigation by smart microswimmers via reinforcement learning. *Phys. Rev. Lett.*, 118(15):158004, 2017.

62. Delphine Collin, Felix Ritort, Christopher Jarzynski, Steven B. Smith, Ignacio Tinoco Jr, and Carlos Bustamante. Verification of the crooks fluctuation theorem and recovery of RNA folding free energies. *Nature*, 437(7056):231–234, 2005.

63. Patrick S. Coppock and James T. Kindt. Atomistic simulations of mixed-lipid bilayers in gel and fluid phases. *Langmuir*, 25(1):352–359, 2009.

64. Thomas M. Cover. *Elements of Information Theory*. John Wiley & Sons, Hoboken, NJ, 1999.

65. Michael Creutz and Andreas Gocksch. Higher-order hybrid Monte Carlo algorithms. *Phys. Rev. Lett.*, 63(1):9, 1989.

66. Gavin E. Crooks. Entropy production fluctuation theorem and the nonequilibrium work relation for free energy differences. *Phys. Rev. E*, 60(3):2721, 1999.

67. James F. Dama, Michele Parrinello, and Gregory A. Voth. Well-tempered metadynamics converges asymptotically. *Phys. Rev. Lett.*, 112(24):240602, 2014.

68. Murray S. Daw and Michael I. Baskes. Semiempirical, quantum mechanical calculation of hydrogen embrittlement in metals. *Phys. Rev. Lett.*, 50(17):1285, 1983.

69. Marta De Toni, Pluton Pullumbi, Francois-Xavier Coudert, and Alain H. Fuchs. Understanding the effect of confinement on the liquid-gas transition: A study of adsorption isotherms in a family of metal–organic frameworks. *J. Phys. Chem. C*, 114(49):21631–21637, 2010.

70. Marco De Vivo, Matteo Masetti, Giovanni Bottegoni, and Andrea Cavalli. Role of molecular dynamics and related methods in drug discovery. *J. Med. Chem.*, 59(9):4035–4061, 2016.

71. Jerome Delhommelle. Rotational viscosity of uniaxial molecules. *Mol. Phys.*, 100(21):3479–3482, 2002.

72. Jerome Delhommelle. Simulations of shear-induced melting in two dimensions. *Phys. Rev. B*, 69(14):144117, 2004.

73. Jerome Delhommelle and Peter T. Cummings. Simulation of friction in nanoconfined fluids for an arbitrarily low shear rate. *Phys. Rev. B*, 72(17):172201, 2005.

74. Jerome Delhommelle, Peter T. Cummings, and Janka Petravic. Conductivity of molten sodium chloride in an arbitrarily weak DC electric field. *J. Chem. Phys.*, 123(11):114505, 2005.

75. Jerome Delhommelle and Denis J. Evans. Comparison of thermostatting mechanisms in NVT and NPT simulations of decane under shear. *J. Chem. Phys.*, 115(1):43–49, 2001.

76. Jerome Delhommelle and Denis J. Evans. Configurational temperature profile in confined fluids. I. Atomic fluid. *J. Chem. Phys.*, 114(14):6229–6235, 2001.

77. Jerome Delhommelle and Denis J. Evans. Poiseuille flow of a micropolar fluid. *Mol. Phys.*, 100(17):2857–2865, 2002.

78. Jerome Delhommelle, Janka Petravic, and Denis J. Evans. On the effects of assuming flow profiles in nonequilibrium simulations. *J. Chem. Phys.*, 119(21):11005–11010, 2003.

79. Jerome Delhommelle, Janka Petravic, and Denis J. Evans. Reexamination of string phase and shear thickening in simple fluids. *Phys. Rev. E*, 68(3):031201, 2003.

80. Christoph Dellago, Peter Bolhuis, and Phillip L. Geissler. Transition path sampling. *Adv. Chem. Phys.*, 123:1–78, 2002.

81. Dario Dell'Arciprete, M. L. Blow, A. T. Brown, F. D. C. Farrell, Juho S. Lintuvuori, A. F. McVey, Davide Marenduzzo, and Wilson C. K. Poon. A growing bacterial colony in two dimensions as an active nematic. *Nat. Commun.*, 9(1):1–9, 2018.

82. Zygmunt S. Derewenda and Peter G. Vekilov. Entropy and surface engineering in protein crystallization. *Acta Crystallogr. D*, 62(1):116–124, 2006.

83. Caroline Desgranges, P. W. Anderson, and Jerome Delhommelle. Classical and quantum many-body effects on the critical properties and thermodynamic regularities of silicon. *J. Phys. Condens. Matt.*, 29(4):045401, 2017.

84. Caroline Desgranges and Jerome Delhommelle. Insights into the molecular mechanism underlying polymorph selection. *J. Am. Chem. Soc.*, 128(47):15104–15105, 2006.

85. Caroline Desgranges and Jerome Delhommelle. Crystallization mechanisms for supercooled liquid Xe at high pressure and temperature: Hybrid Monte Carlo molecular simulations. *Phys. Rev. B*, 77(5):054201, 2008.

86. Caroline Desgranges and Jerome Delhommelle. Molecular simulation of transport in nanopores: Application of the transient-time correlation function formalism. *Phys. Rev. E*, 77(2):027701, 2008.

87. Caroline Desgranges and Jerome Delhommelle. Rheology of liquid FCC metals: Equilibrium and transient-time correlation-function nonequilibrium molecular dynamics simulations. *Phys. Rev. B*, 78(18):184202, 2008.

88. Caroline Desgranges and Jerome Delhommelle. Shear viscosity of liquid copper at experimentally accessible shear rates: Application of the transient-time correlation function formalism. *J. Chem. Phys.*, 128(8):084506, 2008.

89. Caroline Desgranges and Jerome Delhommelle. Accurate determination of normal stress differences via transient-time correlation function–non-equilibrium molecular dynamics (TTCF–NEMD) simulations. *Mol. Simul.*, 35(5):405–408, 2009.

90. Caroline Desgranges and Jerome Delhommelle. Phase equilibria of molecular fluids via hybrid Monte Carlo Wang–Landau simulations: Applications to benzene and n-alkanes. *J. Chem. Phys.*, 130(24):244109, 2009.

91. Caroline Desgranges and Jerome Delhommelle. Evaluation of the grand-canonical partition function using expanded Wang-Landau simulations. I. Thermodynamic properties in the bulk and at the liquid-vapor phase boundary. *J. Chem. Phys.*, 136(18):184107, 2012.

92. Caroline Desgranges and Jerome Delhommelle. Evaluation of the grand-canonical partition function using expanded Wang-Landau simulations. II. Adsorption of atomic and molecular fluids in a porous material. *J. Chem. Phys.*, 136(18):184108, 2012.

93. Caroline Desgranges and Jerome Delhommelle. Evaluation of the grand-canonical partition function using expanded Wang-Landau simulations. III. Impact of combining rules on mixtures properties. *J. Chem. Phys.*, 140(10):104109, 2014.

94. Caroline Desgranges and Jerome Delhommelle. Evaluation of the grand-canonical partition function using expanded Wang-Landau simulations. IV. Performance of many-body force fields and tight-binding schemes for the fluid phases of silicon. *J. Chem. Phys.*, 144(12):124510, 2016.

95. Caroline Desgranges and Jerome Delhommelle. Free energy calculations along entropic pathways. I. Homogeneous vapor-liquid nucleation for atomic and molecular systems. *J. Chem. Phys.*, 145(20):204112, 2016.

96. Caroline Desgranges and Jerome Delhommelle. Free energy calculations along entropic pathways. II. Droplet nucleation in binary mixtures. *J. Chem. Phys.*, 145(23):234505, 2016.

97. Caroline Desgranges and Jerome Delhommelle. Coarse-grained model and boiling point prediction for asphaltene model compounds via HMC-WL simulations. *Energy & Fuels*, 31(10):10699–10705, 2017.

98. Caroline Desgranges and Jerome Delhommelle. Free energy calculations along entropic pathways. III. Nucleation of capillary bridges and bubbles. *J. Chem. Phys.*, 146(18):184104, 2017.

99. Caroline Desgranges and Jerome Delhommelle. Crystal nucleation along an entropic pathway: Teaching liquids how to transition. *Phys. Rev. E*, 98(6):063307, 2018.

100. Caroline Desgranges and Jerome Delhommelle. A new approach for the prediction of partition functions using machine learning techniques. *J. Chem. Phys.*, 149(4):044118, 2018.

101. Caroline Desgranges and Jerome Delhommelle. Unusual crystallization behavior close to the glass transition. *Phys. Rev. Lett.*, 120(11):115701, 2018.

102. Caroline Desgranges and Jerome Delhommelle. Can ordered precursors promote the nucleation of solid solutions? *Phys. Rev. Lett.*, 123(19):195701, 2019.

103. Caroline Desgranges and Jerome Delhommelle. Determination of mixture properties via a combined expanded Wang-Landau simulations-machine learning approach. *Chem. Phys. Lett.*, 715:1–6, 2019.

104. Caroline Desgranges and Jerome Delhommelle. The central role of entropy in adiabatic ensembles and its application to phase transitions in the grand-isobaric adiabatic ensemble. *J. Chem. Phys.*, 153(9):094114, 2020.

105. Caroline Desgranges and Jerome Delhommelle. Ensemble learning of partition functions for the prediction of thermodynamic properties of adsorption in metal–organic and covalent organic frameworks. *J. Phys. Chem. C*, 124(3):1907–1917, 2020.

106. Caroline Desgranges and Jerome Delhommelle. Entropy production in model colloidal suspensions under shear via the fluctuation theorem. *J. Chem. Phys.*, 153(22):224113, 2020.

107. Caroline Desgranges and Jerome Delhommelle. Towards a machine learned thermodynamics: Exploration of free energy landscapes in molecular fluids, biological systems and for gas storage and separation in metal–organic frameworks. *Mol. Syst. Des. Eng.*, 6(1):52–65, 2021.

108. Caroline Desgranges, Melissa Ferrari, Paul M. Chaikin, Stefano Sacanna, Mark E. Tuckerman, and Jerome Delhommelle. Microswimmers under the spotlight: Interplay between agents with different levels of activity. *Soft Matter*, 19: 7334–7342, 2023.

109. Thomas G. Dietterich. Ensemble methods in machine learning. *Multiple Classifier Systems: First International Workshop, MCS 2000*, Cagliari, Italy, pp. 1–15, 2000.

110. Pasquale Digregorio, Demian Levis, Antonio Suma, Leticia F. Cugliandolo, Giuseppe Gonnella, and Ignacio Pagonabarraga. 2D melting and motility induced phase separation in active brownian hard disks and dumbbells. *J. Phys. Conf. Series*, 1163:012073, 2019.

111. Amin Doostmohammadi, Jordi Ignés-Mullol, Julia M. Yeomans, and Francesc Sagués. Active nematics. *Nat. Commun.*, 9(1):1–13, 2018.

112. James W. Dufty and Michael J. Lindenfeld. Nonlinear transport in the Boltzmann limit. *J. Stat. Phys.*, 20:259–301, 1979.

113. Tina Düren, Youn-Sang Bae, and Randall Q. Snurr. Using molecular simulation to characterise metal–organic frameworks for adsorption applications. *Chem. Soc. Rev.*, 38(5):1237–1247, 2009.

114. Krishna Dutta and Andrew Robinsson. *Rabindranath Tagore*. Bloomsbury, London, 1996.

115. Werner Ebeling, Frank Schweitzer, and Benno Tilch. Active brownian particles with energy depots modeling animal mobility. *BioSystems*, 49(1):17–29, 1999.

116. Roger Edberg, Denis J. Evans, and Gary P. Morriss. Constrained molecular dynamics: Simulations of liquid alkanes with a new algorithm. *J. Chem. Phys.*, 84(12):6933–6939, 1986.

117. Hani M. El-Kaderi, Joseph R. Hunt, José L. Mendoza-Cortés, Adrien P. Côté, Robert E. Taylor, Michael O'Keeffe, and Omar M. Yaghi. Designed synthesis of 3D covalent organic frameworks. *Science*, 316(5822):268–272, 2007.

118. Jeffrey R. Errington. Direct calculation of liquid–vapor phase equilibria from transition matrix Monte Carlo simulation. *J. Chem. Phys.*, 118(22):9915–9925, 2003.

119. Jeffrey R. Errington and Athanassios Z. Panagiotopoulos. A new intermolecular potential model for the n-alkane homologous series. *J. Phys. Chem. B*, 103(30):6314–6322, 1999.

120. Jeffrey R. Errington and Vincent K. Shen. Direct evaluation of multicomponent phase equilibria using flat-histogram methods. *J. Chem. Phys.*, 123(16):164103, 2005.

121. Fernando A. Escobedo and Juan J. de Pablo. Monte Carlo simulation of the chemical potential of polymers in an expanded ensemble. *J. Chem. Phys.*, 103(7):2703–2710, 1995.

122. Ilham Essafri, Bappa Ghosh, Caroline Desgranges, and Jerome Delhommelle. Designing, synthesizing, and modeling active fluids. *Phys. Fluids*, 34(7):071301, 2022.

123. Denis J. Evans. On the representatation of orientation space. *Mol. Phys.*, 34(2):317–325, 1977.

124. Denis J. Evans. Response theory as a free-energy extremum. *Phys. Rev. A*, 32(5):2923, 1985.

125. Denis J. Evans, Ezechiel Godert David Cohen, and Gary P. Morriss. Probability of second law violations in shearing steady states. *Phys. Rev. Lett.*, 71(15):2401, 1993.

126. Denis J. Evans and Brad Lee Holian. The Nosé–Hoover thermostat. *J. Chem. Phys.*, 83(8):4069–4074, 1985.

127. Denis J. Evans, William G. Hoover, Bruce H. Failor, Bill Moran, and Anthony J. C. Ladd. Nonequilibrium molecular dynamics via Gauss's principle of least constraint. *Phys. Rev. A*, 28(2):1016, 1983.

128. Denis J. Evans and Debra J. Searles. Equilibrium microstates which generate second law violating steady states. *Phys. Rev. E*, 50(2):1645, 1994.

129. Denis J. Evans and Debra J. Searles. The fluctuation theorem. *Adv. Phys.*, 51(7):1529–1585, 2002.

130. Henry Eyring. The activated complex in chemical reactions. *J. Chem. Phys.*, 3(2):107–115, 1935.

131. Alfred C. K. Farris, Guangjie Shi, Thomas Wüst, and David P. Landau. The role of chain-stiffness in lattice protein models: A replica-exchange Wang-Landau study. *J. Chem. Phys.*, 149(12):125101, 2018.

132. Alan M. Ferrenberg, David P. Landau, and Robert H. Swendsen. Statistical errors in histogram reweighting. *Phys. Rev. E*, 51(5):5092, 1995.

133. Alan M. Ferrenberg and Robert H. Swendsen. New Monte Carlo technique for studying phase transitions. *Phys. Rev. Lett.*, 61(23):2635, 1988.

134. Alan M. Ferrenberg and Robert H. Swendsen. Optimized Monte Carlo data analysis. *Comput. Phys.*, 3(5):101–104, 1989.

135. Laura Filion, Michiel Hermes, Ran Ni, and Marjolein Dijkstra. Crystal nucleation of hard spheres using molecular dynamics, umbrella sampling, and forward flux sampling: A comparison of simulation techniques. *J. Chem. Phys.*, 133(24):244115, 2010.

136. Yaouen Fily, Silke Henkes, and M. Cristina Marchetti. Freezing and phase separation of self-propelled disks. *Soft Matter*, 10(13):2132–2140, 2014.

137. Yaouen Fily and M. Cristina Marchetti. Athermal phase separation of self-propelled particles with no alignment. *Phys. Rev. Lett.*, 108(23):235702, 2012.

138. M. W. Finnis and J. E. Sinclair. A simple empirical N-body potential for transition metals. *Phil. Mag. A*, 50(1):45–55, 1984.

139. Ralph Fowler and Edward A. Guggenheim. *Statistical Thermodynamics*. Cambridge University Press, London, 1939.

140. Giacomo Frangipane, Dario Dell'Arciprete, Serena Petracchini, Claudio Maggi, Filippo Saglimbeni, Silvio Bianchi, Gaszton Vizsnyiczai, Maria Lina Bernardini, and Roberto Di Leonardo. Dynamic density shaping of photokinetic *E. Coli. Elife*, 7:e36608, 2018.

141. Daan Frenkel. Order through entropy. *Nat. Mater.*, 14(1):9, 2014.

142. Daan Frenkel and Anthony J. C. Ladd. New Monte Carlo method to compute the free energy of arbitrary solids. Application to the FCC and HCP phases of hard spheres. *J. Chem. Phys.*, 81(7):3188–3193, 1984.

143. Daan Frenkel and Berend Smit. *Understanding Molecular Simulation: From Algorithms to Applications*. Academic Press, San Diego, CA, 1996.

144. Giovanni Gallavotti. Ergodicity, ensembles, irreversibility in Boltzmann and beyond. *J. Stat. Phys.*, 78:1571–1589, 1995.

145. Giovanni Gallavotti. Chaotic dynamics, fluctuations, nonequilibrium ensembles. *Chaos*, 8(2):384–392, 1998.

146. Giovanni Gallavotti. Ergodicity: A historical perspective. Equilibrium and nonequilibrium. *Eur. Phys. J. H*, 41(3):181–259, 2016.

147. Giovanni Gallavotti and Ezechiel Godert David Cohen. Dynamical ensembles in nonequilibrium statistical mechanics. *Phys. Rev. Lett.*, 74(14):2694, 1995.

148. Giovanni Gallavotti and Ezechiel Godert David Cohen. Dynamical ensembles in stationary states. *J. Stat. Phys.*, 80(5–6):931–970, 1995.

149. Felipe Gándara, Hiroyasu Furukawa, Seungkyu Lee, and Omar M. Yaghi. High methane storage capacity in aluminum metal–organic frameworks. *J. Am. Chem. Soc.*, 136(14):5271–5274, 2014.

150. Chandrima Ganguly and Debasish Chaudhuri. Stochastic thermodynamics of active brownian particles. *Phys. Rev. E*, 88(3):032102, 2013.

151. Georg Ganzenmüller and Philip J. Camp. Applications of Wang-Landau sampling to determine phase equilibria in complex fluids. *J. Chem. Phys.*, 127(15):154504, 2007.

152. Giovanni Garberoglio, Anastasios I. Skoulidas, and J. Karl Johnson. Adsorption of gases in metal-organic materials: Comparison of simulations and experiments. *J. Phys. Chem. B*, 109(27):13094–13103, 2005.

153. Pierre Gaspard. Fluctuation theorem for nonequilibrium reactions. *J. Chem. Phys.*, 120(19):8898–8905, 2004.

154. Carl Friedrich Gau. Über ein neues allgemeines Grundgesetz der Mechanik. *Journal für die reine und angewandte Mathematik*, IV:232–235, 1829.

155. Lev D. Gelb, K. E. Gubbins, R. Radhakrishnan, and M. Sliwinska-Bartkowiak. Phase separation in confined systems. *Rep. Prog. Phys.*, 62(12):1573, 1999.

156. Murray Gell-Mann. *The Quark and the Jaguar: Adventures in the Simple and the Complex*. Macmillan, New York, 1995.

157. Rachel B. Getman, Youn-Sang Bae, Christopher E. Wilmer, and Randall Q. Snurr. Review and analysis of molecular simulations of methane, hydrogen, and acetylene storage in metal–organic frameworks. *Chem. Rev.*, 112(2):703–723, 2012.

158. Cristian Giardina, Jorge Kurchan, and Luca Peliti. Direct evaluation of large-deviation functions. *Phys. Rev. Lett.*, 96(12):120603, 2006.

159. Josiah Willard Gibbs. *Elementary Principles in Statistical Mechanics*. Scribner's, New York, 1902.

160. Francesco Ginelli. The physics of the vicsek model. *Eur. Phys. J. Spec. Top.*, 225(11):2099–2117, 2016.

161. Francesco Ginelli, Fernando Peruani, Marie-Helène Pillot, Hugues Chaté, Guy Theraulaz, and Richard Bon. Intermittent collective dynamics emerge from conflicting imperatives in sheep herds. *Proc. Natl. Acad. Sci. U. S. A.*, 112(41):12729–12734, 2015.

162. Félix Ginot, Isaac Theurkauff, Demian Levis, Christophe Ybert, Lydéric Bocquet, Ludovic Berthier, and Cécile Cottin-Bizonne. Nonequilibrium equation of state in suspensions of active colloids. *Phys. Rev. X.*, 5(1):011004, 2015.

163. Gianpaolo Gobbo, Michael A. Bellucci, Gareth A. Tribello, Giovanni Ciccotti, and Bernhardt L. Trout. Nucleation of molecular crystals driven by relative information entropy. *J. Chem. Theory Comput.*, 14(2):959–972, 2018.

164. Herbert Goldstein, Charles Poole, and John Safko. *Classical Mechanics*. Pearson, London, 2001.

165. Wei Gong, Yi Xie, Thang Duc Pham, Suchetha Shetty, Florencia A. Son, Karam B. Idrees, Zhijie Chen, Haomiao Xie, Yan Liu, Randall Q. Snurr, et al. Creating optimal pockets in a clathrochelate-based metal–organic framework for gas adsorption and separation: Experimental and computational studies. *J. Am. Chem. Soc.*, 144(8):3737–3745, 2022.

166. Ian Goodfellow, Yoshua Bengio, and Aaron Courville. *Deep Learning*. MIT Press, Cambridge, MA, 2016.

167. Leif Goodwin, A. J. Skinner, and D. G. Pettifor. Generating transferable tight-binding parameters: Application to silicon. *Europhys. Lett.*, 9(7):701, 1989.

168. H. W. Graben and John R. Ray. Unified treatment of adiabatic ensembles. *Phys. Rev. A*, 43(8):4100, 1991.

169. H. W. Graben and John R. Ray. Eight physical systems of thermodynamics, statistical mechanics, and computer simulations. *Mol. Phys.*, 80(5):1183–1193, 1993.

170. Jeffery A. Greathouse, Nathan W. Ockwig, Louise J. Criscenti, T. R. Guilinger, Phil Pohl, and Mark D. Allendorf. Computational screening of metal–organic frameworks for large-molecule chemical sensing. *Phys. Chem. Chem. Phys.*, 12(39):12621–12629, 2010.

171. Edward A. Guggenheim. Grand partition functions and so-called "thermodynamic probability". *J. Chem. Phys.*, 7(2):103–107, 1939.

172. Ashley Z. Guo, Emre Sevgen, Hythem Sidky, Jonathan K. Whitmer, Jeffrey A. Hubbell, and Juan J. de Pablo. Adaptive enhanced sampling by force-biasing using neural networks. *J. Chem. Phys.*, 148(13):134108, 2018.

173. Amar Nath Gupta, Abhilash Vincent, Krishna Neupane, Hao Yu, Feng Wang, and Michael T. Woodside. Experimental validation of free-energy-landscape reconstruction from non-equilibrium single-molecule force spectroscopy measurements. *Nat. Phys.*, 7(8):631–634, 2011.

174. J. M. Haile and H. W. Graben. On the isoenthalpic-isobaric ensemble in classical statistical mechanics. *Mol. Phys.*, 40(6):1433–1439, 1980.

175. Sang Soo Han, Hiroyasu Furukawa, Omar M. Yaghi, and William A. Goddard III. Covalent organic frameworks as exceptional hydrogen storage materials. *J. Am. Chem. Soc.*, 130(35):11580–11581, 2008.

176. J. S. Hansen, Peter J. Daivis, and B. D. Todd. Molecular spin in nano-confined fluidic flows. *Microfluid. Nanofluid.*, 6:785–795, 2009.

177. Tajie H. Harris, Edward J. Banigan, David A. Christian, Christoph Konradt, Elia D. Tait Wojno, Kazumi Norose, Emma H. Wilson, Beena John, Wolfgang Weninger, Andrew D. Luster, Andrea J. Liu, and Christopher A. Hunter. Generalized Lévy walks and the role of chemokines in migration of effector CD8+ T cells. *Nature*, 486(7404):545–548, 2012.

178. Walter A. Harrison. *Electronic Structure and the Properties of Solids*. Freeman, San Francisco, CA, 1980.

179. Remco Hartkamp, Stefano Bernardi, and B. D. Todd. Transient-time correlation function applied to mixed shear and elongational flows. *J. Chem. Phys.*, 136(6):064105, 2012.

180. Kumiko Hayashi, Hiroshi Ueno, Ryota Iino, and Hiroyuki Noji. Fluctuation theorem applied to F1-ATPase. *Phys. Rev. Lett.*, 104(21):218103, 2010.

181. Simon Haykin. *Kalman Filtering and Neural Networks*. Wiley, New York, 2001.

182. Robert Hecht-Nielsen. , Chapter III.3: Theory of the backpropagation neural network. In Harry Wechsler (Ed.), *Neural Networks for Perception*, pp. 65–93. Academic Press, Cambridge, MA, 1992.

183. Jerome Henin, Giacomo Fiorin, Christophe Chipot, and Michael L. Klein. Exploring multidimensional free energy landscapes using time-dependent biases on collective variables. *J. Chem. Theory Comput.*, 6(1):35–47, 2010.

184. Terrell L. Hill. *An Introduction to Statistical Thermodynamics*. Dover Books, New York, 1986.

185. Steven Hobday, Roger Smith, and Joe Belbruno. Applications of neural networks to fitting interatomic potential functions. *Model. Simul. Mat. Sci. Eng.*, 7(3):397, 1999.

186. William G. Hoover. Canonical dynamics: Equilibrium phase-space distributions. *Phys. Rev. A*, 31(3):1695, 1985.

187. William G. Hoover. Constant-pressure equations of motion. *Phys. Rev. A*, 34(3):2499, 1986.

188. William G. Hoover, Denis J. Evans, Richard B. Hickman, Anthony J. C. Ladd, William T. Ashurst, and Bill Moran. Lennard-Jones triple-point bulk and shear viscosities. Green-Kubo theory, hamiltonian mechanics, and nonequilibrium molecular dynamics. *Phys. Rev. A*, 22(4):1690, 1980.

189. William G. Hoover, Anthony J. C. Ladd, and Bill Moran. High-strain-rate plastic flow studied via nonequilibrium molecular dynamics. *Phys. Rev. Lett.*, 48(26):1818, 1982.

190. William G. Hoover and Harald A. Posch. Direct measurement of Lyapunov exponents. *Phys. Lett. A*, 113(2):82–84, 1985.

191. William G. Hoover and Francis H. Ree. Melting transition and communal entropy for hard spheres. *J. Chem. Phys.*, 49(8):3609–3617, 1968.

192. Gerhard Hummer and Attila Szabo. Free energy profiles from single-molecule pulling experiments. *Proc. Natl. Acad. Sci. U. S. A.*, 107(50):21441–21446, 2010.

193. Witold Hurewicz and Henry Wallman. *Dimension Theory (PMS-4)*, vol. 4. Princeton University Press, Princeton, NJ, 2015.

194. James T. Hynes and J. M. Deutch. *Nonequilibrium Problems: Projection Operator Techniques*. Academic Press, New York, 1975.

195. Thomas Ihle. Kinetic theory of flocking: Derivation of hydrodynamic equations. *Phys. Rev. E*, 83(3):030901, 2011.

196. Thomas Ihle. Invasion-wave-induced first-order phase transition in systems of active particles. *Phys. Rev. E*, 88(4):040303, 2013.

197. J. H. Irving and John G. Kirkwood. The statistical mechanical theory of transport processes. IV. The equations of hydrodynamics. *J. Chem. Phys.*, 18(6):817–829, 1950.

198. Denis J. Evans and Gary P. Morriss. *Statistical Mechanics of Nonequilbrium Liquids*. Cambridge University Press, Cambridge, 2007.

199. Kevin Maik Jablonka, Daniele Ongari, Seyed Mohamad Moosavi, and Berend Smit. Big-data science in porous materials: Materials genomics and machine learning. *Chem. Rev.*, 120(16):8066–8129, 2020.

200. W. Janke. Histograms and all that. *Computer Simulations of Surfaces and Interfaces*, 114:137–157, 2003.

201. Christopher Jarzynski. Nonequilibrium equality for free energy differences. *Phys. Rev. Lett.*, 78(14):2690, 1997.

202. Owen G. Jepps, Gary Ayton, and Denis J. Evans. Microscopic expressions for the thermodynamic temperature. *Phys. Rev. E*, 62(4):4757, 2000.

203. Nikos Ch Karayiannis, Katerina Foteinopoulou, and Manuel Laso. Entropy-driven crystallization in dense systems of athermal chain molecules. *Phys. Rev. Lett.*, 103(4):045703, 2009.

204. Johannes Kästner. Umbrella sampling. *Wiley Interdiscip. Rev. Comput. Mol. Sci.*, 1(6):932–942, 2011.

205. Kyozi Kawasaki and James D. Gunton. Theory of nonlinear transport processes: Nonlinear shear viscosity and normal stress effects. *Phys. Rev. A*, 8(4):2048, 1973.

206. Felix C. Keber, Etienne Loiseau, Tim Sanchez, Stephen J. DeCamp, Luca Giomi, Mark J. Bowick, M. Cristina Marchetti, Zvonimir Dogic, and Andreas R. Bausch. Topology and dynamics of active nematic vesicles. *Science*, 345(6201):1135–1139, 2014.

207. Seda Keskin and David S. Sholl. Assessment of a metal–organic framework membrane for gas separations using atomically detailed calculations: CO_2, CH_4, N_2, H_2 mixtures in MOF-5. *Ind. Eng. Chem. Res.*, 48(2):914–922, 2009.

208. Vikram Khanna, Jamshed Anwar, Daan Frenkel, Michael F. Doherty, and Baron Peters. Free energies of crystals computed using Einstein crystal with fixed center of mass and differing spring constants. *J. Chem. Phys.*, 154(16):164509, 2021.

209. A. R. V. Koenig, Caroline Desgranges, and Jerome Delhommelle. Adsorption of hydrogen in covalent organic frameworks using expanded Wang–Landau simulations. *Molec. Simul.*, 40(1–3):71–79, 2014.

210. David A. Kofke and Eduardo D. Glandt. Monte Carlo simulation of multicomponent equilibria in a semigrand canonical ensemble. *Mol. Phys.*, 64(6):1105–1131, 1988.

211. Anders Krogh and Jesper Vedelsby. Neural network ensembles, cross validation, and active learning. *Adv. Neur. Inf. Process. Syst.*, 7:231–238, 1995.

212. Ryogo Kubo. Statistical-mechanical theory of irreversible processes. I. General theory and simple applications to magnetic and conduction problems. *J. Phys. Soc. Japan*, 12(6):570–586, 1957.

213. Jorge Kurchan. Fluctuation theorem for stochastic dynamics. *J. Phys. A*, 31(16):3719, 1998.

214. I. Kwon, R. Biswas, C. Z. Wang, K. M. Ho, and C. M. Soukoulis. Transferable tight-binding models for silicon. *Phys. Rev. B*, 49(11):7242, 1994.

215. David Lacoste, A. W. C. Lau, and Kirone Mallick. Fluctuation theorem and large deviation function for a solvable model of a molecular motor. *Phys. Rev. E*, 78(1):011915, 2008.

216. Joao Marcelo Lamim Ribeiro and Pratyush Tiwary. Toward achieving efficient and accurate ligand-protein unbinding with deep learning and molecular dynamics through rave. *J. Chem. Theory Comput.*, 15(1):708–719, 2018.

217. David P. Landau, Shan-Ho Tsai, and M. Exler. A new approach to Monte Carlo simulations in statistical physics: Wang-Landau sampling. *Am. J. Phys.*, 72(10):1294–1302, 2004.

218. Francois A. Lavergne, Hugo Wendehenne, Tobias Bäuerle, and Clemens Bechinger. Group formation and cohesion of active particles with visual perception–dependent motility. *Science*, 364(6435):70–74, 2019.

219. Joel L. Lebowitz and Herbert Spohn. A Gallavotti–Cohen-type symmetry in the large deviation functional for stochastic dynamics. *J. Stat. Phys.*, 95(1–2):333–365, 1999.

220. Yann LeCun, D. Touresky, G. Hinton, and T. Sejnowski. A theoretical framework for back-propagation. In *Proceedings of the 1988 Connectionist Models Summer School*, vol. 1, pp. 21–28, CMU, Pittsburgh, PA, 1988.

221. Byeong-Joo Lee and Michael I. Baskes. Second nearest-neighbor modified embedded-atom-method potential. *Phys. Rev. B*, 62(13):8564, 2000.

222. A. W. Lees and S. F. Edwards. The computer study of transport processes under extreme conditions. *J. Phys. C*, 5(15):1921, 1972.

223. John Edward Lennard-Jones. On the forces between atoms and ions. *Proc. R. Soc. A*, 109(752):584–597, 1925.

224. Gilbert Newton Lewis and Merle Randall. *Thermodynamics and the Free Energy of Chemical Substances*. McGraw-Hill, New York, 1923.

225. Hailian Li, Mohamed Eddaoudi, Michael O'Keeffe, and Omar M. Yaghi. Design and synthesis of an exceptionally stable and highly porous metal-organic framework. *Nature*, 402(6759):276, 1999.

226. Jan Liphardt, Sophie Dumont, Steven B. Smith, Ignacio Tinoco Jr, and Carlos Bustamante. Equilibrium information from nonequilibrium measurements in an experimental test of Jarzynski's equality. *Science*, 296(5574):1832–1835, 2002.

227. Jan Liphardt, Bibiana Onoa, Steven B. Smith, Ignacio Tinoco Jr, and Carlos Bustamante. Reversible unfolding of single RNA molecules by mechanical force. *Science*, 292(5517):733–737, 2001.

228. Yunhua Liu, Dahuan Liu, Qingyuan Yang, Chongli Zhong, and Jianguo Mi. Comparative study of separation performance of COFS and MOFS for $CH_4/CO_2/H_2$ mixtures. *Ind. Eng. Chem. Res.*, 49(6):2902–2906, 2010.

229. Edward N. Lorenz. Deterministic nonperiodic flow. *J. Atmos. Sci.*, 20(2):130–141, 1963.

230. L. Lue, Owen G. Jepps, Jerome Delhommelle, and Denis J. Evans. Configurational thermostats for molecular systems. *Mol. Phys,*, 100(14):2387–2395, 2002.

231. Sheng-Nian Luo, Thomas J. Ahrens, Tahir Çağın, Alejandro Strachan, William A. Goddard III, and Damian C. Swift. Maximum superheating and undercooling: Systematics, molecular dynamics simulations, and dynamic experiments. *Phys. Rev. B*, 68(13):134206, 2003.

232. A. P. Lyubartsev, A. A. Martsinovski, S. V. Shevkunov, and P. N. Vorontsov-Velyaminov. New approach to Monte Carlo calculation of the free energy: Method of expanded ensembles. *J. Chem. Phys.*, 96(3):1776–1783, 1992.

233. Ao Ma and Aaron R. Dinner. Automatic method for identifying reaction coordinates in complex systems. *J. Phys. Chem. B*, 109(14):6769–6779, 2005.

234. James Clerk Maxwell. V. Illustrations of the dynamical theory of gases. Part I: On the motions and collisions of perfectly elastic spheres. *The London, Edinburgh, and Dublin Philosophical Magazine and Journal of Science*, 19(124):19–32, 1860.

235. David MacGowan and Denis J. Evans. Heat and matter transport in binary liquid mixtures. *Phys. Rev. A*, 34(3):2133, 1986.

236. Christian Maes. The fluctuation theorem as a Gibbs property. *J. Stat. Phys.*, 95(1–2):367–392, 1999.

237. Luca Maffioli, Edward R. Smith, James P. Ewen, Peter J. Daivis, Daniele Dini, and B. D. Todd. Slip and stress from low shear rate nonequilibrium molecular dynamics: The transient-time correlation function technique. *J. Chem. Phys.*, 156(18):184111, 2022.

238. Stewart A. Mallory, Chantal Valeriani, and Angelo Cacciuto. An active approach to colloidal self-assembly. *Annu. Rev. Phys. Chem.*, 69:59–79, 2018.

239. Dibyendu Mandal, Katherine Klymko, and Michael R. DeWeese. Entropy production and fluctuation theorems for active matter. *Phys. Rev. Lett.*, 119(25):258001, 2017.

240. Benoit B. Mandelbrot. *The Fractal Geometry of Nature WH Freeman and Company*. WH Freeman and Co., San Francisco, CA, 1982.

241. Niall M. Mangan, J. Nathan Kutz, Steven L. Brunton, and Joshua L. Proctor. Model selection for dynamical systems via sparse regression and information criteria. *Proc. R. Soc. A Math. Phys. Eng. Sci.*, 473(2204):20170009, 2017.

242. M. Cristina Marchetti, Jean-Francois Joanny, Sriram Ramaswamy, Tanniemola B. Liverpool, Jacques Prost, Madan Rao, and R. Aditi Simha. Hydrodynamics of soft active matter. *Rev. Mod. Phys.*, 85(3):1143, 2013.

243. Stefano Martiniani, Paul M. Chaikin, and Dov Levine. Quantifying hidden order out of equilibrium. *Phys. Rev. X*, 9(1):011031, 2019.

244. Glenn J. Martyna, Michael L. Klein, and Mark E. Tuckerman. Nosé–Hoover chains: The canonical ensemble via continuous dynamics. *J. Chem. Phys.*, 97(4):2635–2643, 1992.

245. Glenn J. Martyna, Mark E. Tuckerman, Douglas J. Tobias, and Michael L. Klein. Explicit reversible integrators for extended systems dynamics. *Mol. Phys.*, 87(5):1117–1157, 1996.

246. Anthony F. Masters and Thomas Maschmeyer. Zeolites–from curiosity to cornerstone. *Micropor. Mesopor. Mat.*, 142(2–3):423–438, 2011.

247. Nobuyuki Matubayasi and Masaru Nakahara. Reversible molecular dynamics for rigid bodies and hybrid Monte Carlo. *J. Chem. Phys.*, 110(7):3291–3301, 1999.

248. James Clerk Maxwell. On Boltzmann's Theorem on the average distribution of energy in a system of material points. In *The Scientific Papers of J.C. Maxwell: Volume 2*, pp. 713–741. Cambridge University Press, Cambridge, 1879.

249. Stephen L. Mayo, Barry D. Olafson, and William A. Goddard. Dreiding: A generic force field for molecular simulations. *J. Phys. Chem.*, 94(26):8897–8909, 1990.

250. Oleg A. Mazyar, Guoai Pan, and Clare McCabe. Transient time correlation function calculation of the viscosity of a molecular fluid at low shear rates: A comparison of stress tensors. *Mol. Phys.*, 107(14):1423–1429, 2009.

251. Warren S. McCulloch and Walter Pitts. A logical calculus of the ideas immanent in nervous activity. *Bull. Math. Biophys.*, 5(4):115–133, 1943.

252. Michael T. McManus, Christian P. Petersen, Brian B. Haines, Jianzhu Chen, and Phillip A. Sharp. Gene silencing using micro-RNA designed hairpins. *RNA*, 8(6):842–850, 2002.

253. Alan D. McNaught, Andrew Wilkinson, et al. *Compendium of Chemical Terminology*, vol. 1669. Blackwell Science, Oxford, 1997.

254. J. Liam McWhirter. The stability of planar Couette flow simulated by molecular dynamics. *J. Chem. Phys.*, 118(6):2824–2836, 2003.

255. B. Mehlig, D. W. Heermann, and B. M. Forrest. Hybrid Monte Carlo method for condensed-matter systems. *Phys. Rev. B*, 45(2):679, 1992.

256. J. Mei, J. W. Davenport, and G. W. Fernando. Analytic embedded-atom potentials for FCC metals: Application to liquid and solid copper. *Phys. Rev. B*, 43(6):4653, 1991.

257. Evert Jan Meijer, Daan Frenkel, Richard A. LeSar, and Anthony J. C. Ladd. Location of melting point at 300K of nitrogen by Monte Carlo simulation. *J. Chem. Phys.*, 92(12):7570–7575, 1990.

258. Simone Melchionna, Giovanni Ciccotti, and Brad Lee Holian. Hoover NPT dynamics for systems varying in shape and size. *Mol. Phys.*, 78(3):533–544, 1993.

259. Jose L. Mendoza-Cortes, William A. Goddard III, Hiroyasu Furukawa, and Omar M. Yaghi. A covalent organic framework that exceeds the DOE 2015 volumetric target for H2 uptake at 298K. *J. Phys. Chem. Lett.*, 3(18):2671–2675, 2012.

260. Nicholas Metropolis, Arianna W. Rosenbluth, Marshall N. Rosenbluth, Augusta H. Teller, and Edward Teller. Equation of state calculations by fast computing machines. *J. Chem. Phys.*, 21(6):1087–1092, 1953.

261. Shriyaa Mittal and Diwakar Shukla. Recruiting machine learning methods for molecular simulations of proteins. *Mol. Simul.*, 44(11):891–904, 2018.

262. Hazime Mori. Transport, collective motion, and brownian motion. *Prog. Theor. Phys.*, 33(3):423–455, 1965.

263. Gary P. Morriss and Denis J. Evans. Application of transient correlation functions to shear flow far from equilibrium. *Phys. Rev. A*, 35(2):792, 1987.

264. M. Müller and W. Paul. Measuring the chemical potential of polymer solutions and melts in computer simulations. *J. Chem. Phys.*, 100(1):719–724, 1994.

265. Gorka Munoz-Gil, Giovanni Volpe, Miguel Angel Garcia-March, Erez Aghion, Aykut Argun, Chang Beom Hong, Tom Bland, Stefano Bo, J. Alberto Conejero, Nicolás Firbas, et al. Objective comparison of methods to decode anomalous diffusion. *Nat. Commun.*, 12(1):6253, 2021.

266. Alan L. Myers and Peter A. Monson. Adsorption in porous materials at high pressure: Theory and experiment. *Langmuir*, 18(26):10261–10273, 2002.

267. Cesare Nardini, Étienne Fodor, Elsen Tjhung, Frédéric Van Wijland, Julien Tailleur, and Michael E. Cates. Entropy production in field theories without time-reversal symmetry: Quantifying the non-equilibrium character of active matter. *Phys. Rev. X*, 7(2):021007, 2017.

268. Afshin Eskandari Nasrabad, Rozita Laghaei, and U. K. Deiters. Prediction of the thermophysical properties of pure neon, pure argon, and the binary mixtures neon-argon and argon-krypton by Monte Carlo simulation using ab initio potentials. *J. Chem. Phys.*, 121(13):6423–6434, 2004.

269. Frank Noé, Gianni De Fabritiis, and Cecilia Clementi. Machine learning for protein folding and dynamics. *Curr. Opin. Struct. Biol.*, 60:77–84, 2020.

270. Shuichi Nosé. A molecular dynamics method for simulations in the canonical ensemble. *Mol. Phys.*, 52(2):255–268, 1984.

271. Shuichi Nosé. A unified formulation of the constant temperature molecular dynamics methods. *J. Chem. Phys.*, 81(1):511–519, 1984.

272. Shuichi Nose. Constant-temperature molecular dynamics. *J. Phys. Condens. Matter*, 2(S):SA115, 1990.

273. Gerardo Odriozola and Marcelo Lozada-Cassou. Entropy driven self-assembly in charged lock–key particles. *J. Phys. Chem. B*, 120(26):5966–5974, 2016.

274. Ahmad K. Omar, Zhen-Gang Wang, and John F. Brady. Microscopic origins of the swim pressure and the anomalous surface tension of active matter. *Phys. Rev. E*, 101(1):012604, 2020.

275. Lars Onsager. Reciprocal relations in irreversible processes. I. *Phys. Rev.*, 37(4):405, 1931.

276. Lars Onsager. Reciprocal relations in irreversible processes. II. *Phys. Rev.*, 38(12):2265, 1931.

277. Jeremie Palacci, Stefano Sacanna, Asher Preska Steinberg, David J. Pine, and Paul M. Chaikin. Living crystals of light-activated colloidal surfers. *Science*, 339(6122):936–940, 2013.

278. Siddharth Paliwal, Jeroen Rodenburg, René van Roij, and Marjolein Dijkstra. Chemical potential in active systems: Predicting phase equilibrium from bulk equations of state? *New J. Phys.*, 20(1):015003, 2018.

279. Andrew S. Paluch, Vincent K. Shen, and Jeffrey R. Errington. Comparing the use of Gibbs ensemble and grand-canonical transition-matrix Monte Carlo methods to determine phase equilibria. *Ind. Eng. Chem. Res.*, 47(13):4533–4541, 2008.

280. Guoai Pan and Clare McCabe. Prediction of viscosity for molecular fluids at experimentally accessible shear rates using the transient time correlation function formalism. *J. Chem. Phys.*, 125(19):194527, 2006.

281. Athanassios Z. Panagiotopoulos. Direct determination of phase coexistence properties of fluids by Monte Carlo simulation in a new ensemble. *Mol. Phys.*, 61(4):813–826, 1987.

282. Athanassios Z. Panagiotopoulos, Nicholas Quirke, M. Stapleton, and Dominic J. Tildesley. Phase equilibria by simulation in the Gibbs ensemble: Alternative derivation, generalization and application to mixture and membrane equilibria. *Mol. Phys.*, 63(4):527–545, 1988.

283. Sanghyun Park, Fatemeh Khalili-Araghi, Emad Tajkhorshid, and Klaus Schulten. Free energy calculation from steered molecular dynamics simulations using Jarzynski's equality. *J. Chem. Phys.*, 119(6):3559–3566, 2003.

284. Michele Parrinello and Aneesur Rahman. Polymorphic transitions in single crystals: A new molecular dynamics method. *J. Appl. Phys.*, 52(12):7182–7190, 1981.

285. Leopold Alexander Pars. *A Treatise on Analytical Dynamics*. Heineman, London, 1968.

286. Drew F. Parsons and David R. M. Williams. Globule transitions of a single homopolymer: A Wang-Landau Monte Carlo study. *Phys. Rev. E*, 74(4):041804, 2006.

287. Baron Peters and Bernhardt L. Trout. Obtaining reaction coordinates by likelihood maximization. *J. Chem. Phys.*, 125(5):054108, 2006.

288. Janka Petravic and Jerome Delhommelle. Conductivity of molten sodium chloride and its supercritical vapor in strong DC electric fields. *J. Chem. Phys.*, 118(16):7477–7485, 2003.

289. Pablo M. Piaggi, Omar Valsson, and Michele Parrinello. Enhancing entropy and enthalpy fluctuations to drive crystallization in atomistic simulations. *Phys. Rev. Lett.*, 119(1):015701, 2017.

290. Patrick Pietzonka, Eva Zimmermann, and Udo Seifert. Fine-structured large deviations and the fluctuation theorem: Molecular motors and beyond. *EPL-Europhys. Lett.*, 107(2):20002, 2014.

291. David J. Pine, Jerry P. Gollub, John F. Brady, and Alexander M. Leshansky. Chaos and threshold for irreversibility in sheared suspensions. *Nature*, 438(7070):997–1000, 2005.

292. Len M. Pismen. Dynamics of defects in an active nematic layer. *Phys. Rev. E*, 88(5):050502, 2013.

293. James M. Polson, Emmanuel Trizac, S. Pronk, and Daan Frenkel. Finite-size corrections to the free energies of crystalline solids. *J. Chem. Phys.*, 112(12):5339–5342, 2000.

294. Dirk Porezag, Th Frauenheim, Th Köhler, Gotthard Seifert, and R. Kaschner. Construction of tight-binding-like potentials on the basis of density-functional theory: Application to carbon. *Phys. Rev. B*, 51(19):12947, 1995.

295. Jeffrey J. Potoff and Athanassios Z. Panagiotopoulos. Surface tension of the three-dimensional Lennard-Jones fluid from histogram-reweighting Monte Carlo simulations. *J. Chem. Phys.*, 112(14):6411–6415, 2000.

296. Jeffrey J. Potoff and Jörn Ilja Siepmann. Vapor–liquid equilibria of mixtures containing alkanes, carbon dioxide, and nitrogen. *AIChE J.*, 47(7):1676–1682, 2001.

297. William H. Press, Saul A. Teukolsky, William T. Vetterling, and Brian P. Flannery. *Numerical Recipes 3rd Edition: The Art of Scientific Computing*. Cambridge University Press, Cambridge, 2007.

298. Ilya Prigogine and Isabelle Stengers. *The End of Certainty*. Simon and Schuster, New York, 1997.

299. Elias Putzig, Gabriel S. Redner, Arvind Baskaran, and Aparna Baskaran. Instabilities, defects, and defect ordering in an overdamped active nematic. *Soft Matter*, 12(17):3854–3859, 2016.

300. Mithun Radhakrishna, Sumit Sharma, and Sanat K. Kumar. Enhanced Wang–Landau sampling of adsorbed protein conformations. *J. Chem. Phys.*, 136(11):03B609, 2012.

301. Sriram Ramaswamy. Active matter. *J. Stat. Mech. Theory Exp.*, 2017(5):054002, 2017.

302. Padinhateeri Ranjith, David Lacoste, Kirone Mallick, and Jean-Francois Joanny. Nonequilibrium self-assembly of a filament coupled to ATP/GTP hydrolysis. *Biophys. J.*, 96(6):2146–2159, 2009.

303. John R. Ray and H. W. Graben. Fourth adiabatic ensemble. *J. Chem. Phys.*, 93(6):4296–4298, 1990.

304. John R. Ray, H. W. Graben, and J. M. Haile. A new adiabatic ensemble with particle fluctuations. *J. Chem. Phys.*, 75(8):4077–4079, 1981.

305. John R. Ray and Ralph J. Wolf. Monte Carlo simulations at constant chemical potential and pressure. *J. Chem. Phys.*, 98(3):2263–2267, 1993.

306. Nitzan Razin. Entropy production of an active particle in a box. *Phys. Rev. E*, 102(3):030103, 2020.

307. Gabriel S. Redner, Michael F. Hagan, and Aparna Baskaran. Structure and dynamics of a phase-separating active colloidal fluid. *Phys. Rev. Lett.*, 110(5):055701, 2013.

308. Sunghan Ro, Buming Guo, Aaron Shih, Trung V. Phan, Robert H. Austin, Dov Levine, Paul M. Chaikin, and Stefano Martiniani. Model-free measurement of local entropy production and extractable work in active matter. *Phys. Rev. Lett.*, 129(22):220601, 2022.

309. Jeroen Rodenburg, Marjolein Dijkstra, and René van Roij. Van't Hoff's law for active suspensions: The role of the solvent chemical potential. *Soft Matter*, 13(47):8957–8963, 2017.

310. Jutta Rogal, Elia Schneider, and Mark E. Tuckerman. Neural-network-based path collective variables for enhanced sampling of phase transformations. *Phys. Rev. Lett.*, 123(24):245701, 2019.

311. Pawel Romanczuk, Markus Bär, Werner Ebeling, Benjamin Lindner, and Lutz Schimansky-Geier. Active brownian particles. *Eur. Phys. J. Spec. Top.*, 202(1):1–162, 2012.

312. L. Rondoni and Ezechiel Godert David Cohen. Gibbs entropy and irreversible thermodynamics. *Nonlinearity*, 13(6):1905, 2000.

313. Frank Rosenblatt. The perceptron: A probabilistic model for information storage and organization in the brain. *Psychol. Rev.*, 65(6):386, 1958.

314. Jörg Rösgen, B. Montgomery Pettitt, John Perkyns, and David Wayne Bolen. Statistical thermodynamic approach to the chemical activities in two-component solutions. *J. Phys. Chem. B*, 108(6):2048–2055, 2004.

315. Samuel H. Rudy, Steven L. Brunton, Joshua L. Proctor, and J. Nathan Kutz. Data-driven discovery of partial differential equations. *Sci. Adv.*, 3(4):e1602614, 2017.

316. David Ruelle. A measure associated with Axiom-A attractors. *Am. J. Math.*, 98:619–654, 1976.

317. David Ruelle. Conversations on nonequilibrium physics with an extraterrestrial. *Phys. Today*, 57(5):48–53, 2004.

318. Hans Henrik Rugh. Dynamical approach to temperature. *Phys. Rev. Lett.*, 78(5):772, 1997.

319. Stuart Russell and Peter Norvig. *Artificial Intelligence: A Modern Approach*, 4th edition. Pearson, Hoboken, NJ, 2021.

320. Adal Sabri, Xinran Xu, Diego Krapf, and Matthias Weiss. Elucidating the origin of heterogeneous anomalous diffusion in the cytoplasm of mammalian cells. *Phys. Rev. Lett.*, 125(5):058101, 2020.

321. Otto F. Sankey and Roland E. Allen. Atomic forces from electronic energies via the Hellmann-Feynman theorem, with application to semiconductor (110) surface relaxation. *Phys. Rev. B*, 33(10):7164, 1986.

322. Markus Schöberl, Nicholas Zabaras, and Phaedon-Stelios Koutsourelakis. Predictive collective variable discovery with deep Bayesian models. *J. Chem. Phys.*, 150(2):024109, 2019.

323. Erwin Schrödinger. *Statistical Thermodynamics: A Course of Seminar Lecture.* Cambridge University Press, Cambridge, 1952.

324. Debra J. Searles and Denis J. Evans. Ensemble dependence of the transient fluctuation theorem. *J. Chem. Phys.*, 113(9):3503–3509, 2000.

325. Daniel T. Seaton, Stefan Schnabel, David P. Landau, and Michael Bachmann. From flexible to stiff: Systematic analysis of structural phases for single semiflexible polymers. *Phys. Rev. Lett.*, 110(2):028103, 2013.

326. Udo Seifert. Stochastic thermodynamics, fluctuation theorems and molecular machines. *Rep. Prog. Phys.*, 75(12):126001, 2012.

327. Claude E. Shannon. A mathematical theory of communication. *Bell Syst. Tech. J.*, 27(3):379–423, 1948.

328. Claude E. Shannon. Von Neumann's contributions to automata theory. *Bull. Am. Math. Soc.*, 64(3):123–129, 1958.

329. Hythem Sidky, Wei Chen, and Andrew L. Ferguson. Machine learning for collective variable discovery and enhanced sampling in biomolecular simulation. *Mol. Phys.*, 118(5):e1737742, 2020.

330. Jonathan Tammo Siebert, Janina Letz, Thomas Speck, and Peter Virnau. Phase behavior of active brownian disks, spheres, and dimers. *Soft Matter*, 13(5):1020–1026, 2017.

331. Jörn Iija Siepmann, Sami Karaborni, and Berend Smit. Simulating the critical behaviour of complex fluids. *Nature*, 365(6444):330–332, 1993.

332. Jörn Ilja Siepmann and Daan Frenkel. Configurational bias Monte Carlo: A new sampling scheme for flexible chains. *Mol. Phys.*, 75(1):59–70, 1992.

333. Yakov G. Sinai. Gibbs measures in ergodic theory. *Russ. Math. Surv.*, 27(4):21, 1972.

334. Anastasios I. Skoulidas and David S. Sholl. Self-diffusion and transport diffusion of light gases in metal–organic framework materials assessed using molecular dynamics simulations. *J. Phys. Chem. B*, 109(33):15760–15768, 2005.

335. John C. Slater and George F. Koster. Simplified LCAO method for the periodic potential problem. *Phys. Rev.*, 94(6):1498, 1954.

336. Alexandre P. Solon, Hugues Chaté, and Julien Tailleur. From phase to microphase separation in flocking models: The essential role of nonequilibrium fluctuations. *Phys. Rev. Lett.*, 114(6):068101, 2015.

337. Alexandre P. Solon, Joakim Stenhammar, Michael E. Cates, Yariv Kafri, and Julien Tailleur. Generalized thermodynamics of motility-induced phase separation: Phase equilibria, laplace pressure, and change of ensembles. *New J. Phys.*, 20(7):075001, 2018.

338. Thomas Speck, Julian Bialké, Andreas M. Menzel, and Hartmut Löwen. Effective Cahn-Hilliard equation for the phase separation of active brownian particles. *Phys. Rev. Lett.*, 112(21):218304, 2014.

339. Paul J. Steinhardt, David R. Nelson, and Marco Ronchetti. Bond-orientational order in liquids and glasses. *Phys. Rev. B*, 28(2):784, 1983.

340. Joakim Stenhammar, Davide Marenduzzo, Rosalind J. Allen, and Michael E. Cates. Phase behaviour of active brownian particles: The role of dimensionality. *Soft Matter*, 10(10):1489–1499, 2014.

341. Joakim Stenhammar, Adriano Tiribocchi, Rosalind J. Allen, Davide Marenduzzo, and Michael E. Cates. Continuum theory of phase separation kinetics for active brownian particles. *Phys. Rev. Lett.*, 111(14):145702, 2013.

342. Joakim Stenhammar, Raphael Wittkowski, Davide Marenduzzo, and Michael E. Cates. Light-induced self-assembly of active rectification devices. *Sci. Adv.*, 2(4):e1501850, 2016.

343. Mark J. Stevens and Mark O. Robbins. Simulations of shear-induced melting and ordering. *Phys. Rev. E*, 48(5):3778, 1993.

344. Frank H. Stillinger and Thomas A. Weber. Computer simulation of local order in condensed phases of silicon. *Phys. Rev. B*, 31(8):5262, 1985.

345. Yangzesheng Sun, Robert F. DeJaco, and Jörn Ilja Siepmann. Deep neural network learning of complex binary sorption equilibria from molecular simulation data. *Chem. Sci.*, 10(16):4377–4388, 2019.

346. A. P. Sutton and J. Chen. Long-range Finnis–Sinclair potentials. *Phil. Mag. Lett.*, 61(3):139–146, 1990.

347. Adam D. Swetnam and Michael P. Allen. Improving the Wang–Landau algorithm for polymers and proteins. *J. Comput. Chem.*, 32(5):816–821, 2011.

348. William C. Swope, Hans C. Andersen, Peter H. Berens, and Kent R. Wilson. A computer simulation method for the calculation of equilibrium constants for the formation of physical clusters of molecules: Application to small water clusters. *J. Chem. Phys.*, 76(1):637–649, 1982.

349. Julien Tailleur and Michael E. Cates. Statistical mechanics of interacting run-and-tumble bacteria. *Phys. Rev. Lett.*, 100(21):218103, 2008.

350. Sho C. Takatori and John F. Brady. Towards a thermodynamics of active matter. *Phys. Rev. E*, 91(3):032117, 2015.

351. Sho C. Takatori and John F. Brady. Forces, stresses and the (thermo?) dynamics of active matter. *Curr. Opin. Colloid Interface*, 21:24–33, 2016.

352. Sho C. Takatori, Wen Yan, and John F. Brady. Swim pressure: Stress generation in active matter. *Phys. Rev. Lett.*, 113(2):028103, 2014.

353. Mark P. Taylor, Wolfgang Paul, and Kurt Binder. Phase transitions of a single polymer chain: A Wang–Landau simulation study. *J. Chem. Phys.*, 131(11):114907, 2009.

354. Mark P. Taylor, Christopher Vinci, and Ryogo Suzuki. Effects of macromolecular crowding on the folding of a polymer chain: A Wang–Landau simulation study. *J. Chem. Phys.*, 153(17):174901, 2020.

355. Pieter Rein ten Wolde and Daan Frenkel. Computer simulation study of gas–liquid nucleation in a Lennard-Jones system. *J. Chem. Phys.*, 109(22):9901–9918, 1998.

356. Pieter Rein Ten Wolde, Maria J. Ruiz-Montero, and Daan Frenkel. Numerical evidence for BCC ordering at the surface of a critical FCC nucleus. *Phys. Rev. Lett.*, 75(14):2714, 1995.

357. J. Tersoff. Empirical interatomic potential for carbon, with applications to amorphous carbon. *Phys. Rev. Lett.*, 61(25):2879, 1988.

358. J. Tersoff. Modeling solid-state chemistry: Interatomic potentials for multicomponent systems. *Phys. Rev. B*, 39(8):5566, 1989.

359. Ryan V. Thaner, Youngeun Kim, Ting I. N..G. Li, Robert J. Macfarlane, SonBinh T. Nguyen, Monica Olvera de la Cruz, and Chad A. Mirkin. Entropy-driven crystallization behavior in DNA-mediated nanoparticle assembly. *Nano Lett.*, 15(8):5545–5551, 2015.

360. Elsen Tjhung, Cesare Nardini, and Michael E. Cates. Cluster phases and bubbly phase separation in active fluids: Reversal of the Ostwald process. *Phys. Rev. X.*, 8(3):031080, 2018.

361. B. D. Todd. Application of transient-time correlation functions to nonequilibrium molecular-dynamics simulations of elongational flow. *Phys. Rev. E*, 56(6):6723, 1997.

362. B. D. Todd, Denis J. Evans, and Peter J. Daivis. Pressure tensor for inhomogeneous fluids. *Phys. Rev. E*, 52(2):1627, 1995.

363. Richard C. Tolman. On the establishment of grand canonical distributions. *Phys. Rev.*, 57(12):1160, 1940.

364. Glenn M. Torrie and John P. Valleau. Nonphysical sampling distributions in Monte Carlo free-energy estimation: Umbrella sampling. *J. Comput. Phys.*, 23(2):187–199, 1977.

365. Shoichi Toyabe and Masaki Sano. Nonequilibrium fluctuations in biological strands, machines, and cells. *J. Phys. Soc. Jpn*, 84(10):102001, 2015.

366. Shoichi Toyabe, Takahiro Watanabe-Nakayama, Tetsuaki Okamoto, Seishi Kudo, and Eiro Muneyuki. Thermodynamic efficiency and mechanochemical coupling of F1-ATPase. *Proc. Natl. Acad. Sci. U. S. A.*, 108(44):17951–17956, 2011.

367. Karl P. Travis, Peter J. Daivis, and Denis J. Evans. Thermostats for molecular fluids undergoing shear flow: Application to liquid chlorine. *J. Chem. Phys.*, 103(24):10638–10651, 1995.

368. Karl P. Travis and Denis J. Evans. Molecular spin in a fluid undergoing poiseuille flow. *Phys. Rev. E*, 55(2):1566, 1997.

369. Karl P. Travis, B. D. Todd, and Denis J. Evans. Departure from Navier–Stokes hydrodynamics in confined liquids. *Phys. Rev. E*, 55(4):4288, 1997.

370. Benjamin Trefz, Jonathan Tammo Siebert, Thomas Speck, Kurt Binder, and Peter Virnau. Estimation of the critical behavior in an active colloidal system with vicsek-like interactions. *J. Chem. Phys.*, 146(7):074901, 2017.

371. Mark E. Tuckerman, Bruce J. Berne, and Glenn J. Martyna. Reversible multiple time scale molecular dynamics. *J. Chem. Phys.*, 97(3):1990–2001, 1992.

372. Philippe Ungerer, Christele Beauvais, Jérôme Delhommelle, Anne Boutin, Bernard Rousseau, and Alain H. Fuchs. Optimization of the anisotropic united atoms intermolecular potential for N-alkanes. *J. Chem. Phys.*, 112(12):5499–5510, 2000.

373. Irais Valencia-Jaime, Caroline Desgranges, and Jerome Delhommelle. Viscosity of a highly compressed methylated alkane via equilibrium and nonequilibrium molecular dynamics simulations. *Chem. Phys. Lett.*, 719:103–109, 2019.

374. Adri C. T. Van Duin, Siddharth Dasgupta, Francois Lorant, and William A. Goddard. Reaxff: A reactive force field for hydrocarbons. *J. Phys. Chem. A*, 105(41):9396–9409, 2001.

375. Titus S. Van Erp and Peter G. Bolhuis. Elaborating transition interface sampling methods. *J. Comput. Phys.*, 205(1):157–181, 2005.

376. Tim Van Erven and Peter Harremos. Rényi divergence and Kullback-Leibler divergence. *IEEE Trans. Inf. Theory*, 60(7):3797–3820, 2014.

377. Ramses Van Zon and Ezechiel Godert David Cohen. Extension of the fluctuation theorem. *Phys. Rev. Lett.*, 91(11):110601, 2003.

378. Natan B. Vargaftik, Yurii K. Vinoradov, and Vadim S. Yargin. *Handbook of Physical Properties of Liquids and Gases*. Begell House, New York, 1996.

379. Carlos Vega and Eva G. Noya. Revisiting the Frenkel-Ladd method to compute the free energy of solids: The Einstein molecule approach. *J. Chem. Phys.*, 127(15):154113, 2007.

380. Loup Verlet. Computer "experiments" on classical fluids. I. Thermodynamical properties of Lennard-Jones molecules. *Phys. Rev.*, 159(1):98, 1967.

381. Gatien Verley, Kirone Mallick, and David Lacoste. Modified fluctuation-dissipation theorem for non-equilibrium steady states and applications to molecular motors. *EPL-Europhys. Lett.*, 93(1):10002, 2011.

382. Tamás Vicsek, András Czirók, Eshel Ben-Jacob, Inon Cohen, and Ofer Shochet. Novel type of phase transition in a system of self-driven particles. *Phys. Rev. Lett.*, 75(6):1226, 1995.

383. Peter Virnau, M. Müller, Luis González MacDowell, and Kurt Binder. Phase diagrams of hexadecane–CO_2 mixtures from histogram-reweighting Monte Carlo. *Comp. Phys. Comm.*, 147(1–2):378–381, 2002.

384. William M. Visscher. Transport processes in solids and linear-response theory. *Phys. Rev. A*, 10(6):2461, 1974.

385. Giorgio Volpe, Sylvain Gigan, and Giovanni Volpe. Simulation of the active brownian motion of a microswimmer. *Am. J. Phys.*, 82(7):659–664, 2014.

386. David J. Wales. Exploring energy landscapes. *Annu. Rev. Phys.*, 69:401–425, 2018.

387. Peter J. Waller, Felipe Gandara, and Omar M. Yaghi. Chemistry of covalent organic frameworks. *Acc. Chem. Res.*, 48(12):3053–3063, 2015.

388. Jessica M. Walter, Derek Greenfield, Carlos Bustamante, and Jan Liphardt. Light-powering *Escherichia Coli* with proteorhodopsin. *Proc. Natl. Acad. Sci. U. S. A.*, 104(7):2408–2412, 2007.

389. Krista S. Walton, Andrew R. Millward, David Dubbeldam, Houston Frost, John J. Low, Omar M. Yaghi, and Randall Q. Snurr. Understanding inflections and steps in carbon dioxide adsorption isotherms in metal–organic frameworks. *J. Am. Chem. Soc.*, 130(2):406–407, 2008.

390. Fugao Wang and David P. Landau. Efficient, multiple-range random walk algorithm to calculate the density of states. *Phys. Rev. Lett.*, 86(10):2050, 2001.

391. Stephen Whitelam. Hierarchical assembly may be a way to make large information-rich structures. *Soft Matter*, 11(42):8225–8235, 2015.

392. Edmund Taylor Whittaker. *A Treatise on the Analytical Dynamics of Particles and Rigid Bodies*. Cambridge University Press, Cambridge, 1964.

393. Collin D. Wick, Marcus G. Martin, and Jörn Ilja Siepmann. Transferable potentials for phase equilibria. 4. United-atom description of linear and branched alkenes and alkylbenzenes. *J. Phys. Chem. B*, 104(33):8008–8016, 2000.

394. Thomas Wüst and David P. Landau. The HP model of protein folding: A challenging testing ground for Wang–Landau sampling. *Comp. Phys. Comm.*, 179(1–3):124–127, 2008.

395. Omar M. Yaghi, Michael O'Keeffe, Nathan W. Ockwig, and Hee K. Chae. Reticular synthesis and the design of new materials. *Nature*, 423(6941):705, 2003.

396. Tomoji Yamada and Kyozi Kawasaki. Nonlinear effects in the shear viscosity of critical mixtures. *Prog. Theor. Phys.*, 38(5):1031–1051, 1967.

397. Yuguang Yang and Michael A. Bevan. Cargo capture and transport by colloidal swarms. *Sci. Adv.*, 6(4):eaay7679, 2020.

398. Yuguang Yang, Michael A. Bevan, and Bo Li. Efficient navigation of colloidal robots in an unknown environment via deep reinforcement learning. *Adv. Intell. Syst.*, 2(1):1900106, 2020.

399. Yuguang Yang, Michael A. Bevan, and Bo Li. Micro/nano motor navigation and localization via deep reinforcement learning. *Adv. Theory Simul.*, 3(6):2000034, 2020.

400. He-Peng Zhang, Avraham Beer, E.-L. Florin, and Harry L. Swinney. Collective motion and density fluctuations in bacterial colonies. *Proc. Natl. Acad. Sci. U. S. A.*, 107(31):13626–13630, 2010.

401. Jianli Zhang, Junyan Yang, Yuanxing Zhang, and Michael A. Bevan. Controlling colloidal crystals via morphing energy landscapes and reinforcement learning. *Sci. Adv.*, 6(48):eabd6716, 2020.

402. Linfeng Zhang, Han Wang, and E. Weinan. Reinforced dynamics for enhanced sampling in large atomic and molecular systems. *J. Chem. Phys.*, 148(12):124113, 2018.

403. Pan Zhang, Lin Shen, and Weitao Yang. Solvation free energy calculations with quantum mechanics/molecular mechanics and machine learning models. *J. Phys. Chem. B*, 123(4):901–908, 2018.

404. Ren Zhang, Bongjoon Lee, Christopher M. Stafford, Jack F. Douglas, Andrey V. Dobrynin, Michael R. Bockstaller, and Alamgir Karim. Entropy-driven segregation of polymer-grafted nanoparticles under confinement. *Proc. Natl. Acad. Sci. U. S. A.*, 114(10):2462–2467, 2017.

405. Zhenxia Zhao, Zhong Li, and Y. S. Lin. Adsorption and diffusion of carbon dioxide on metal-organic framework (MOF-5). *Ind. Eng. Chem. Res.*, 48(22):10015–10020, 2009.

406. Robert Zwanzig. Time-correlation functions and transport coefficients in statistical mechanics. *Ann. Rev. Phys. Chem.*, 16(1):67–102, 1965.

407. Robert Zwanzig. *Nonequilibrium Statistical Mechanics*. Oxford University Press, New York, 2001.

Index